Mathematical Physics Studies

The series publishes original research monographs on contemporary mathematical physics. The focus is on important recent developments at the interface of Mathematics, and Mathematical and Theoretical Physics. These will include, but are not restricted to: application of algebraic geometry, D-modules and symplectic geometry, category theory, number theory, low-dimensional topology, mirror symmetry, string theory, quantum field theory, noncommutative geometry, operator algebras, functional analysis, spectral theory, and probability theory.

More information about this series at http://www.springer.com/series/6316

Fumio Hiai

Quantum f-Divergences in von Neumann Algebras

Reversibility of Quantum Operations

Fumio Hiai
Abiko, Chiba, Japan

ISSN 0921-3767 ISSN 2352-3905 (electronic)
Mathematical Physics Studies
ISBN 978-981-33-4201-9 ISBN 978-981-33-4199-9 (eBook)
https://doi.org/10.1007/978-981-33-4199-9

This Springer imprint is published by the registered company Springer Nature Singapore Pte Ltd.
The registered company address is: 152 Beach Road, #21-01/04 Gateway East, Singapore 189721, Singapore

To the memory of Dénes Petz (1953–2018)

Preface

After I wrote in 1981 the joint paper [61] with M. Tsukada and M. Ohya on sufficiency of von Neumann subalgebras and the relative entropy in a specialized situation, I wanted to extend the sufficiency notion to general von Neumann subalgebras. But I felt some difficulty in doing that because of the lack of the conditional expectation onto a general von Neumann subalgebra. In the meantime, I was shocked by D. Petz' remarkable paper [105] in 1986, where he settled the problem mentioned above by making use of the generalized conditional expectation due to L. Accardi and C. Cecchini [1] in 1982. In his 1988 paper [106] Petz followed the same idea to introduce the notion of sufficient (or reversible) channels (more precisely, unital normal 2-positive maps) between von Neumann algebras, and he further collaborated with A. Jenčová on the subject in [68, 69].

After the joint work of the 2011 paper [63], I thought that it would be nice to extend the materials for [63] to the von Neumann algebra setting, including a renewal of Petz' theory on sufficiency and reversibility in von Neumann algebras. But that was not undertaken due to my laziness before I looked at the papers [20, 66], where the sandwiched Rényi divergence, a new Rényi divergence extensively developed in recent years, was extended to the von Neumann algebra setting. Those papers strongly motivated me to develop a general theory of quantum f-divergences in von Neumann algebras with applications to the sufficiency and reversibility problem. I planned to first write three papers on three different types of f-divergences (the standard f-divergences, the maximal f-divergences, and the measured f-divergences) in von Neumann algebras. The first two papers were published in [54, 55], which are surveyed in Chaps. 2 and 4 of this monograph. However, I wrote the third one as Chap. 5 instead of publishing a separate paper, so Chap. 5 of the monograph is new. In Chaps. 6–8, the main part of the monograph, I revisit the work [68, 105, 106, 108] with slight extensions and some modifications.

Acknowledgments I would like to thank Milán Mosonyi, Anna Jenčová, and Hideki Kosaki. I greatly enjoyed joint work with Mosonyi in [58, 63]. Many results and proofs in this monograph are modeled on those in [58, 63] in the finite-dimensional case. I thank Mosonyi also for inviting me to lectures for an intensive course [56] in April 2019 at the Department of Mathematical Analysis, Budapest University of Technology and Economics. Appendix A of this monograph is a short extract from those lecture notes. I learned from Jenčová important points in Sects. 3.2 and 3.3, in particular, I am grateful to her for permitting me to include Lemma 3.5, which is due to her. Discussions with Kosaki in recent joint work [57] on operator connections of unbounded positive operators and positive forms were helpful in writing Appendix D.

I am grateful to Masayuki Nakamura (Springer Japan) for his constant support, without which it would not have been possible for me to write this monograph.

This work was supported in part by JSPS KAKENHI Grant Number JP17K05266.

Chiba, Japan
October 2020

Fumio Hiai

Contents

Chapter 1
Introduction

The notion of quantum divergences has played a significant role in quantum information, which defines important quantum quantities to discriminate between states of a quantum system. A quantum system is mathematically described, in most cases, by an operator algebra $\mathcal{A}$ on a Hilbert space, either finite-dimensional or infinite-dimensional, and a quantum divergence is generally given as a function $S(\psi\|\varphi)$ of two states or more generally, two positive linear functionals ψ and φ on $\mathcal{A}$. This monograph is aimed at presenting a comprehensive survey of quantum f-divergences and showing their significant role in the reversibility problem of quantum operations in the general von Neumann algebra setting.

A general and rigorous framework of quantum information should be that of von Neumann algebras, while, in the present circumstances, quantum information literature is mostly presented in the finite-dimensional setting or the matrix setting. Indeed, the reader can consult the textbooks [50, 100, 109, 135] for broad and intensive developments of quantum information in the last quarter century. As widely believed, the von Neumann algebra framework is the most suitable in studying non-commutative (= quantum) probability and integration from the pure mathematical side [32, 99, 118] and also in developing quantum statistical mechanics from the mathematical physics side [22, 23]. The idea provides a good motivation to study quantum information in the von Neumann algebra setting. C^*-algebras, a more abstract class of operator algebras, are useful as well in studying quantum physics, but discussions in C^*-algebras can be, in many cases, reduced to those in von Neumann algebras by taking the GNS (Gelfand–Naimark–Segal) construction associated with a relevant state on a C^*-algebra. Furthermore, several well-developed machines are at our disposal in theory of von Neumann algebras such as, starting from the structure theory based on the Tomita–Takesaki theory, the standard forms, the relative modular operators, the non-commutative L^p-spaces, and so on. Those basics of von Neumann algebras are briefly surveyed, for the convenience of the reader, in Appendix A of the monograph as preliminaries for quantum information in the von Neumann algebra setting.

F. Hiai, *Quantum f-Divergences in von Neumann Algebras*,
Mathematical Physics Studies, https://doi.org/10.1007/978-981-33-4199-9_1

In the first half of the monograph we present a comprehensive survey of quantum f-divergences, which are introduced for normal positive linear functionals ψ, φ on a von Neumann algebra M with a parametrization of convex functions (often operator convex functions) f on $(0, \infty)$. In Chaps. 2–5 we deal with three different types of quantum f-divergences. First, the standard f-divergences $S_f(\psi\|\varphi)$ discussed in Chap. 2 are defined by making use of the relative modular operators due to Araki [11], which are a special case of quasi-entropies introduced by Petz [103, 104]. The most notable example (when $f(t) = t\log t$) is the relative entropy $D(\psi\|\varphi)$ having a long history as the quantum version of the Kullback–Leibler divergence in classical theory, first introduced by Umegaki [134] in semifinite von Neumann algebras and extended by Araki [10, 11] to the general von Neumann algebra case. The content of Chap. 2 is a survey from [54], and a main point there is a variational expression (see Theorem 2.5) of $S_f(\psi\|\varphi)$ extending Kosaki's expression [79] of the relative entropy. See [54] for the proofs of the results given in Chap. 2, and more detailed discussions in the finite-dimensional case are found in [58, 63].

The maximal f-divergences $\widehat{S}_f(\psi\|\varphi)$ are discussed in Chap. 4. When specialized to the finite-dimensional case, those f-divergences are defined as the trace value of the operator perspective $D_\varphi^{1/2} f(D_\varphi^{-1/2} D_\psi D_\varphi^{1/2}) D_\varphi^{1/2}$ of the density operators D_ψ, D_φ, whose detailed discussions in the finite-dimensional case are found in [58, 92]. A special example of the maximal f-divergence (when $f(t) = t\log t$) is a variant of the relative entropy introduced by Belavkin and Staszewski [17], denoted by $D_{\mathrm{BS}}(\psi\|\varphi)$. The content of Chap. 4 is a survey from [55], where, apart from the definition of $\widehat{S}_f(\psi\|\varphi)$, the main results are the integral expression (see Theorem 4.8) and the variational expression (see Theorem 4.17) in terms of reverse tests introduced following Matsumoto's idea in [92].

Assume that f is an operator convex function on $(0, \infty)$. Among many properties of the quantum f-divergence S_f (resp., $\widehat{S}_f$), the most important is the monotonicity property (also called the DPI, the data-processing inequality)

$$S_f(\psi\circ\gamma\|\varphi\circ\gamma) \le S_f(\psi\|\varphi), \qquad \widehat{S}_f(\psi\circ\gamma\|\varphi\circ\gamma) \le \widehat{S}_f(\psi\|\varphi) \tag{1.1}$$

for any unital normal Schwarz (resp., simply positive) map $\gamma : N \to M$ between von Neumann algebras N, M (in particular, normal completely positive maps or quantum channels). Here, a linear map $\gamma : N \to M$ is a Schwarz map if $\gamma(x^*x) \ge \gamma(x)^*\gamma(x)$ for all $x \in N$. In Chap. 5 we introduce the measured f-divergences $S_f^{\mathrm{meas}}(\psi\|\varphi)$ by the supremum of $S_f(\mathcal{M}(\psi)\|\mathcal{M}(\varphi)) := \sum_{j=1}^n \varphi(A_j) f(\psi(A_j)/\varphi(A_j))$, the classical f-divergence of $\mathcal{M}(\psi) := (\psi(A_j))$ and $\mathcal{M}(\varphi) = (\varphi(A_j))$ over all measurements (or quantum-classical channels) $\mathcal{M} = (A_j)_{1\le j\le n}$ in M. The projectively measured f-divergence $S_f^{\mathrm{pr}}(\psi\|\varphi)$ is also defined to be the supremum taken with restriction to projective measurements in M. The measured f-divergences S_f^{meas} and S_f^{pr} are also defined by using $\widehat{S}_f$ in place of S_f because $\widehat{S}_f(\psi\|\varphi) = S_f(\psi\|\varphi)$ for $\psi, \varphi \in M_*^+$ if M is an abelian von Neumann algebra. By construction we have the relation

$$S_f^{\mathrm{pr}} \le S_f^{\mathrm{meas}} \le S_f \le \widehat{S}_f.$$

A more intrinsic explanation of the above relation is given at the end of Sect. 5.1. For the measured f-divergences S_f^{meas} and S_f^{pr}, we discuss in Chap. 5 their basic properties, variational expressions, and the existence of optimal measurements. This part of the measured f-divergences in the von Neumann algebra setting is new, while discussions in the finite-dimensional case are found in [58, 91]. So the presentation of Chap. 5 is mostly self-contained including proofs.

Together with the relative entropy, among the most used quantum divergences is the α-Rényi divergence (or the α-Rényi relative entropy) $D_\alpha(\psi\|\varphi)$ with parameter $\alpha \in (0,\infty)\setminus\{1\}$, which is essentially the standard f-divergence for $f(t) = t^\alpha$ (for $\alpha > 1$) or $-t^\alpha$ (for $0 < \alpha < 1$). In recent years it has also been widely known that another variant of the α-Rényi divergence $\widetilde{D}_\alpha(\psi\|\varphi)$ with parameter $\alpha \in [1/2,\infty)\setminus\{1\}$, called the sandwiched α-Rényi divergence, is equally useful in quantum information, in particular, in quantum state discrimination, see [96, 97] for example. In the finite-dimensional case with the density operators D_ψ, D_φ, the divergences D_α and $\widetilde{D}_\alpha$ are given by

$$D_\alpha(\psi\|\varphi) := \frac{1}{\alpha-1}\log\frac{\mathrm{Tr}\, D_\psi^\alpha D_\varphi^{1-\alpha}}{\psi(1)},$$

$$\widetilde{D}_\alpha(\psi\|\varphi) := \frac{1}{\alpha-1}\log\frac{\mathrm{Tr}\big(D_\varphi^{\frac{1-\alpha}{2\alpha}} D_\psi D_\varphi^{\frac{1-\alpha}{2\alpha}}\big)^\alpha}{\psi(1)}.$$

The monograph [130] is a good source for Rényi-type quantum divergences and their mathematical backgrounds in the finite-dimensional setting. The sandwiched Rényi divergence $\widetilde{D}_\alpha$ has recently been extended to the von Neumann algebra setting by Berta, Scholz and Tomamichel [20] and Jenčová [66, 67], while the description of D_α in von Neumann algebras is found in [54]. Chapter 3 is a survey from those articles. In particular, Sect. 3.3 contains a proof of the fact that the two definitions of $\widetilde{D}_\alpha$ in [20, 66] are equivalent, which is different from that given in [66]. Moreover, yet another equivalent definition of $\widetilde{D}_\alpha$ was given in [67].

In Chaps. 6–8, the second half of the monograph, we develop the reversibility theory for quantum operations between von Neumann algebras via quantum f-divergences. Let $\gamma : N \to M$ be a unital normal 2-positive map, and let $\psi,\varphi \in M_*^+$. We say that γ is reversible for $\{\psi,\varphi\}$ if there exists a unital normal 2-positive map $\beta : M \to N$ such that

$$\psi\circ\gamma\circ\beta = \psi \quad \text{and} \quad \varphi\circ\gamma\circ\beta = \varphi.$$

When M_0 is a von Neumann subalgebra of M, we say that M_0 is sufficient for $\{\psi,\varphi\}$ if there exists a unital normal 2-positive map $\beta : M \to M_0$ such that $\psi\circ\beta = \psi$ and $\varphi\circ\beta = \varphi$. Here is a small historical remark on sufficiency in von Neumann algebras. The study of the subject was initiated by Umegaki [133, 134] around 1960 (rather soon after the emergence of non-commutative probability and integration [32, 118]), and the idea was further discussed later in [61, 62]. In these papers, the sufficiency of a subalgebra M_0 was defined by the existence of the common conditional expectation from M onto M_0 for $\{\psi, \varphi\}$, so discussions were necessarily quite limited because of lack of the conditional expectation onto a general von Neumann subalgebra. But a breakthrough was provided by Petz' paper [105] in 1986 as referred to shortly.

If γ is reversible for $\{\psi, \varphi\}$, then the double use of the monotonicity inequality (1.1) gives

$$S_f(\psi \circ \gamma \| \varphi \circ \gamma) = S_f(\psi \| \varphi) \tag{1.2}$$

for any standard f-divergence S_f with an operator convex function f. Similarly, if a subalgebra M_0 is sufficient for $\{\psi, \varphi\}$, then

$$S_f(\psi|_{M_0} \| \varphi|_{M_0}) = S_f(\psi \| \varphi). \tag{1.3}$$

The reverse direction of the above is our concern of Chap. 6, that is, we are interested in the problem whether equality (1.2) (resp., (1.3)) with a finite value implies the reversibility of γ (resp., the sufficiency of M_0) for $\{\psi, \varphi\}$. The problem was formerly studied by Petz [105, 106] (also in [101, Chap. 9]) and by Jenčová and Petz [68, 69], where the treated quantum divergences were the relative entropy and the transition probability (or equivalently, the standard f-divergence for $f(t) = -t^{1/2}$). It was assumed in [105, 106] that $\psi, \varphi \in M_*^+$ are faithful, while the assumption of ψ being faithful was removed in [68]. Further detailed discussions are found in [63] (also [65]) on the reversibility in the finite-dimensional case via standard f-divergences under a mild assumption on f. Other discussions on sufficiency in von Neumann algebras are found in [89, 90] for example.

In Sects. 6.1–6.4 we revisit the sufficiency and reversibility results in [68, 105, 106] with slight improvements and furthermore obtain the reversibility result (see Theorem 6.19) via general standard f-divergences under the assumption that the support of μ_f has a limit point in $(0, \infty)$, where μ_f is the representing measure of the integral expression of an operator convex function f. (Note that μ_f is supported on the whole $(0, \infty)$ when $f(t) = t \log t$ for the relative entropy and when $f(t) = -t^{1/2}$ for the transition probability.) In recent papers [66, 67] Jenčová proved the reversibility theorems via the sandwiched Rényi divergences in von Neumann algebras, which are briefly reported in Sect. 6.5.

A significant point in the reversibility theorems in Chap. 6 is that when a unital normal 2-positive map $\gamma : N \to M$ is reversible for $\{\psi, \varphi\}$, we have a reverse map $\beta = \gamma_\varphi^* : M \to N$ that is canonically defined as a certain dual map of γ with respect to φ. The map γ_φ^* is often called Petz' recovery map and its construction described in Sect. 6.1 is originally due to Accardi and Cecchini [1].

A more extended form of reversibility (or sufficiency) is the approximate reversibility for a sequence of quantum operations $\beta_k : M_k \to M$ (or quantum channels with input M and outputs M_k). When $\beta_k : M_k \to M$ ($k \in \mathbb{N}$) are unital normal 2-positive maps between von Neumann algebras, we say that $(\beta_k : M_k \to M)_{k \in \mathbb{N}}$ is approximately reversible for $\{\psi, \varphi\}$ in M_*^+ if there exist unital normal

2-positive maps $\beta_k : M \to M_k$ ($k \in \mathbb{N}$) such that

$$\psi \circ \alpha_k \circ \beta_k \longrightarrow \psi \quad \text{and} \quad \varphi \circ \alpha_k \circ \beta_k \longrightarrow \varphi \tag{1.4}$$

in the $\sigma(M_*, M)$-topology. Then a natural problem is to characterize the approximate reversibility by the approximation of quantum f-divergences

$$\lim_k S_f(\psi \circ \alpha_k \| \varphi \circ \alpha_k) = S_f(\psi \| \varphi) < +\infty \tag{1.5}$$

for a certain f.

In a quantum mechanical system we have final data by measurements (or quantum-classical channels). From this point of view it is reasonable for us to consider the reversibility for $\{\psi, \varphi\}$ via a measurement channel, that is, to find unital normal positive maps $\alpha : \mathcal{A} \to M$ and $\beta : M \to \mathcal{A}$ with a commutative von Neumann algebra $\mathcal{A}$ such that $\psi = \psi \circ \alpha \circ \beta$ and $\varphi = \varphi \circ \alpha \circ \beta$. It is also meaningful to consider the approximate reversibility via measurement channels, that is, to find unital normal positive maps $\alpha_k : \mathcal{A}_k \to M$ and $\beta_k : M \to \mathcal{A}_k$ ($k \in \mathbb{N}$) with commutative $\mathcal{A}_k$ satisfying (1.4). In the case of measurement channels (i.e., M_k's are commutative) the approximation for S_f in (1.5) is equivalently stated as

$$S_f^{\mathrm{meas}}(\psi \| \varphi) = S_f(\psi \| \varphi) < +\infty.$$

The approximate reversibility problem was studied by Petz [108]. It was shown in [108] that if the approximate reversibility via measurements as in (1.4) (with commutative M_k's) is satisfied for faithful ψ, φ, then ψ, φ commute. This means that the approximate reversibility via measurements holds only when ψ, φ are in the classical position. In Chap. 7 we extend this result to ψ, φ with $s(\psi) \le s(\varphi)$ (see Theorems 7.10 and 7.11). Furthermore, it was shown in [108] that when $\delta\varphi \le \psi \le \delta^{-1}\varphi$ for some $\delta > 0$ and M_k's are general von Neumann algebras, condition (1.5) for $S_f = D$, the relative entropy, implies the approximate reversibility in (1.4). We extend this result to S_f with a general f satisfying the same support condition as in Theorem 6.19 (see Theorem 7.8).

In the proofs of the reversibility and the approximate reversibility theorems in Chaps. 6 and 7, a main ingredient is to obtain the (approximate) preservation of Connes' cocycle derivatives, as done in Sects. 6.3 and 7.1. Both Araki's relative modular operator $\Delta_{\psi,\varphi}$ ([11]) and Connes' cocycle derivative $[D\psi : D\varphi]_t$ ([26]) are certain types of the Radon–Nikodym derivative. In the finite-dimensional $M = B(\mathcal{H})$ case with the density operators D_ψ, D_φ, those are given as $\Delta_{\psi,\varphi}(X) = D_\psi X D_\varphi^{-1}$ for $X \in B(\mathcal{H})$ (where $M = B(\mathcal{H})$ is represented by the left multiplication on $B(\mathcal{H})$ with the Hilbert–Schmidt inner product) and $[D\psi : D_\varphi]_t = D_\psi^{it} D_\varphi^{-it}$ for $t \in \mathbb{R}$. So $\Delta_{\psi,\varphi}$ and $[D\psi : D\varphi]_t$ are essentially in the same vein (roughly speaking, $[D\psi : D\varphi]_t$ is the it-power of $\Delta_{\psi,\varphi}$). A concise account of Connes' cocycle derivatives is included in Sect. A.7 for the convenience of the reader.

The reversibility problem may be also considered for the maximal f-divergence $\widehat{S}_f$, in place of S_f discussed in Sect. 6.4. But it is known [58, Example 4.8] that even in the finite-dimensional case, the equality case (with a finite value) in the monotonicity inequality for $\widehat{S}_f$ in (1.1) does not necessarily imply the reversibility of γ for $\{\psi, \varphi\}$. Even though the situation is negative in that way, it is meaningful to characterize the preservation

$$\widehat{S}_f(\psi \circ \gamma \| \varphi \circ \gamma) = \widehat{S}_f(\psi \| \varphi) \tag{1.6}$$

under a unital normal positive map $\gamma : N \to M$, which was discussed in [58] in the finite-dimensional case. Our concern in Chap. 8 is to extend the discussions in [58] to the von Neumann algebra case. Among many equivalent conditions, we show that for any $\psi, \varphi \in M_*^+$, if (1.6) holds with a finite value for some nonlinear operator convex function f, then the same equality holds for all operator convex functions f (see Theorem 8.4). In the course of discussions, we are led to consider operator connections (developed by Kubo and Ando [85]) for positive functionals in M_*^+. The idea is interesting on its own, so a somewhat detailed account of operator connections of elements in M_*^+ is given in Appendix D separately.

The rest of the monograph contains four appendices, which are used in the main body. In the long Appendix A we give concise accounts of selected topics of von Neumann algebras, including the Tomita–Takesaki theory, the standard forms, the relative modular operators, Haagerup's L^p-spaces, Connes' cocycle derivatives, and so on. These might be beneficial for the reader who is not familiar with von Neumann algebras. Appendix B contains a few preliminaries on positive self-adjoint operators. Appendix C is concerned with operator convex functions on $(0, 1)$, which turn up in the integral expression of Theorem 4.8 and may be of independent interest. Finally, Appendix D is concerned with operator connections of normal positive functionals, which is a variant of Kubo and Ando's theory in [85].

Chapter 2
Standard f-Divergences

2.1 Definition

Let M be a general von Neumann algebra, and M_*^+ be the positive cone of the predual M_* consisting of normal positive linear functionals on M. Basics of von Neumann algebras are given in Sect. A.1. Throughout this monograph, we consider M in its *standard form*

$$(M, \mathcal{H}, J, \mathcal{P}),$$

that is, M is faithfully represented on a Hilbert space $\mathcal{H}$ with a conjugate-linear involution J and a self-dual cone $\mathcal{P}$ called the *natural positive cone*, satisfying properties (a)–(d) given in Sect. A.3. Any von Neumann algebra has a standard form, which is unique up to unitary conjugation, see Theorem A.19.

Let $\psi, \varphi \in M_*^+$, whose vector representatives in $\mathcal{P}$ are denoted by Ψ, Φ respectively, so that

$$\psi(x) = \langle \Psi, x\Psi \rangle, \qquad \varphi(x) = \langle \Phi, x\Phi \rangle, \qquad x \in M.$$

The M-support projection $s(\varphi) = s_M(\varphi) \in M$ and the M'-support projection $s_{M'}(\varphi) \in M'$ are the orthogonal projections onto $\overline{M'\Phi}$ (the closure of $M'\Phi := \{x'\Phi : x' \in M'\}$) and onto $\overline{M\Phi}$ respectively, and similarly for $s(\psi) = s_M(\psi)$, $s_{M'}(\psi)$. We have the *relative modular operator* $\Delta_{\psi,\varphi}$ introduced in [11], which is described in Sect. A.4. The support projection of $\Delta_{\psi,\varphi}$ is $s_M(\psi)s_{M'}(\varphi)$. We write the spectral decomposition of $\Delta_{\psi,\varphi}$ as

$$\Delta_{\psi,\varphi} = \int_{[0,\infty)} t \, dE_{\psi,\varphi}(t). \tag{2.1}$$

F. Hiai, *Quantum f-Divergences in von Neumann Algebras*,
Mathematical Physics Studies, https://doi.org/10.1007/978-981-33-4199-9_2

Assume that $f : (0, \infty) \to \mathbb{R}$ is a convex function. The limits

$$f(0^+) := \lim_{t \searrow 0} f(t), \qquad f'(\infty) := \lim_{t \to \infty} \frac{f(t)}{t}$$

exist in $(-\infty, +\infty]$. Below we will understand the expression $bf(a/b)$ for $a = 0$ or $b = 0$ in the following way:

$$bf(a/b) := \begin{cases} f(0^+)b & \text{for } a = 0,\ b \geq 0, \\ f'(\infty)a & \text{for } a > 0,\ b = 0, \end{cases} \tag{2.2}$$

where we use the convention that $(+\infty)0 := 0$ and $(+\infty)a := +\infty$ for $a > 0$. In particular, we set $0f(0/0) = 0$.

The next definition is a specialization of the *quasi-entropy* introduced in [77, 103] with modifications.

Definition 2.1 For each $\varphi, \psi \in M_*^+$ and f as above, we define the *standard f-divergence* $S_f(\psi\|\varphi)$ of ψ, φ by

$$S_f(\psi\|\varphi) := \int_{(0,\infty)} f(t)\, d\|E_{\psi,\varphi}(t)\Phi\|^2 + f(0^+)\varphi(1 - s_M(\psi)) + f'(\infty)\psi(1 - s_M(\varphi)), \tag{2.3}$$

where we note that the integral above is on $(0, \infty)$ instead of $[0, \infty)$.

The integral in (2.3) may be written as $\langle\Phi, f(\Delta_{\psi,\varphi})\Phi\rangle$ in terms of the self-adjoint operator $f(\Delta_{\psi,\varphi}) := \int_{(0,\infty)} f(t)\, dE_{\psi,\varphi}(t)$ on $s_M(\psi)s_{M'}(\varphi)\mathcal{H}$, although, to be more precise, it should be understood in the sense of a lower-bounded form (see [114]). From the convexity of f, there are $a, b \in \mathbb{R}$ such that $f(t) \geq a + bt$, $t > 0$. Then expression (2.3) is well defined with values in $(-\infty, +\infty]$ since

$$\begin{aligned} \int_{(0,\infty)} f(t)\, d\|E_{\psi,\varphi}(t)\Phi\|^2 &\geq \int_{(0,\infty)} (a + bt)\, d\|E_{\psi,\varphi}(t)\Phi\|^2 \\ &= a\|s_M(\psi)s_{M'}(\varphi)\Phi\|^2 + b\|\Delta_{\psi,\varphi}^{1/2}\Phi\|^2 \\ &= a\|s_M(\psi)\Phi\|^2 + b\|s_M(\varphi)\Psi\|^2 \\ &= a\varphi(s_M(\psi)) + b\psi(s_M(\varphi)) > -\infty \end{aligned}$$

thanks to $J\Delta_{\psi,\varphi}^{1/2}\Phi = s_M(\varphi)\Psi$. By the above computation and (2.3) we also see that

$$S_{a+bt}(\psi\|\varphi) = a\varphi(1) + b\psi(1). \tag{2.4}$$

Examples 2.2 The following are the special cases of abelian M (classical case) and $M = B(\mathcal{H})$.

(1) Let $M = L^\infty(X, \mu)$ be an abelian von Neumann algebra, where $(X, \mathcal{X}, \mu)$ is a σ-finite measure space. For every $\psi, \varphi \in L^1(X, \mu)_+ \cong M_*^+$, since $\Delta_{\psi,\varphi}$ is the multiplication of $1_{\{\varphi>0\}}(\psi/\varphi)$ (see Examples A.14 (1) and A.23 (1)), we have

$$S_f(\psi\|\varphi) = \int_{\{\psi>0\}\cap\{\varphi>0\}} \varphi f(\psi/\varphi)\, d\mu + f(0^+) \int_{\{\psi=0\}} \varphi\, d\mu + f'(\infty) \int_{\{\varphi=0\}} \psi\, d\mu,$$

which equals the classical f-divergence $S_f(\psi|\varphi) = \int_X \psi f(\psi/\varphi)\, d\mu$ under the convention (2.2).

(2) Let $M = B(\mathcal{H})$ be a factor of type I, where $\mathcal{H}$ is an arbitrary Hilbert space. The standard form of $B(\mathcal{H})$ is described in Example A.14 (2). For every $\psi, \varphi \in B(\mathcal{H})_*^+$ the relative modular operators $\Delta_{\psi,\varphi}$ on $C_2(\mathcal{H})$ (the Hilbert space consisting of Hilbert–Schmidt operators) is given in Example A.23 (2) in terms of the spectral decompositions $D_\psi = \sum_{a>0} a P_a$ and $D_\varphi = \sum_{b>0} b Q_b$ of the density operators of ψ, φ. From (A.15) the definition of $S_f(\psi\|\varphi)$ in (2.3) is rewritten as

$$\begin{aligned} &S_f(\psi\|\varphi) \\ &= \sum_{a,b>0} b f(ab^{-1}) \mathrm{Tr}\, P_a Q_b + f(0^+) \mathrm{Tr}(I - D_\psi^0) D_\varphi + f'(\infty) \mathrm{Tr}\, D_\psi (I - D_\varphi^0), \end{aligned}$$

where D_φ^0 is the support projection of D_φ. The above coincides with an expression in [58, Proposition 3.2] when $\dim \mathcal{H} < \infty$.

The next proposition summarizes simple properties of $S_f(\psi\|\varphi)$.

Proposition 2.3 ([54]) *Let* $\psi, \varphi \in M_*^+$.

(1) Invariance: *If* $\alpha : N \to M$ *is a* $*$*-isomorphism between von Neumann algebras, then*

$$S_f(\psi \circ \alpha \| \varphi \circ \alpha) = S_f(\psi\|\varphi).$$

(2) *In the case* $\psi = 0$ *or* $\varphi = 0$ *or* $\psi = \varphi$,

$$S_f(0\|\varphi) = f(0^+)\varphi(1), \quad S_f(\psi\|0) = f'(\infty)\psi(1), \quad S_f(\varphi\|\varphi) = f(1)\varphi(1).$$

(3) Homogeneity: *For every* $\lambda \in [0, \infty)$,

$$S_f(\lambda\psi\|\lambda\varphi) = \lambda S_f(\psi\|\varphi).$$

(4) Additivity: *Let $M = M_1 \oplus M_2$ be the direct sum of von Neumann algebras M_1 and M_2. If $\psi_i, \varphi_i \in (M_i)_*^+$ for $i = 1, 2$, then*

$$S_f(\psi_1 \oplus \psi_2 \| \varphi_1 \oplus \varphi_2) = S_f(\psi_1 \| \varphi_1) + S_f(\psi_2 \| \varphi_2).$$

(5) Transpose: *Let $\widetilde{f}$ be the* transpose *of f defined by*

$$\widetilde{f}(t) := tf(t^{-1}), \qquad t \in (0, \infty).$$

Then $S_f(\psi \| \varphi) = S_{\widetilde{f}}(\varphi \| \psi)$. (Note that $\widetilde{f}(0^+) = f'(\infty)$ and $\widetilde{f}'(\infty) = f(0^+)$.)

Example 2.4 The relative entropy is the most used quantum divergence, which is the standard f-divergence for $f(t) = t \log t$ on $(0, \infty)$ with the transpose $\widetilde{f}(t) = -\log t$. Since $f(0^+) = 0$ and $f'(\infty) = +\infty$, the *relative entropy* $D(\psi \| \varphi)$ of $\psi, \varphi \in M_*^+$ [10, 11] is

$$D(\psi \| \varphi) = S_{t \log t}(\psi \| \varphi) = \begin{cases} \int_{(0,\infty)} t \log t \, d\|E_{\psi,\varphi}(t)\Phi\|^2 & \text{if } s(\psi) \le s(\varphi), \\ +\infty & \text{otherwise,} \end{cases}$$

or $D(\psi \| \varphi) = S_{-\log t}(\varphi \| \psi)$. When $M = B(\mathcal{H})$, $D(\psi \| \varphi)$ is *Umegaki's relative entropy* [134], whose alternative familiar expression is

$$D(\psi \| \varphi) = \begin{cases} \mathrm{Tr}\, D_\psi (\log D_\psi - \log D_\varphi) & \text{if } D_\psi^0 \le D_\varphi^0, \\ +\infty & \text{otherwise,} \end{cases}$$

where D_ψ, D_φ are as in Example 2.2 (2).

2.2 Variational Expression of Standard f-Divergences

We extend the variational expression of the relative entropy given in [79] to standard f-divergences. The extended expression will be quite useful to verify various properties of standard f-divergences. Throughout this section, we assume that a function $f : (0, \infty) \to \mathbb{R}$ is *operator convex*, i.e., the operator inequality

$$f(\lambda A + (1 - \lambda)B) \le \lambda f(A) + (1 - \lambda) f(B), \qquad 0 \le \lambda \le 1$$

holds for every invertible $A, B \in B(\mathcal{H})_+$ of any $\mathcal{H}$. Also, a function $h : (0, \infty) \to \mathbb{R}$ is said to be *operator monotone* if $A \le B \implies h(A) \le h(B)$ for every invertible $A, B \in B(\mathcal{H})_+$ of any $\mathcal{H}$. It is well-known that an operator monotone function h on $(0, \infty)$ is automatically *operator concave* (i.e., $-h$ is operator convex). For general theory on operator monotone and operator convex functions, see, e.g., [21, 51]. Note here [58, Proposition A.1] that if f is operator convex on $(0, \infty)$, then so is the transpose $\widetilde{f}(t) = tf(t^{-1})$.

Recall [87] (also [39, Theorem 5.1] for a more general form) that any operator convex function f on $(0, \infty)$ has the integral expression

$$f(t) = a + b(t-1) + c(t-1)^2 + \int_{[0,\infty)} \frac{(t-1)^2}{t+s}\, d\mu(s), \quad t \in (0, \infty), \tag{2.5}$$

where $a, b, c \in \mathbb{R}$ with $c \geq 0$ and μ is a positive measure on $[0, \infty)$ satisfying $\int_{[0,\infty)} (1+s)^{-1}\, d\mu(s) < +\infty$ (moreover, a, b, c and μ are uniquely determined). Letting $d := \mu(\{0\}) \geq 0$ we also write

$$f(t) = a + b(t-1) + c(t-1)^2 + d\, \frac{(t-1)^2}{t} + \int_{(0,\infty)} \frac{(t-1)^2}{t+s}\, d\mu(s), \quad t \in (0, \infty).$$

One can easily verify that

$$f(0^+) = a - b + c + (+\infty)d + \int_{(0,\infty)} s^{-1}\, d\mu(s),$$

$$f'(\infty) = b + (+\infty)c + d + \int_{(0,\infty)} d\mu(s).$$

For each $n \in \mathbb{N}$ we define

$$f_n(t) := a + b(t-1) + c\, \frac{n(t-1)^2}{t+n} + d\, \frac{(t-1)^2}{t + (1/n)} + \int_{[1/n,n]} \frac{(t-1)^2}{t+s}\, d\mu(s), \quad t \in (0, \infty). \tag{2.6}$$

Then f_n is operator convex on $(0, \infty)$, and we have

$$f_n(0^+) = a - b + c + nd + \int_{[1/n,n]} s^{-1}\, d\mu(s) < +\infty, \tag{2.7}$$

$$f_n'(\infty) = b + nc + d + \int_{[1/n,n]} d\mu(s) < +\infty, \tag{2.8}$$

$$f_n(0^+) \nearrow f(0^+), \quad f_n'(\infty) \nearrow f'(\infty), \quad f_n(t) \nearrow f(t) \tag{2.9}$$

as $n \to \infty$ for all $t \in (0, \infty)$. Moreover, for every $\psi, \varphi \in M_*^+$ we have

$$S_{f_n}(\psi\|\varphi) \nearrow S_f(\psi\|\varphi) \qquad \text{as } n \nearrow \infty.$$

Next, we define

$$h_n(t) := \int_{(0,\infty)} \frac{t(1+s)}{t+s}\, d\nu_n(s), \qquad t \in [0, \infty),$$

where ν_n is a finite positive measure supported on $[1/n, n]$ given by

$$d\nu_n(s) := c(1+n)\delta_n + d(1+n)\delta_{1/n} + 1_{[1/n,n]}(s)\frac{1+s}{s}\, d\mu(s) \tag{2.10}$$

with the point masses δ_n at n and $\delta_{1/n}$ at $1/n$. Then h_n is an operator monotone function on $[0, \infty)$, and f_n is written as

$$f_n(t) = f_n(0^+) + f_n'(\infty)t - h_n(t), \qquad t \in (0, \infty). \tag{2.11}$$

Now, let L be a subspace of M containing 1, and assume that L is dense in M with respect to the strong* operator topology. Our variational expression of $S_f(\psi\|\varphi)$ is given as follows:

Theorem 2.5 ([54]) *Let f be an operator convex function on $(0, \infty)$. For each $n \in \mathbb{N}$ let $f_n(0^+)$, $f_n'(\infty)$ and ν_n be given in* (2.7), (2.8) *and* (2.10), *respectively. Then for every $\psi, \varphi \in M_*^+$,*

$$\begin{aligned} S_f(\psi\|\varphi) = \sup_{n\in\mathbb{N}} \sup_{x(\cdot)} \Bigg[& f_n(0^+)\varphi(1) + f_n'(\infty)\psi(1) \\ & - \int_{[1/n,n]} \{\varphi((1-x(s))^*(1-x(s))) + s^{-1}\psi(x(s)x(s)^*)\}(1+s)\, d\nu_n(s)\Bigg], \end{aligned} \tag{2.12}$$

where the second supremum is taken over all L-valued (finitely many values) step functions $x(\cdot)$ on $(0, \infty)$.

To prove this, note from (2.11) and (2.4) that

$$\begin{aligned} S_{f_n}(\psi\|\varphi) &= f_n(0^+)\varphi(1) + f_n'(\infty)\psi(1) + S_{-h_n}(\psi\|\varphi) \\ &= f_n(0^+)\varphi(1) + f_n'(\infty)\psi(1) - \int_{(0,\infty)} h_n(t)\, d\|E_{\psi,\varphi}(t)\Phi\|^2. \end{aligned}$$

Since $h_n(0) = h_n'(\infty) = 0$, we can apply [79, Theorem 2.2] to the above integral term, so we have

$$S_{f_n}(\psi\|\varphi) = \sup_{x(\cdot)}\Bigg[f_n(0^+)\varphi(1) + f_n'(\infty)\psi(1) - \int_{[1/n,n]}\{\varphi((1-x(s))^*(1-x(s))) + s^{-1}\psi(x(s)x(s)^*)\}(1+s)\,d\nu_n(s)\Bigg].$$

Taking $\sup_n$ of both sides of the above yields (2.12).

Example 2.6 Consider $f(t) = -\log t$, whose integral expression in (2.5) is

$$-\log t = -(t-1) + \int_{(0,\infty)} \frac{(t-1)^2}{(t+s)(1+s)^2}\,ds.$$

Hence, in this case,

$$a = c = d = 0, \quad b = -1, \quad d\mu(s) = \frac{1}{(1+s)^2}\,ds, \quad f(0^+) = +\infty, \quad f'(\infty) = 0.$$

Moreover, compute

$$f_n(0^+) = 1 + \int_{1/n}^{n} \frac{1}{s(1+s)^2}\,ds = \frac{2}{n+1} + \log n,$$

$$f_n'(\infty) = -1 + \int_{1/n}^{n} \frac{1}{(1+s)^2}\,ds = -\frac{2}{n+1},$$

$$d\nu_n(s) = 1_{[1/n,n]}(s)\frac{1}{s(1+s)}\,ds.$$

For every $\psi, \varphi \in M_*^+$ the *relative entropy* $D(\varphi\|\psi)$ is $S_{t\log t}(\varphi\|\psi) = S_{-\log t}(\psi\|\varphi)$ (see Example 2.4), for which one can write expression (2.12) as

$$D(\varphi\|\psi) = \sup_{n\in\mathbb{N}}\sup_{x(\cdot)}\Bigg[\varphi(1)\log n + (\varphi(1) - \psi(1))\frac{2}{n+1} - \int_{[1/n,n]}\{\varphi((1-x(s))^*(1-x(s))) + s^{-1}\psi(x(s)x(s)^*)\}s^{-1}\,ds\Bigg].$$

This expression resembles the variational expression

$$D(\varphi\|\psi) = \sup_{n\in\mathbb{N}} \sup_{x(\cdot)} \Bigg[\varphi(1)\log n - \int_{[1/n,\infty)} \{\varphi((1-x(s))^*(1-x(s))) + s^{-1}\psi(x(s)x(s)^*)\} s^{-1}\, ds \Bigg]$$

in [79, Theorem 3.2].

2.3 Properties of Standard f-Divergences

Most of the important properties of standard f-divergences can immediately be verified from the variational expression in Theorem 2.5, similarly to [79] where the relative entropy was treated. In the rest of this section, assume that f is an operator convex function on $(0, \infty)$.

Theorem 2.7 ([54]) *Let $\psi, \varphi, \psi_i, \varphi_i \in M_*^+$ for $i = 1, 2$.*

(i) Joint lower semicontinuity: *The map $(\psi, \varphi) \in M_*^+ \times M_*^+ \mapsto S_f(\psi\|\varphi) \in (-\infty, +\infty]$ is jointly lower semicontinuous in the $\sigma(M_*, M)$-topology.*

(ii) Joint convexity: *The map in (i) is jointly convex and jointly subadditive, i.e., for every $\psi_i, \varphi_i \in M_*^+$, $1 \le i \le k$,*

$$S_f\Bigg(\sum_{i=1}^k \psi_i \Bigg\| \sum_{i=1}^k \varphi_i\Bigg) \le \sum_{i=1}^k S_f(\psi_i\|\varphi_i).$$

(iii) *If $f(0^+) \le 0$ and $\varphi_1 \le \varphi_2$, then $S_f(\psi\|\varphi_1) \ge S_f(\psi\|\varphi_2)$. Also, if $f'(\infty) \le 0$ and $\psi_1 \le \psi_2$, then $S_f(\psi_1\|\varphi) \ge S_f(\psi_2\|\varphi)$.*

(iv) Monotonicity: *Let N be another von Neumann algebra and $\gamma : N \to M$ be a unital positive linear map that is normal (i.e., if $\{x_\alpha\}$ is an increasing net in M_+ with $x_\alpha \nearrow x \in M_+$, then $\gamma(x_\alpha) \nearrow \gamma(x)$) and is a Schwarz map (i.e., $\gamma(x^*x) \ge \gamma(x)^*\gamma(x)$ for all $x \in N$). Then*

$$S_f(\psi\circ\gamma\|\varphi\circ\gamma) \le S_f(\psi\|\varphi). \tag{2.13}$$

In particular, if N is a unital von Neumann subalgebra of M, then

$$S_f(\psi|_N\|\varphi|_N) \le S_f(\psi\|\varphi). \tag{2.14}$$

(v) Peierls–Bogolieubov inequality:

$$S_f(\psi\|\varphi) \ge \varphi(1) f(\psi(1)/\varphi(1)). \tag{2.15}$$

Assume that f is non-linear and $\psi, \varphi \neq 0$. Then equality holds in (2.15) if and only if $\psi = (\psi(1)/\varphi(1))\varphi$.

(vi) Strict positivity: *Assume that f is non-linear with $f(1) = 0$ and $\psi(1) = \varphi(1) > 0$. Then $S_f(\psi\|\varphi) \geq 0$, and $S_f(\psi\|\varphi) = 0$ if and only if $\psi = \varphi$.*

(vii) Martingale convergence: *If $\{M_\alpha\}$ is an increasing net of unital von Neumann subalgebras of M such that $\left(\bigcup_\alpha M_\alpha\right)'' = M$, then*

$$S_f(\psi|_{M_\alpha}\|\varphi|_{M_\alpha}) \nearrow S_f(\psi\|\varphi).$$

For each $\varphi \in M_*^+$ and any projection $e \in M$, we write $e\varphi e$ for the restriction of φ to the reduced von Neumann algebra eMe. The next corollary is easily seen from properties in Theorem 2.7.

Corollary 2.8

(1) *If $e \in M$ is a projection such that $s_M(\psi), s_M(\varphi) \leq e$, then*

$$S_f(\psi\|\varphi) = S_f(e\psi e\|e\varphi e).$$

(2) *If $\psi_i, \varphi_i \in M_*^+$, $i = 1, 2$, and $s_M(\psi_1) \vee s_M(\varphi_1) \perp s_M(\psi_2) \vee s_M(\varphi_2)$, then*

$$S_f(\psi_1 + \psi_2\|\varphi_1 + \varphi_2) = S_f(\psi_1\|\varphi_1) + S_f(\psi_2\|\varphi_2).$$

(This is a stronger version of Proposition 2.3 (4).)

(3) *If $\omega_1, \omega_2 \in M_*^+$ and $S_f(\omega_1\|\omega_2) < +\infty$, then for every $\psi, \varphi \in M_*^+$,*

$$S_f(\psi\|\varphi) = \lim_{\varepsilon \searrow 0} S_f(\psi + \varepsilon\omega_1\|\varphi + \varepsilon\omega_2).$$

In particular, for every $\psi, \varphi, \omega \in M_*^+$,

$$S_f(\psi\|\varphi) = \lim_{\varepsilon \searrow 0} S_f(\psi + \varepsilon\omega\|\varphi + \varepsilon\omega). \tag{2.16}$$

The relative entropy $D(\psi\|\varphi)$ has the scaling property

$$D(\lambda\psi\|\mu\varphi) = \lambda D(\psi\|\varphi) + \lambda\psi(1)\log\frac{\lambda}{\mu}, \qquad \lambda, \mu > 0,$$

from which it is clear that if $D(\psi\|\varphi) < +\infty$ then $D(\lambda\psi\|\mu\varphi) < +\infty$ for all $\lambda, \mu > 0$. The next proposition says that this holds for any standard f-divergence, which will be used in the proof of Theorem 7.11 of Sect. 7.2.

Proposition 2.9 *Let $\psi, \varphi \in M_*^+$. If $S_f(\psi\|\varphi) < +\infty$, then $S_f(\lambda\psi\|\mu\varphi) < +\infty$ and $S_f(\lambda\psi + \mu\varphi\|\varphi) < +\infty$ for all $\lambda, \mu > 0$.*

Proof Since $S_f(\lambda\psi\|\lambda\mu) = \lambda S_f(\psi\|\varphi)$, we may assume $\mu = 1$ for the first assertion. Since $\Delta_{\lambda\psi,\varphi} = \lambda\Delta_{\psi,\varphi}$, we have

$$\Delta_{\lambda\psi,\varphi} = \int_{[0,\infty)} \lambda t\, dE_{\psi,\varphi}(t) = \int_{[0,\infty)} t\, dE_{\psi,\varphi}(\lambda^{-1}t).$$

Note that

$$\begin{aligned} S_f(\psi\|\varphi) &= \int_{(0,\infty)} f(t)\, d\|E_{\psi,\varphi}(t)\Phi\|^2 \\ &\quad + f(0^+)\varphi(1 - s(\psi)) + f'(+\infty)\psi(1 - s(\varphi)) < +\infty, \\ S_f(\lambda\psi\|\varphi) &= \int_{(0,\infty)} f(t)\, d\|E_{\psi,\varphi}(\lambda^{-1}t)\Phi\|^2 \\ &\quad + f(0^+)\varphi(1 - s(\psi)) + f'(+\infty)\lambda\psi(1 - s(\varphi)) \\ &= \int_{(0,\infty)} f(\lambda t)\, d\|E_{\psi,\varphi}(t)\Phi\|^2 \\ &\quad + f(0^+)\varphi(1 - s(\psi)) + f'(+\infty)\lambda\psi(1 - s(\varphi)), \end{aligned}$$

so it suffices to prove that

$$\int_{(0,\infty)} f(\lambda t)\, d\|E_{\psi,\varphi}(t)\Phi\|^2 < +\infty. \tag{2.17}$$

One can write $f(t) = f_0(t) + at + b$ for $t > 0$ with an operator convex function f_0 on $(0, \infty)$ such that $f_0(1) = 0$ and $f_0(t) \ge 0$ for all $t > 0$. So we may assume that $f(1) = 0$ and $f(t) \ge 0$ for all $t > 0$.

Assume that $\lambda > 1$. Note that $g(t) := \frac{f(t)-f(1)}{t-1}$ (where $g(1) = f'(1) = 0$) is operator monotone and so operator concave on $(0, \infty)$, see [51, Corollaries 2.7.8 and 2.5.4]. Hence for every $t > 1$ one has $\frac{g(t)}{t-1} \ge \frac{g(\lambda t)}{\lambda t-1}$ so that

$$f(\lambda t) \le \left(\frac{\lambda t - 1}{t - 1}\right)^2 f(t), \qquad t > 1.$$

Since $\left(\frac{\lambda t-1}{t-1}\right)^2 \le 2\lambda^2$ for all sufficiently large t, one can choose a $c > 0$ such that

$$f(\lambda t) \le 2\lambda^2 f(t) + c, \qquad t \ge 1,$$

and so

$$\int_{[1,\infty)} f(\lambda t)\, d\|E_{\psi,\varphi}(t)\Phi\|^2 \le \int \bigl\{2\lambda^2 f(t) + c\bigr\}\, d\|E_{\psi,\varphi}(t)\Phi\|^2 < +\infty.$$

Moreover, for every $t \in (0, \lambda^{-1})$ one has $\frac{f(t)}{t-1} \le \frac{f(\lambda t)}{\lambda t - 1}$ so that

$$f(\lambda t) \le \frac{1-\lambda t}{1-t} f(t) \le f(t), \qquad 0 < t < \lambda^{-1}.$$

Hence one can choose a $d > 0$ such that

$$f(\lambda t) \le f(t) + d, \qquad 0 < t \le 1,$$

and so

$$\int_{(0,1]} f(\lambda t)\, d\|E_{\psi,\varphi}(t)\Phi\|^2 \le \int_{(0,1]} \{f(t) + d\}\, d\|E_{\psi,\varphi}(t)\Phi\|^2 < +\infty.$$

Therefore, (2.17) follows.

Next, assume that $0 < \lambda < 1$. One can apply the above argument to the transpose $\widetilde{f}(t) := t f(t^{-1})$ and $\lambda^{-1} > 1$. Hence one can choose $c', d' > 0$ such that

$$\widetilde{f}(\lambda^{-1} t) \le 2\lambda^{-2} \widetilde{f}(t) + c', \qquad t \ge 1,$$
$$\widetilde{f}(\lambda^{-1} t) \le \tilde{f}(t) + d', \qquad 0 < t \le 1.$$

By replacing t with t^{-1} these are rewritten as

$$f(\lambda t) \le 2\lambda^{-1} f(t) + c'\lambda t, \qquad 0 < t \le 1,$$
$$f(\lambda t) \le \lambda f(t) + d'\lambda t, \qquad t \ge 1,$$

from which (2.17) follows as in the above case where $\lambda > 1$.

For the second assertion we may assume by the first that $\lambda + \mu = 1$. Then Theorem 2.7 (ii) gives $S_f(\lambda\psi + \mu\varphi\|\varphi) \le \lambda S_f(\psi\|\varphi) + \mu S_f(\varphi\|\varphi) < +\infty$.

The next continuity property is not included in the martingale convergence in Theorem 2.7, because eMe is not a unital von Neumann subalgebra of M.

Proposition 2.10 ([54]) *Let $\{e_\alpha\}$ be an increasing net of projections in M such that $e_\alpha \nearrow 1$. Then for every $\psi, \varphi \in M_*^+$,*

$$\lim_\alpha S_f(e_\alpha \psi e_\alpha \| e_\alpha \varphi e_\alpha) = S_f(\psi\|\varphi).$$

When $f \ge 0$ in Proposition 2.10, from the monotonicity of S_f we see that $S_f(e_\alpha \psi e_\alpha \| e_\alpha \varphi e_\alpha)$ is increasing as $e_\alpha \nearrow 1$. But this is not the case unless $f \ge 0$.

Chapter 3
Rényi Divergences and Sandwiched Rényi Divergences

3.1 Rényi Divergences

Let M be a general von Neumann algebra given in a standard form $(M, \mathcal{H}, J, \mathcal{P})$ as in Chap. 2. The notion of Rényi divergences $D_\alpha(\psi\|\varphi)$ for $\psi, \varphi \in M_*^+$ where $\alpha \in [0, \infty) \setminus \{1\}$ is defined as follows:

Definition 3.1 Let $\psi, \varphi \in M_*^+$. Since the vector representative Φ of φ is in $\mathcal{D}(\Delta_{\psi,\varphi}^{1/2})$, the domain of $\Delta_{\psi,\varphi}^{1/2}$, note that $\Phi \in \mathcal{D}(\Delta_{\psi,\varphi}^{\alpha/2})$ for any $\alpha \in [0, 1]$. Define the quantities $Q_\alpha(\psi\|\varphi)$ for $\alpha \geq 0$ as follows: When $0 \leq \alpha < 1$,

$$Q_\alpha(\psi\|\varphi) := \|\Delta_{\psi,\varphi}^{\alpha/2}\Phi\|^2 \in [0, \infty), \tag{3.1}$$

and when $\alpha > 1$,

$$Q_\alpha(\psi\|\varphi) := \begin{cases} \|\Delta_{\psi,\varphi}^{\alpha/2}\Phi\|^2 & \text{if } s_M(\psi) \leq s_M(\varphi) \text{ and } \Phi \in \mathcal{D}(\Delta_{\psi,\varphi}^{\alpha/2}), \\ +\infty & \text{otherwise.} \end{cases} \tag{3.2}$$

Moreover, when $\alpha = 1$, define $Q_1(\psi\|\varphi) := \psi(1)$.

Assume that $\psi \neq 0$. For each $\alpha \in [0, \infty) \setminus \{1\}$ the *α-Rényi divergence* (or *α-Rényi relative entropy*) $D_\alpha(\psi\|\varphi)$ is defined by

$$D_\alpha(\psi\|\varphi) := \frac{1}{\alpha - 1} \log \frac{Q_\alpha(\psi\|\varphi)}{\psi(1)}. \tag{3.3}$$

In particular, $Q_0(\alpha\|\varphi) = \varphi(s_M(\psi))$ and $D_0(\psi\|\varphi) = -\log\left[\varphi(s_M(\psi))/\psi(1)\right]$.

F. Hiai, *Quantum f-Divergences in von Neumann Algebras*,
Mathematical Physics Studies, https://doi.org/10.1007/978-981-33-4199-9_3

Define convex functions f_α on $[0, \infty)$ by

$$f_\alpha(t) := \begin{cases} t^\alpha & \text{if } \alpha \ge 1, \\ -t^\alpha & \text{if } 0 < \alpha < 1. \end{cases}$$

Then for every $\psi, \varphi \in M_*^+$, $Q_\alpha(\psi\|\varphi)$ is given as

$$Q_\alpha(\psi\|\varphi) = \begin{cases} S_{f_\alpha}(\psi\|\varphi) & \text{if } \alpha \ge 1, \\ -S_{f_\alpha}(\psi\|\varphi) & \text{if } 0 < \alpha < 1. \end{cases}$$

Indeed, when $0 < \alpha < 1$, since $f_\alpha(0) = f'_\alpha(\infty) = 0$, by (3.1),

$$S_{f_\alpha}(\psi\|\varphi) = \int_{(0,\infty)} (-t^\alpha)\, d\|E_{\psi,\varphi}(t)\Phi\|^2 = -Q_\alpha(\psi\|\varphi).$$

When $\alpha = 1$, by (2.4), $S_{f_1}(\psi\|\varphi) = \psi(1) = Q_1(\psi\|\varphi)$. When $\alpha > 1$, since $f_\alpha(0) = 0$ and $f'_\alpha(\infty) = +\infty$, by (3.2),

$$S_{f_\alpha}(\psi\|\varphi) = \int_{(0,\infty)} t^\alpha\, d\|E_{\psi,\varphi}(t)\Phi\|^2 + (+\infty)\psi(1 - s_M(\varphi)) = Q_\alpha(\psi\|\varphi).$$

Therefore, $Q_\alpha(\psi\|\varphi)$ is essentially a standard f-divergence and so $D_\alpha(\psi\|\varphi)$ is a variant of standard f-divergences.

Properties of Q_α and D_α are summarized in the next theorem, some of which are also found in, e.g., [20, 52, 88, 94, 95] though mostly in the finite-dimensional situation. Many of them are shown based on Theorem 2.7 applied to S_{f_α}.

Theorem 3.2 ([54]) *Let $\psi, \varphi \in M_*^+$ with $\psi \ne 0$.*

(1) *Unless $s_M(\psi) \perp s_M(\varphi)$, $Q_\alpha(\psi\|\varphi) > 0$ for all $\alpha \ge 0$ and the function $\alpha \in [0, \infty) \mapsto \log Q_\alpha(\psi\|\varphi)$ is convex.*
(2) Limit values: *The limit $D_1(\psi\|\varphi) := \lim_{\alpha\nearrow 1} D_\alpha(\psi\|\varphi)$ exists and*

$$D_1(\psi\|\varphi) = \frac{D(\psi\|\varphi)}{\psi(1)},$$

where $D(\psi\|\varphi)$ is the relative entropy. If $D_\alpha(\psi\|\varphi) < +\infty$ for some $\alpha > 1$, then $\lim_{\alpha\searrow 1} D_\alpha(\psi\|\varphi) = D_1(\psi\|\varphi)$. Moreover, if $\psi(1) = \varphi(1)$, then $\lim_{\alpha\searrow 0} \frac{1}{\alpha} D_\alpha(\psi\|\varphi) = D_1(\varphi\|\psi)$.
(3) *The function $\alpha \in [0, \infty) \mapsto D_\alpha(\psi\|\varphi)$ is monotone increasing.*
(4) Joint lower semicontinuity: *For every $\alpha \in [0, 2]$, the map*

$$(\psi, \varphi) \in (M_*^+ \setminus \{0\}) \times M_*^+ \longmapsto D_\alpha(\psi\|\varphi) \in (-\infty, +\infty]$$

is jointly lower semicontinuous in the $\sigma(M_*, M)$-topology.

(5) *The map* $(\psi,\varphi)\in M_*^+\times M_*^+\mapsto Q_\alpha(\psi\|\varphi)$ *is jointly concave and jointly superadditive for* $0\le\alpha\le 1$, *and jointly convex and jointly subadditive for* $1\le\alpha\le 2$. *Hence, when* $0\le\alpha\le 1$, $D_\alpha(\psi\|\varphi)$ *is jointly convex on* $\{(\psi,\varphi)\in M_*^+\times M_*^+:\psi(1)=c\}$ *for any fixed* $c>0$.

(6) *When* $0\le\alpha\le 2$, *the map* $\varphi\in M_*^+\mapsto D_\alpha(\psi\|\varphi)$ *is convex for any fixed* $\psi\in M_*^+$ *with* $\psi\ne 0$.

(7) *Let* $\psi_i,\varphi_i\in M_*^+$ *for* $i=1,2$. *If* $0\le\alpha<1$, $\psi_1\le\psi_2$ *and* $\varphi_1\le\varphi_2$, *then* $Q_\alpha(\psi_1\|\varphi_1)\le Q_\alpha(\psi_2\|\varphi_2)$. *If* $1\le\alpha\le 2$ *and* $\varphi_1\le\varphi_2$, *then* $Q_\alpha(\psi\|\varphi_1)\ge Q_\alpha(\psi\|\varphi_2)$. *Hence, if* $\varphi_1\le\varphi_2$, *then* $D_\alpha(\psi\|\varphi_1)\ge D_\alpha(\psi\|\varphi_2)$ *for all* $\alpha\in[0,2]$.

(8) Monotonicity: *For each* $\alpha\in[0,2]$, $D_\alpha(\psi\|\varphi)$ *is monotone under unital normal Schwarz maps, i.e.,*

$$D_\alpha(\psi\circ\gamma\|\varphi\circ\gamma)\le D_\alpha(\psi\|\varphi)$$

for any unital normal Schwarz map $\gamma:N\to M$ *as in Theorem 2.7 (iv).*

(9) Strict positivity: *Let* $\alpha\in(0,\infty)$ *and* $\psi,\varphi\ne 0$. *The inequality*

$$D_\alpha(\psi\|\varphi)\ge\log\frac{\psi(1)}{\varphi(1)} \tag{3.4}$$

holds, and equality holds in (3.4) *if and only if* $\psi=(\psi(1)/\varphi(1))\varphi$. *If* $\psi(1)=\varphi(1)$, *then* $D_\alpha(\psi\|\varphi)\ge 0$, *and* $D_\alpha(\psi\|\varphi)=0$ *if and only if* $\psi=\varphi$.

Proof We prove (6) only since all other items are from [54, Proposition 5.3]. In view of (5) we may assume that $1<\alpha\le 2$, and let $\psi\ (\ne 0),\varphi_1,\varphi_2\in M_*^+$ with respective vector representatives Ψ,Φ_1,Φ_2. Let $0<\lambda<1$ and $\varphi:=\lambda\varphi_1+(1-\lambda)\varphi_2$. What we need to prove is

$$Q_\alpha(\psi\|\varphi)\le Q_\alpha(\psi\|\varphi_1)^\lambda Q_\alpha(\psi\|\varphi_2)^{1-\lambda}. \tag{3.5}$$

Since $Q_\alpha(\psi\|\varphi_i)>0$ $(i=1,2)$ from (1) and (3.2), we may assume that $Q_\alpha(\psi\|\varphi_i)<+\infty$ for $i=1,2$, so $s(\psi)\le s(\varphi_i)$ and $\Phi_i\in\mathcal{D}(\Delta_{\psi,\varphi_i}^{\alpha/2})$. By Proposition A.22 of Sect. A.4 note that $s(\Delta_{\psi,\varphi_i})=s(\psi)Js(\varphi_i)J$ and $\Delta_{\varphi_i,\psi}^{-1}=J\Delta_{\psi,\varphi_i}J$, where $\Delta_{\varphi_i,\psi}^{-1}$ is the inverse with restriction to the support. Hence $s(\Delta_{\psi,\varphi_i})\ge s(\psi)Js(\psi)J$ so that $\Psi\in s(\Delta_{\psi,\varphi_i})\mathcal{H}$. Since $\Delta_{\psi,\varphi_i}^{1/2}\Phi_i=\Psi$, it follows from Proposition B.2 that $\Psi\in\mathcal{D}(\Delta_{\psi,\varphi_i}^{(\alpha-1)/2})$. Hence we find that $\Psi\in\mathcal{D}(\Delta_{\varphi_i,\psi}^{(1-\alpha)/2})$ (see Proposition B.2 again) and

$$Q_\alpha(\psi\|\varphi_i)=\|\Delta_{\psi,\varphi_i}^{\alpha/2}\Phi_i\|^2=\|\Delta_{\psi,\varphi_i}^{(\alpha-1)/2}\Psi\|^2=\|\Delta_{\varphi_i,\psi}^{(1-\alpha)/2}\Psi\|^2,\qquad i=1,2.$$

Since Proposition A.22 (5) gives

$$\Delta_{\varphi,\psi}=\Delta_{\lambda\varphi_1,\psi}\dot{+}\Delta_{(1-\lambda)\varphi_2,\psi}=\lambda\Delta_{\varphi_1,\psi}\dot{+}(1-\lambda)\Delta_{\varphi_2,\psi},$$

it follows from Proposition B.10 that

$$\|\Delta_{\varphi,\psi}^{(1-\alpha)/2}\Psi\|^2 \le \|\Delta_{\varphi_1,\psi}^{(1-\alpha)/2}\Psi\|^{2\lambda}\|\Delta_{\varphi_2,\psi}^{(1-\alpha)/2}\Psi\|^{2(1-\lambda)} = Q_\alpha(\psi\|\varphi_1)^\lambda Q_\alpha(\psi\|\varphi_2)^{1-\lambda}.$$

As above, note that $\Delta_{\varphi,\psi}^{-1} = J\Delta_{\psi,\varphi}J$ and $\Psi \in s(\Delta_{\psi,\varphi})\mathcal{H}$ (thanks to $s(\psi) \le s(\varphi_i) \le s(\varphi)$). Therefore, we have $\|\Delta_{\psi,\varphi}^{(\alpha-1)/2}\Psi\| = \|\Delta_{\varphi,\psi}^{(1-\alpha)/2}\Psi\| < +\infty$. Since $\Psi = \Delta_{\psi,\varphi}^{1/2}\Phi$, this implies by Proposition B.2 again that $\Phi \in \mathcal{D}(\Delta_{\psi,\varphi}^{\alpha/2})$ and (3.5) follows.

Remarks 3.3

(1) In Theorem 3.2 (2) the assumption that $D_\alpha(\psi\|\varphi) < +\infty$ for some $\alpha > 1$ is essential to have $\lim_{\alpha\searrow 1} D_\alpha(\psi\|\varphi) = D_1(\psi\|\varphi)$ in the infinite-dimensional case.
(2) Theorem 3.2 (6) settles a question raised in [54, Remark 5.4 (3)].
(3) The convexity of Q_α for $1 \le \alpha \le 2$ in Theorem 3.2 (5) cannot extend to $\alpha > 2$ even in the finite-dimensional case and in separate arguments. This implies that the monotonicity property of D_α in (8) fails to hold for $\alpha > 2$, because the monotonicity of Q_α under unital CP maps yields its joint convexity, see Sect. 6.1 for the definition of CP (completely positive) maps. Also, the monotone decreasing of $\varphi \mapsto Q_\alpha(\psi\|\varphi)$ for $1 \le \alpha \le 2$ in (7) cannot extend to $\alpha > 2$.

3.2 Description of Rényi Divergences

When $M = B(\mathcal{H})$ on a finite-dimensional Hilbert space $\mathcal{H}$, the main term Q_α of the Rényi divergence D_α in (3.3) has a simple expression

$$Q_\alpha(\psi\|\varphi) = \mathrm{Tr}\big(D_\psi^\alpha D_\varphi^{1-\alpha}\big) \tag{3.6}$$

in terms of the density operators D_ψ, D_φ, where $D_\varphi^{1-\alpha}$ is defined with restriction to the support of φ. In this section we give a similar expression of Q_α in the general von Neumann algebra case.

It is convenient for us to work in the framework of *Haagerup's L^p-spaces* $L^p(M)$, which are constructed inside the space $\widetilde{N}$ of τ-measurable operators affiliated with the crossed product $N := M \rtimes_\sigma \mathbb{R}$ of M by the modular automorphism group σ_t ($t \in \mathbb{R}$). Here note that N is a semifinite von Neumann algebra with the canonical trace. See the beginning of Sect. A.5 for operators affiliated with N and Sect. A.6 for Haagerup's L^p-spaces. Also, recall that $M_* \cong L^1(M)$ by an order-isomorphic linear bijection $\varphi \in M_* \mapsto h_\varphi \in L^1(M)$, the functional tr on $L^1(M)$ is defined by $\mathrm{tr}(h_\varphi) = \varphi(1)$ for $\varphi \in M_*$, and the standard form of M is given as

$$\left(M, L^2(M), J = {}^*, L^2(M)_+\right),$$

where $x \in M$ (resp., $JxJ \in M'$) acts as the left (resp., the right) multiplication by x on $L^2(M)$. In particular, for any $\varphi \in M_*^+$ the support projection $s(\varphi) = s_M(\varphi)$ is equal to $s(h_\varphi)$ and it acts as the left multiplication on $L^2(M)$, while the M'-support projection $s_{M'}(\varphi) = Js(\varphi)J$ does so as the right multiplication by $s(\varphi)$.

Let $\psi, \varphi \in M_*^+$ with corresponding $h_\psi, h_\varphi \in L^1(M)_+$. For each $p \geq 0$ define

$$\mathcal{D}_p(\psi, \varphi) := \left\{\xi \in L^2(M) : h_\psi^p \xi s(\varphi) = \eta h_\varphi^p \text{ for some } \eta \in L^2(M)s(\varphi)\right\}. \tag{3.7}$$

We prove that the domain $\mathcal{D}(\Delta_{\psi,\varphi}^p)$ coincides with $\mathcal{D}_p(\psi, \varphi)$ for any $p \geq 0$. The next lemma was first observed by Kosaki [75] (see also [54, Lemma A.3]).

Lemma 3.4 *For every $\psi, \varphi \in M_*^+$ and $0 \leq p \leq 1/2$,*

$$\mathcal{D}(\Delta_{\psi,\varphi}^p) = \mathcal{D}_p(\psi, \varphi).$$

Moreover, if $\xi, \eta \in L^2(M)$ are given as in the RHS of (3.7)*, then*

$$\Delta_{\psi,\varphi}^p \xi = \Delta_{\psi,\varphi}^p (\xi s(\varphi)) = \eta. \tag{3.8}$$

The next extension to the whole $p \geq 0$ is due to Jenčová.[1]

Lemma 3.5 *The results of Lemma 3.4 hold for all $p \geq 0$.*

Proof First, note that $\eta \in L^2(M)s(\varphi)$ in the RHS of (3.7) is uniquely determined for each $\xi \in \mathcal{D}_p(\psi, \varphi)$. Also, recall that $s(\Delta_{\psi,\varphi}) = s(\psi)Js(\varphi)J$ by Proposition A.22 (1), so the first equality in (3.8) is clear. Define

$$\mathcal{D}_\infty(\psi, \varphi) := \left\{\xi \in L^2(M) : t \in \mathbb{R} \mapsto \Delta_{\psi,\varphi}^{it}\xi \in L^2(M) \text{ extends to an entire function}\right\}.$$

By a familiar regularization technique with Gaussian kernels (which will be also used in the last part of the proof here), it is seen that $\mathcal{D}_\infty(\psi, \varphi)$ is dense in $L^2(M)$ and is a core of $\Delta_{\psi,\varphi}^p$ for any $p \geq 0$. Let $\xi \in \mathcal{D}_\infty(\psi, \varphi)$, and show that for every $p \geq 0$, $\xi \in \mathcal{D}_p(\psi, \varphi)$ and

$$(\Delta_{\psi,\varphi}^p \xi) h_\varphi^p = h_\psi^p \xi s(\varphi). \tag{3.9}$$

[1] This is a private communication "A remark on the relative modular operator" (2017, September). The author is grateful to A. Jenčová for permitting her proof to be included here.

When $0 \le p \le 1/2$, (3.9) holds by Lemma 3.4. For every $p > 1/2$ write $p = k/2 + p_0$ with $k \in \mathbb{N}$ and $p_0 \in [0, 1/2)$. Using Lemma 3.4 and Proposition B.2 repeatedly, one has

$$\begin{aligned}(\Delta_{\psi,\varphi}^p \xi)h_\varphi^p &= (\Delta_{\psi,\varphi}^{1/2}(\Delta_{\psi,\varphi}^{p-1/2}\xi))h_\varphi^{1/2}h_\varphi^{p-1/2} = h_\psi^{1/2}(\Delta_{\psi,\varphi}^{p-1/2}\xi)h_\varphi^{p-1/2} \\ &= h_\psi^{1/2}(\Delta_{\psi,\varphi}^{1/2}(\Delta_{\psi,\varphi}^{p-1}\xi))h_\varphi^{1/2}h_\varphi^{p-1} = h_\psi(\Delta_{\psi,\varphi}^{p-1}\xi)h_\varphi^{p-1} \\ &= \cdots = h_\psi^{k/2}(\Delta_{\psi,\varphi}^{p_0}\xi)h_\varphi^{p_0} = h_\psi^p \xi s(\varphi).\end{aligned}$$

Therefore, (3.9) has been shown for any $p \ge 0$, which implies that $\mathcal{D}_\infty(\psi, \varphi) \subset \mathcal{D}_p(\psi, \varphi)$ and (3.8) hold.

Now, let $p > 1/2$ and let T_p be the operator with the domain $\mathcal{D}_p(\psi, \varphi)$ defined by $T_p\xi := \eta$, which is clearly a linear operator on $L^2(M)$. By the previous paragraph note that $T_p\xi = \Delta_{\psi,\varphi}^p\xi$ for all $\xi \in \mathcal{D}_\infty(\psi, \varphi)$. We show that T_p is a closed operator with a core $\mathcal{D}_\infty(\psi, \varphi)$. Then the result follows since $\mathcal{D}_\infty(\psi, \varphi)$ is also a core of $\Delta_{\psi,\varphi}^p$.

So let $\{\xi_n\}$ be a sequence in $\mathcal{D}_p(\psi, \varphi)$ such that $\xi_n \to \xi$ and $T_p\xi_n \to \eta$ in $L^2(M)$. Then by [129, Chap. II, Proposition 26], $\xi_n \to \xi$ and $T_p\xi_n \to \eta$ in $\widetilde{N}$. Since $\widetilde{N}$ is a complete Hausdorff topological $*$-algebra with the measure topology (see [129, Chap. I, Theorem 28]), one has

$$h_\psi^p \xi s(\varphi) = \lim_n h_\psi^p \xi_n s(\varphi) = \lim_n (T_p\xi_n)h_\varphi^p = \eta h_\varphi^p$$

in $\widetilde{N}$, so that $\xi \in \mathcal{D}_p(\psi, \varphi)$ and $T_p\xi = \eta$. Hence T_p is closed. To show that $\mathcal{D}_\infty(\psi, \varphi)$ is a core of T_p, let $\xi \in \mathcal{D}_p(\psi, \varphi)$ and $\eta = T_p\xi$. Here, since $s(h_\psi) = s(\psi)$, one can assume that $\xi = s(\psi)\xi$. For $n \in \mathbb{N}$ set

$$\xi_n := \sqrt{\frac{n}{\pi}} \int_{-\infty}^{\infty} e^{-nt^2} \Delta_{\psi,\varphi}^{it}\xi \, dt + \xi(1 - s(\varphi)),$$

$$\eta_n := \sqrt{\frac{n}{\pi}} \int_{-\infty}^{\infty} e^{-nt^2} \Delta_{\psi,\varphi}^{it}\eta \, dt.$$

Then $\xi_n \in \mathcal{D}_\infty(\psi, \varphi)$ and $\xi_n \to \xi$, $\eta_n \to \eta$ in $L^2(M)$. From the formula (A.18) of Sect. A.6 it follows that

$$h_\psi^p \Delta_{\psi,\varphi}^{it}\xi = h_\psi^{it} h_\psi^p \xi h_\varphi^{-it} = h_\psi^{it} \eta h_\varphi^p h_\varphi^{-it} = (\Delta_{\psi,\varphi}^{it}\eta)h_\varphi^p, \qquad t \in \mathbb{R}. \tag{3.10}$$

For each n, to see that $T_p\xi_n = \eta_n$, one can take sequences $\{\xi_{n,k}\}_{k=1}^{\infty}$, $\{\eta_{n,k}\}_{k=1}^{\infty}$ of Riemann sums

$$\xi_{n,k} := \sqrt{\frac{n}{\pi}} \sum_{l=1}^{m_k} \big(t_l^{(k)} - t_{l-1}^{(k)}\big) e^{-n\left(t_l^{(k)}\right)^2} \Delta_{\psi,\varphi}^{it_l^{(k)}} \xi + \xi(1-s(\varphi)),$$

$$\eta_{n,k} := \sqrt{\frac{n}{\pi}} \sum_{l=1}^{m_k} \big(t_l^{(k)} - t_{l-1}^{(k)}\big) e^{-n\left(t_l^{(k)}\right)^2} \Delta_{\psi,\varphi}^{it_l^{(k)}} \eta,$$

with $-\infty < t_0^{(k)} < t_1^{(k)} < \cdots < t_{m_k}^{(k)} < \infty$, such that $\|\xi_{n,k} - \xi_n\|_2 \to 0$ and $\|\eta_{n,k} - \eta_n\|_2 \to 0$ as $k \to \infty$. Then it follows from (3.10) that

$$h_\psi^p \xi_n s(\varphi) = \lim_k h_\psi^p \xi_{n,k} s(\varphi) = \lim_k \eta_{n,k} h_\varphi^p = \eta_n h_\varphi^p$$

in $\widetilde{N}$, so that $T_p\xi_n = \eta_n$ for all n. Hence $\mathcal{D}_\infty(\psi,\varphi)$ is a core of T_p, as desired.

The following theorem provides an explicit description of Q_α (hence the Rényi divergence $D_\alpha(\psi\|\varphi)$) in terms of $h_\psi, h_\varphi \in L^1(M)_+$.

Theorem 3.6 *Let $\psi, \varphi \in M_*^+$.*

(1) *When $0 \le \alpha < 1$,*

$$Q_\alpha(\psi\|\varphi) = \operatorname{tr}(h_\psi^\alpha h_\varphi^{1-\alpha}). \tag{3.11}$$

(2) *When $s(\psi) \le s(\varphi)$ and $\alpha > 1$, the following conditions are equivalent:*

  (i) $h_\varphi^{1/2} \in \mathcal{D}(\Delta_{\psi,\varphi}^{\alpha/2})$;

  (ii) $h_\psi^{1/2} \in \mathcal{D}(\Delta_{\psi,\varphi}^{(\alpha-1)/2})$;

  (iii) *there exists an $\eta \in L^2(M)s(\varphi)$ such that $h_\psi^{\alpha/2} = \eta h_\varphi^{(\alpha-1)/2}$.*

  If the above conditions hold, then η in (iii) is unique and $Q_\alpha(\psi\|\varphi) = \|\eta\|_2^2$.

Proof

(1) For $0 < \alpha < 1$, since $h_\psi^{\alpha/2} h_\varphi^{1/2} = (h_\psi^{\alpha/2} h_\varphi^{(1-\alpha)/2}) h_\varphi^{\alpha/2}$, we have by (3.8)

$$\begin{aligned} Q_\alpha(\psi\|\varphi) &= \|\Delta_{\psi,\varphi}^{\alpha/2} h_\varphi^{1/2}\|^2 = \|h_\psi^{\alpha/2} h_\varphi^{(1-\alpha)/2}\|^2 \\ &= \operatorname{tr}(h_\varphi^{(1-\alpha)/2} h_\psi^\alpha h_\varphi^{(1-\alpha)/2}) = \operatorname{tr}(h_\psi^\alpha h_\varphi^{1-\alpha}). \end{aligned}$$

(2) Assume that $s(\psi) \le s(\varphi)$ and $\alpha > 1$. Since $h_\varphi^{1/2} \in \mathcal{D}(\Delta_{\psi,\varphi}^{1/2})$ and $\Delta_{\psi,\varphi}^{1/2} h_\varphi^{1/2} = h_\psi^{1/2} s(\varphi) = h_\psi^{1/2}$, it follows from Proposition B.2 that (i) $\iff$ (ii) and in this case $\Delta_{\psi,\varphi}^{\alpha/2} h_\varphi^{1/2} = \Delta_{\psi,\varphi}^{(\alpha-1)/2} h_\psi^{1/2}$. Hence Lemma 3.5 with $p = (\alpha-1)/2$ implies

that (ii) $\iff$ (iii) and in this case $Q_\alpha(\psi\|\varphi) = \|\eta\|_2^2$. The uniqueness of η in (iii) is obvious.

Remark 3.7 In the above (2), if $h_\psi^{\alpha/2} = \eta h_\varphi^{(\alpha-1)/2}$ with $\eta \in L^2(M)s(\varphi)$, then we may write $\eta = h_\psi^{\alpha/2} h_\varphi^{(1-\alpha)/2}$ in a formal sense, so that

$$Q_\alpha(\psi\|\varphi) = \|h_\psi^{\alpha/2} h_\varphi^{(1-\alpha)/2}\|^2 = \mathrm{tr}(h_\psi^{\alpha/2} h_\varphi^{1-\alpha} h_\psi^{\alpha/2}) = \mathrm{tr}(h_\psi^\alpha h_\varphi^{1-\alpha}),$$

which is the same expression as in (1). The form is the same as (3.6) in the finite-dimensional case, with tr instead of the usual trace and h_ψ, h_φ instead of the density operators.

The next corollary supplements Theorem 3.2 (4).

Corollary 3.8 *When $0 \le \alpha < 1$, the map $(\psi, \varphi) \in (M_*^+ \setminus \{0\}) \times M_*^+ \mapsto D_\alpha(\psi\|\varphi) \in \mathbb{R}$ is jointly continuous in the norm topology.*

Proof By (3.11) it suffices to show that $(\psi, \varphi) \in M_*^+ \times M_*^+ \mapsto h_\psi^\alpha h_\varphi^{1-\alpha} \in L^1(M)$ is continuous. Let $\psi_n, \psi, \varphi_n, \varphi \in M_*^+$ be such that $\|\psi_n - \psi\|_1 \to 0$ and $\|\varphi_n - \varphi\|_1 \to 0$. Then

$$\begin{aligned}
\|h_{\psi_n}^\alpha h_{\varphi_n}^{1-\alpha} - h_\psi^\alpha d_\varphi^{1-\alpha}\|_1 &\le \|(h_{\psi_n}^\alpha - h_\psi^\alpha) h_{\varphi_n}^{1-\alpha}\|_1 + \|h_\psi^\alpha (h_{\varphi_n}^{1-\alpha} - h_\varphi^{1-\alpha})\|_1 \\
&\le \|h_{\psi_n}^\alpha - h_\psi^\alpha\|_{1/\alpha} \|h_{\varphi_n}^{1-\alpha}\|_{1/(1-\alpha)} \\
&\qquad + \|h_\psi^\alpha\|_{1/\alpha} \|h_{\varphi_n}^{1-\alpha} - h_\varphi^{1-\alpha}\|_{1/(1-\alpha)} \\
&\le \|h_{\psi_n} - h_\varphi\|_1^\alpha \|h_{\varphi_n}\|_1^{1-\alpha} + \|h_\psi\|_1^\alpha \|h_{\varphi_n} - h_\varphi\|_1^{1-\alpha} \\
&\longrightarrow 0,
\end{aligned}$$

where the second inequality is Hölder's inequality for Haagerup's L^p-norms (see Theorem A.38) and the third follows from Kosaki's generalized Powers–Størmer inequality given in the appendix of [59].

3.3 Sandwiched Rényi Divergences

When $\mathcal{H}$ is finite-dimensional, for every $\psi, \varphi \in B(\mathcal{H})_*^+$ ($\psi \ne 0$) with the density operators D_ψ, D_φ and every $\alpha \in (0, \infty) \setminus \{1\}$, the *$\alpha$-sandwiched Rényi divergence* introduced in [98, 136] is

$$\widetilde{D}_\alpha(\psi\|\varphi) := \frac{1}{\alpha - 1} \log \frac{\widetilde{Q}_\alpha(\psi\|\varphi)}{\psi(1)},$$

where

$$\widetilde{Q}_\alpha(\psi\|\varphi) := \begin{cases} \mathrm{Tr}\Big(D_\varphi^{\frac{1-\alpha}{2\alpha}} D_\psi D_\varphi^{\frac{1-\alpha}{2\alpha}}\Big)^\alpha & \text{if } 0 < \alpha < 1 \text{ or } s(\psi) \le s(\varphi), \\ +\infty & \text{otherwise.} \end{cases} \tag{3.12}$$

These sandwiched variants of Rényi divergences have been in active consideration with applications, e.g., [16, 38, 96, 97]. A more general notion of the *α-z-Rényi relative entropies* was first introduced in [64, Sec. 4.3.3] and further discussed in [15]. In this section we will survey the α-sandwiched Rényi divergences in von Neumann algebras, recently developed in [20, 66, 67].

Definition 3.9 ([66]) Let $\psi, \varphi \in M_*^+$ with $\psi \neq 0$, and $1 < \alpha < \infty$. The *α-sandwiched Rényi divergence* due to Jenčová is

$$\widetilde{D}_\alpha^{(\mathrm{J})}(\psi\|\varphi) := \frac{1}{\alpha-1}\log\frac{\widetilde{Q}_\alpha^{(\mathrm{J})}(\psi\|\varphi)}{\psi(1)},$$

where

$$\widetilde{Q}_\alpha^{(\mathrm{J})}(\psi\|\varphi) := \begin{cases} \|h_\psi\|_{\alpha,\varphi}^\alpha & \text{if } s(\psi) \le s(\varphi) \text{ and } h_\psi \in L^\alpha(M,\varphi), \\ +\infty & \text{otherwise.} \end{cases}$$

Here, $L^\alpha(M,\varphi) = L^\alpha(M,\varphi)_{\eta=1/2}$ is *Kosaki's L^α-space* in the symmetric case and $\|\cdot\|_{\alpha,\varphi}$ is the norm on $L^\alpha(M,\varphi)$, more precisely, $\|\cdot\|_{\alpha,\varphi,\eta}$ with $\eta = 1/2$, see Remark A.62 of Sect. A.8. (Here we put the subscript φ to specify the dependence on it.)

An alternative definition of the sandwiched divergence in von Neumann algebras given in [20] is based on *Araki and Masuda's L^p-norm* [13]. So here we briefly recall the definition in [13]. Let $\psi, \varphi \in M_*^+$. For $2 \le p \le \infty$, Araki and Masuda's L^p-norm of the vector representative $h_\psi^{1/2}$ with respect to φ is

$$\|h_\psi^{1/2}\|_{p,\varphi} := \begin{cases} \sup\Big\{\big\|\Delta_{\omega,\varphi}^{\frac{1}{2}-\frac{1}{p}} h_\psi^{1/2}\big\| : \omega \in M_*^+,\ \omega(1) = 1\Big\} & \text{if } s(\psi) \le s(\varphi), \\ +\infty & \text{otherwise.} \end{cases} \tag{3.13}$$

For $1 \le p < 2$, Araki and Masuda's L^p-norm of $h_\psi^{1/2}$ is

$$\|h_\psi^{1/2}\|_{p,\varphi} := \inf\Big\{\big\|\Delta_{\omega,\varphi}^{\frac{1}{2}-\frac{1}{p}} h_\psi^{1/2}\big\| : \omega \in M_*^+,\ \omega(1) = 1,\ s(\omega) \ge s(\psi)\Big\}, \tag{3.14}$$

where $\|\Delta_{\omega,\psi}^{\frac{1}{2}-\frac{1}{p}} h_\psi^{1/2}\|$ means $+\infty$ if $h_\psi^{1/2}$ is not in $\mathcal{D}(\Delta_{\omega,\psi}^{\frac{1}{2}-\frac{1}{p}})$. (Note that the above p-norm of vector representatives can be defined under any $*$-representation of M on any Hilbert space $\mathcal{H}$, while here we use the standard representation on Haagerup's $L^2(M)$.)

Definition 3.10 ([20]) For every $\psi, \varphi \in M_*^+$ with $\psi \neq 0$ and $\alpha \in [1/2, \infty) \setminus \{1\}$, the *$\alpha$-sandwiched Rényi divergence* due to Berta–Scholz–Tomamichel (called the *Araki–Masuda divergence* in [20]) is

$$\widetilde{D}_\alpha^{(\mathrm{BST})}(\psi\|\varphi) := \frac{1}{\alpha-1}\log\frac{\widetilde{Q}_\alpha^{(\mathrm{BST})}(\psi\|\varphi)}{\psi(1)}, \quad \text{where } \widetilde{Q}_\alpha^{(\mathrm{BST})}(\psi\|\varphi) := \|h_\psi^{1/2}\|_{2\alpha,\varphi}^{2\alpha}.$$

The next theorem is a main result of this section, which was proved by Jenčová. The proof below is different from that in [66], see also Remark 3.20 (1) below.

Theorem 3.11 ([66]) *Let $\psi, \varphi \in M_*^+$ with $\psi \neq 0$.*

(1) *For every $\alpha \in (1, \infty)$ we have*

$$\widetilde{Q}_\alpha^{(\mathrm{BST})}(\psi\|\varphi) = \widetilde{Q}_\alpha^{(\mathrm{J})}(\psi\|\varphi). \tag{3.15}$$

Hence

$$\widetilde{D}_\alpha^{(\mathrm{BST})}(\psi\|\varphi) = \widetilde{D}_\alpha^{(\mathrm{J})}(\psi\|\varphi).$$

(2) *For every $\alpha \in [1/2, 1)$ we have*

$$\widetilde{Q}_\alpha^{(\mathrm{BST})}(\psi\|\varphi) = \mathrm{tr}\Big(h_\varphi^{\frac{1-\alpha}{2\alpha}} h_\psi h_\varphi^{\frac{1-\alpha}{2\alpha}}\Big)^\alpha. \tag{3.16}$$

Hence

$$\widetilde{D}_\alpha^{(\mathrm{BST})}(\psi\|\varphi) = \frac{1}{\alpha-1}\log\frac{\mathrm{tr}\big(h_\varphi^{\frac{1-\alpha}{2\alpha}} h_\psi h_\varphi^{\frac{1-\alpha}{2\alpha}}\big)^\alpha}{\psi(1)}.$$

To prove the theorem, we first give two lemmas.

Lemma 3.12 *Assume that $1 < \alpha < \infty$.*

(1) *$\widetilde{D}_\alpha^{(\mathrm{J})}(\psi\|\varphi)$ is jointly lower semicontinuous in $\psi, \varphi \in M_*^+$ in the norm topology.*
(2) *$\widetilde{Q}_\alpha^{(\mathrm{J})}(\psi\|\varphi)$ is jointly convex in $\psi, \varphi \in M_*^+$.*
(3) *$\widetilde{Q}_\alpha^{(\mathrm{BST})}(\psi\|\varphi)$ is jointly convex in $\psi, \varphi \in M_*^+$.*

Proof See [66] for the proofs of (1) and (2).

(3) Consider a unital $*$-homomorphism (hence CP) $\gamma : M \to M \oplus M$ given by $\gamma(x) := x \oplus x$. For $\psi_i, \varphi_i \in M_*^+$ $(i = 1, 2)$ and $\lambda \in (0, 1)$ set $\psi := \lambda\psi_1 \oplus$

$(1-\lambda)\psi_2$ and $\varphi := \lambda\varphi_1 \oplus (1-\lambda)\varphi_2$. Since $\psi\circ\gamma = \lambda\psi_1 + (1-\lambda)\psi_2$ and $\varphi\circ\gamma = \lambda\varphi_1 + (1-\lambda)\varphi_2$, from the monotonicity property of $\widetilde{D}_\alpha^{(\mathrm{BST})}$ (equivalent to that of $\widetilde{Q}_\alpha^{(\mathrm{BST})}$) given in [20, Theorem 14] it follows that

$$\widetilde{Q}_\alpha^{(\mathrm{BST})}(\lambda\psi_1 + (1-\lambda)\psi_2 \| \lambda\varphi_1 + (1-\lambda)\varphi_2) \le \widetilde{Q}_\alpha^{(\mathrm{BST})}(\psi\|\varphi).$$

So it suffices to prove that

$$\widetilde{Q}_\alpha^{(\mathrm{BST})}(\psi\|\varphi) = \lambda\widetilde{Q}_\alpha^{(\mathrm{BST})}(\psi_1\|\varphi_1) + (1-\lambda)\widetilde{Q}_\alpha^{(\mathrm{BST})}(\psi_2\|\varphi_2). \tag{3.17}$$

Note that $L^2(M\oplus M) = L^2(M)\oplus L^2(M)$ and

$$h_\psi^{1/2} = \lambda^{1/2}h_{\psi_1}^{1/2} \oplus (1-\lambda)^{1/2}h_{\psi_2}^{1/2}, \quad h_\varphi^{1/2} = \lambda^{1/2}h_{\varphi_1}^{1/2} \oplus (1-\lambda)^{1/2}h_{\varphi_2}^{1/2}.$$

Let $p := 2\alpha \in (2,\infty)$. For every $\omega \in (M\oplus M)_*^+$ with $\omega(1) = 1$, we can write $\omega = t\omega_1 \oplus (1-t)\omega_2$, where $\omega_i \in M_*^+$, $\omega_i(1) = 1$ and $0 \le t \le 1$. Since

$$\Delta_{\omega,\varphi} = \Delta_{t\omega_1,\lambda\varphi_1} \oplus \Delta_{(1-t)\omega_2,(1-\lambda)\varphi_2} = \frac{t}{\lambda}\,\Delta_{\omega_1,\varphi_1} \oplus \frac{1-t}{1-\lambda}\,\Delta_{\omega_2,\varphi_2},$$

we have

$$\|\Delta_{\omega,\varphi}^{\frac12-\frac1p}h_\psi^{1/2}\| = \Big\{t^{1-\frac2p}\lambda^{\frac2p}\|\Delta_{\omega_1,\varphi_1}^{\frac12-\frac1p}h_{\psi_1}^{1/2}\|^2 + (1-t)^{1-\frac2p}(1-\lambda)^{\frac2p}\|\Delta_{\omega_2,\varphi_2}^{\frac12-\frac1p}h_{\psi_2}^{1/2}\|^2\Big\}^{1/2}.$$

Therefore,

$$\begin{aligned}
&\|h_\psi^{1/2}\|_{p,\varphi}\\
&= \sup_{t,\omega_1,\varphi_2}\Big\{t^{1-\frac2p}\lambda^{\frac2p}\|\Delta_{\omega_1,\varphi_1}^{\frac12-\frac1p}h_{\psi_1}^{1/2}\|^2 + (1-t)^{1-\frac2p}(1-\lambda)^{\frac2p}\|\Delta_{\omega_2,\varphi_2}^{\frac12-\frac1p}h_{\psi_2}^{1/2}\|^2\Big\}^{1/2}\\
&\qquad\qquad (\text{the sup is over } 0\le t\le 1,\ \omega_1,\omega_2 \in M_*^+,\ \omega_1(1) = \omega_2(1) = 1)\\
&= \sup_{0\le t\le 1}\Big\{t^{1-\frac2p}\lambda^{\frac2p}\|h_{\psi_1}^{1/2}\|_{p,\varphi_1}^2 + (1-t)^{1-\frac2p}(1-\lambda)^{\frac2p}\|h_{\psi_2}^{1/2}\|_{p,\varphi_2}^2\Big\}^{1/2}\\
&= \Big\{\lambda\|h_{\psi_1}^{1,2}\|_{p,\varphi_1}^p + (1-\lambda)\|h_{\psi_2}^{1/2}\|_{p,\varphi_2}^p\Big\}^{1/p},
\end{aligned}$$

so that $\|h_\psi^{1/2}\|_{p,\varphi}^p = \lambda\|h_{\psi_1}^{1/2}\|_{p,\varphi_1}^p + (1-\lambda)\|h_{\psi_2}^{1/2}\|_{p,\varphi_2}^p$, which is (3.17).

Lemma 3.13 *Let $\alpha \in (1,\infty)$ and $\psi,\varphi \in M_*^+$. Then:*

(1) $\widetilde{D}_\alpha^{(\mathrm{J})}(\psi\|\varphi) = \lim_{\varepsilon\searrow 0}\widetilde{D}_\alpha^{(\mathrm{J})}(\psi\|\varphi+\varepsilon\psi)$, *and*
(2) $\widetilde{D}_\alpha^{(\mathrm{BST})}(\psi\|\varphi) = \lim_{\varepsilon\searrow 0}\widetilde{D}_\alpha^{(\mathrm{BST})}(\psi\|\varphi+\varepsilon\psi)$ *if $s(\psi)\le s(\varphi)$.*

Proof

(1) It is clear that $\widetilde{Q}_\alpha^{(\mathrm{J})}(\psi\|\psi) = 1$. Hence by Lemma 3.12 (2),

$$\widetilde{Q}_\alpha^{(\mathrm{J})}(\psi\|(\varphi+\varepsilon\psi)/(1+\varepsilon)) \le \frac{\widetilde{Q}_\alpha^{(\mathrm{J})}(\psi\|\varphi)+\varepsilon}{1+\varepsilon}.$$

Moreover, as easily seen and noted in [66, (14)],

$$\widetilde{D}_\alpha^{(\mathrm{J})}(\psi\|(\varphi+\varepsilon\psi)/(1+\varepsilon)) = \widetilde{D}_\alpha^{(\mathrm{J})}(\psi\|\varphi+\varepsilon\psi) + \log(1+\varepsilon).$$

Therefore,

$$\limsup_{\varepsilon\searrow 0} \widetilde{D}_\alpha^{(\mathrm{J})}(\psi\|\varphi+\varepsilon\psi) = \limsup_{\varepsilon\searrow 0} \widetilde{D}_\alpha^{(\mathrm{J})}(\psi\|(\varphi+\varepsilon\psi)/(1+\varepsilon)) \le \widetilde{D}_\alpha^{(\mathrm{J})}(\psi\|\varphi).$$

On the other hand, by Lemma 3.12 (1), $\widetilde{D}_\alpha^{(\mathrm{J})}(\psi\|\varphi) \le \liminf_{\varepsilon\searrow 0} \widetilde{D}_\alpha^{(\mathrm{J})}(\psi\|\varphi+\varepsilon\psi)$.

(2) The scaling property of $\widetilde{D}_\alpha^{(\mathrm{BST})}$ is the same as [66, (14)] for $\widetilde{D}_\alpha^{(\mathrm{J})}$. Hence as in the above proof of (1), by Lemma 3.12 (3) we have

$$\limsup_{\varepsilon\searrow 0} \widetilde{D}_\alpha^{(\mathrm{BST})}(\psi\|\varphi+\varepsilon\psi) \le \widetilde{D}_\alpha^{(\mathrm{BST})}(\psi\|\varphi).$$

So it suffices to prove that

$$\widetilde{D}_\alpha^{(\mathrm{BST})}(\psi\|\varphi) \le \liminf_{\varepsilon\searrow 0} \widetilde{D}_\alpha^{(\mathrm{BST})}(\psi\|\varphi+\varepsilon\psi),$$

that is, for $p := 2\alpha \in (2,\infty)$,

$$\|h_\psi^{1/2}\|_{p,\varphi} \le \liminf_{\varepsilon\searrow 0} \|h_\psi^{1/2}\|_{p,\varphi+\varepsilon\psi}. \tag{3.18}$$

Let $\varepsilon_n \searrow 0$ be arbitrary and let $\omega \in M_*^+$. By [11, Lemma 4.1] we have

$$\big(1+\Delta_{\omega,\varphi+\varepsilon_n\psi}^{1/2}\big)^{-1} Js(\varphi)J \longrightarrow (1+\Delta_{\omega,\varphi}^{1/2})^{-1} Js(\varphi)J \quad \text{strongly}.$$

Since $s(\psi) \le s(\varphi)$, both of $\Delta_{\omega,\varphi+\varepsilon_n\psi}$ and $\Delta_{\omega,\varphi}$ have the support projection $s(\omega)Js(\varphi)J$ ($\le Js(\varphi)J$). Hence the above strong convergence means that

$$\big(1+\Delta_{\omega,\varphi+\varepsilon_n\psi}^{1/2}\big)^{-1} \longrightarrow \big(1+\Delta_{\omega,\varphi}^{1/2}\big)^{-1} \quad \text{strongly}. \tag{3.19}$$

Moreover, since $s(\varphi + \varepsilon_1\psi) = s(\varphi)$, we find by Proposition A.22 (4) of Sect. A.4 that $\Delta_{\omega,\varphi+\varepsilon_1\psi} \le \Delta_{\omega,\varphi+\varepsilon_2\psi} \le \cdots$. Hence $\Delta^{1/2}_{\omega,\varphi+\varepsilon_1\psi} \le \Delta^{1/2}_{\omega,\varphi+\varepsilon_2\psi} \le \cdots$ by Lemma B.7 of Appendix B. By this and (3.19) we can apply Proposition B.9 to $A_n = \Delta^{1/2}_{\omega,\varphi+\varepsilon_n\psi}$; then we have

$$\big\|\Delta_{\omega,\varphi}^{\frac{1}{2}-\frac{1}{p}} h_\psi^{1/2}\big\|^2 = \lim_{n\to\infty} \big\|\Delta_{\omega,\varphi+\varepsilon_n\psi}^{\frac{1}{2}-\frac{1}{p}} h_\psi^{1/2}\big\|^2,$$

which implies (3.18) in view of (3.13).

Proof (Theorem 3.11)

(1) If $s(\psi) \not\le s(\varphi)$, then both sides of (3.15) are $+\infty$ by Definitions 3.9 and 3.10. Hence by Lemma 3.13 we may prove equality (3.15) in the case where $\psi \le \lambda\varphi$ for some $\lambda > 0$. Then by Lemma A.24, $h_\psi^{1/2} = Ah_\varphi^{1/2}$ for some $A \in s(\varphi)Ms(\varphi)$. Let $p := 2\alpha \in (2, \infty)$ and $\omega \in M_*^+$ with $\omega(1) = 1$. From (A.19) of Sect. A.6 one has

$$\Delta_{\omega,\varphi}^{\frac{1}{2}-\frac{1}{p}} h_\psi^{1/2} = \Delta_{\omega,\varphi}^{\frac{1}{2}-\frac{1}{p}} Ah_\varphi^{1/2} = h_\omega^{\frac{1}{2}-\frac{1}{p}} Ah_\varphi^{\frac{1}{p}}$$

so that

$$\big\|\Delta_{\omega,\varphi}^{\frac{1}{2}-\frac{1}{p}} h_\psi^{1/2}\big\|^2 = \mathrm{tr}\Big(h_\omega^{\frac{1}{2}-\frac{1}{p}} Ah_\varphi^{\frac{2}{p}} A^* h_\omega^{\frac{1}{2}-\frac{1}{p}}\Big) = \mathrm{tr}(h_\omega^{(p-2)/p} Ah_\varphi^{2/p} A^*). \tag{3.20}$$

Apply Hölder's inequality (Theorem A.38) to $h_\omega^{(p-2)/p} \in L^{p/(p-2)}(M)$ and $Ah_\varphi^{2/p}A^* \in L^{p/2}(M)$ to obtain

$$\big|\mathrm{tr}(h_\omega^{(p-2)/p} Ah_\varphi^{2/p} A^*)\big| \le \|Ah_\varphi^{2/p} A^*\|_{p/2} \tag{3.21}$$

thanks to $\|h_\omega^{(p-2)/p}\|_{p/(p-2)} = (\mathrm{tr}\, h_\omega)^{(p-2)/p} = 1$. On the other hand, set $h := (Ah_\varphi^{2/p}A^*)^{p/2}/\|Ah_\varphi^{2/p}A^*\|_{p/2}^{p/2}$. Then $h \in L^1(M)_+$ and $\mathrm{tr}(h) = 1$, so that $h = h_{\omega_0}$ for some $\omega_0 \in M_*^+$ with $\omega_0(1) = 1$. We then have

$$\mathrm{tr}\big(h_{\omega_0}^{(p-2)/p} Ah_\varphi^{2/p} A^*\big) = \frac{\mathrm{tr}\big((Ah_\varphi^{2/p}A^*)^{p/2}\big)}{\mathrm{tr}\big((Ah_\varphi^{2/p}A^*)^{p/2}\big)^{(p-2)/p}} = \|Ah_\varphi^{2/p} A^*\|_{p/2}. \tag{3.22}$$

Combining (3.20)–(3.22) yields $\|h_\psi^{1/2}\|_{p,\varphi}^2 = \|Ah_\varphi^{2/p}A^*\|_{p/2}$ so that

$$\|h_\psi^{1/2}\|_{2\alpha,\varphi}^{2\alpha} = \|Ah_\varphi^{1/\alpha}A^*\|_\alpha^\alpha = \big\|h_\varphi^{\frac{1}{2\alpha}} A^*Ah_\varphi^{\frac{1}{2\alpha}}\big\|_\alpha^\alpha. \tag{3.23}$$

Now define

$$h_0 := \left(h_\varphi^{\frac{1}{2\alpha}} A^* A h_\varphi^{\frac{1}{2\alpha}}\right)^\alpha \in L^1(M),$$

for which we have, with $1/\alpha + 1/\beta = 1$,

$$h_\varphi^{\frac{1}{2\beta}} h_0^{1/\alpha} h_\varphi^{\frac{1}{2\beta}} = h_\varphi^{\frac{1}{2\alpha}+\frac{1}{2\beta}} A^* A h_\varphi^{\frac{1}{2\alpha}+\frac{1}{2\beta}} = h_\varphi^{1/2} A^* A h_\varphi^{1/2} = h_\psi.$$

This implies that $h_\psi \in L^\alpha(M, \varphi)$ by Theorem A.61 (in the $\eta = 1/2$ case) and

$$\|h_\psi\|_{\alpha,\varphi} = (\operatorname{tr} h_0)^{1/\alpha} = \left\|h_\varphi^{\frac{1}{2\alpha}} A^* A h_\varphi^{\frac{1}{2\alpha}}\right\|_\alpha. \tag{3.24}$$

By (3.23) and (3.24) we thus obtain $\|h_\psi^{1/2}\|_{2\alpha,\varphi}^{2\alpha} = \|h_\psi\|_{\alpha,\varphi}^\alpha$, which yields (3.15) by Definitions 3.9 and 3.10.

(2) Assume that $1/2 \le \alpha < 1$. First, note that the RHS of (3.16) is well defined since $\left(h_\varphi^{\frac{1-\alpha}{2\alpha}} h_\psi h_\varphi^{\frac{1-\alpha}{2\alpha}}\right)^\alpha$ is in $L^1(M)$. Since $\Delta_{\omega,\varphi}^{-1} = J\Delta_{\varphi,\omega}J$ for every $\omega, \varphi \in M_*^+$ by Proposition A.22 (3), the definition of $\|h_\psi^{1/2}\|_{2\alpha,\varphi}$ in (3.14) is rewritten as

$$\|h_\psi^{1/2}\|_{2\alpha,\varphi} = \inf\left\{\left\|\Delta_{\varphi,\omega}^{\frac{1}{2\alpha}-\frac{1}{2}} h_\psi^{1/2}\right\| : \omega \in M_*^+,\ \omega(1) = 1,\ s(\omega) \ge s(\psi)\right\}.$$

Using Lemma 3.4 we find that

$$\begin{aligned}\|h_\psi^{1/2}\|_{2\alpha,\varphi}^2 = \inf\Big\{\|\eta\|_2^2 : \omega \in M_*^+,\ \omega(1) = 1,\ s(\omega) \ge s(\psi),\\ \eta \in L^2(M),\ h_\varphi^{\frac{1-\alpha}{2\alpha}} h_\psi^{1/2} = \eta h_\omega^{\frac{1-\alpha}{2\alpha}}\Big\},\end{aligned} \tag{3.25}$$

where we note that $h_\psi^{1/2} s(\omega) = h_\psi^{1/2}$ (thanks to $s(\omega) \ge s(\psi)$), $\eta h_\omega^{\frac{1-\alpha}{2\alpha}} = \eta s(\omega) h_\omega^{\frac{1-\alpha}{2\alpha}}$ and $\|\eta s(\omega)\|_2^2 \le \|\eta\|_2^2$. Now let $p := \alpha$ and $q := \frac{\alpha}{1-\alpha}$ so that $1/p - 1/q = 1$. By letting $b := h_\omega^{\frac{1-\alpha}{\alpha}} \in L^q(M)_+$ expression (3.25) can be rewritten as

$$\begin{aligned}\|h_\psi^{1/2}\|_{2\alpha,\varphi}^2 = \inf\Big\{\|\eta\|_2^2 : b \in L^q(M)_+,\ \|b\|_q = 1,\ s(b) \ge s(\psi),\\ \eta \in L^2(M),\ h_\varphi^{\frac{1-\alpha}{2\alpha}} h_\psi^{1/2} = \eta b^{1/2}\Big\}.\end{aligned}$$

Furthermore, let $a := \left|h_\varphi^{\frac{1-\alpha}{2\alpha}} h_\psi^{1/2}\right|^2$ and $h_\varphi^{\frac{1-\alpha}{2\alpha}} h_\psi^{1/2} = v a^{1/2}$ be the polar decomposition where $v \in M$ is a partial isometry with $v^* v = s(a) \le s(\psi)$.

Since $va^{1/2} = \eta b^{1/2} \iff a^{1/2} = v^*\eta b^{1/2} \iff va^{1/2} = vv^*\eta b^{1/2}$ and $\|\eta\|_2 \ge \|v^*\eta\|_2 = \|vv^*\eta\|_2$, it follows that

$$\|h_\psi^{1/2}\|_{2\alpha,\varphi}^2 = \inf\big\{\|\eta\|_2^2 : b \in L^q(M)_+,\ \|b\|_q = 1,\ s(b) \ge s(\psi),\\ \eta \in L^2(M),\ a^{1/2} = \eta b^{1/2}\big\}. \tag{3.26}$$

Now we show that the RHS of (3.26) is equal to $\|a\|_p$. For this we may assume that $a \ne 0$ (the result is immediate when $a = 0$). If b and η are as in the above RHS, then Hölder's inequality (Theorem A.38) implies that

$$\|a\|_p^{1/2} = \|a^{1/2}\|_{2p} = \|\eta b^{1/2}\|_{2p} \le \|\eta\|_2\|b^{1/2}\|_{2q} = \|\eta\|_2\|b\|_q^{1/2} = \|\eta\|_2.$$

Furthermore, set $\eta := \|a\|_p^{p/2q}a^{p/2}$ and $b := \|a\|_p^{-p/q}a^{p/q}$; then it is immediate to check that $\eta \in L^2(M)$, $b \in L^q(M)_+$, $\|b\|_q = 1$, $a^{1/2} = \eta b^{1/2}$ and $\|\eta\|_2^2 = \|a\|_p$. Note that $s(\eta) = s(b) = s(a) \le s(\psi)$. Choose a $b_0 \in L^q(M)_+$ with $s(b_0) = s(\psi) - s(a)$ and for $\varepsilon > 0$ set $\eta_\varepsilon := \|b + \varepsilon b_0\|_q^{1/2}\eta$ and $b_\varepsilon := \|b + \varepsilon b_0\|_q^{-1}(b + \varepsilon b_0)$. Then $s(b_\varepsilon) = s(\psi)$, $a^{1/2} = \eta_\varepsilon b_\varepsilon^{1/2}$ and $\|\eta_\varepsilon\|_2^2 \to \|\eta\|_2^2 = \|a\|_p$ as $\varepsilon \searrow 0$. Therefore, the RHS of (3.26) is $\|a\|_p$ so that

$$\|h_\psi^{1/2}\|_{2\alpha,\varphi}^2 = \|a\|_p = \Big[\mathrm{tr}\Big(h_\varphi^{\frac{1-\alpha}{2\alpha}} h_\psi h_\varphi^{\frac{1-\alpha}{2\alpha}}\Big)^\alpha\Big]^{1/\alpha},$$

which is (3.16) by Definition 3.10 when $1/2 \le \alpha < 1$.

In view of Theorem 3.11 we simply denote the α-sandwiched Rényi divergence by

$$\widetilde{D}_\alpha(\psi\|\varphi) = \frac{1}{\alpha - 1}\log\frac{\widetilde{Q}_\alpha(\psi\|\varphi)}{\psi(1)} \tag{3.27}$$

for $\psi, \varphi \in M_*^+$ and $\alpha \in [1/2, \infty) \setminus \{1\}$, where

$$\widetilde{Q}_\alpha(\psi\|\varphi) = \begin{cases} \widetilde{Q}_\alpha^{(\mathrm{BST})}(\psi\|\varphi) = \widetilde{Q}_\alpha^{(\mathrm{J})}(\psi\|\varphi) & \text{if } 1 < \alpha < \infty, \\ \widetilde{Q}_\alpha^{(\mathrm{BST})}(\psi\|\varphi) = \mathrm{tr}\Big(h_\varphi^{\frac{1-\alpha}{2\alpha}} h_\psi h_\varphi^{\frac{1-\alpha}{2\alpha}}\Big)^\alpha & \text{if } 1/2 \le \alpha < 1. \end{cases} \tag{3.28}$$

Remark 3.14 The expression for $1/2 \le \alpha < 1$ in (3.28) has a complete resemblance to the matrix case in (3.12). For $1 < \alpha < \infty$, if $h_\psi \in L^\alpha(M, \varphi)$ (Kosaki's L^α-space in the symmetric case), then there exists an $x \in (s(\varphi)L^\alpha(M)s(\varphi))_+$ such that $h_\psi = h_\varphi^{\frac{\alpha-1}{2\alpha}} x h_\varphi^{\frac{\alpha-1}{2\alpha}}$ and $\|h_\psi\|_{\alpha,\varphi} = \|x\|_\alpha$. Hence we may write

$$\widetilde{Q}_\alpha(\psi\|\varphi) = \mathrm{tr}\, x^\alpha = \mathrm{tr}\Big(h_\varphi^{\frac{1-\alpha}{2\alpha}} h_\psi h_\varphi^{\frac{1-\alpha}{2\alpha}}\Big)^\alpha$$

in a formal sense, which is the same expression as (3.12) again.

Remark 3.15 In particular, when $\alpha = 1/2$, note that

$$\widetilde{Q}_{1/2}(\psi\|\varphi) = F(\psi, \varphi), \qquad \widetilde{D}_{1/2}(\psi\|\varphi) = -2\log\frac{F(\psi, \varphi)}{\psi(1)}, \tag{3.29}$$

where $F(\psi, \varphi) := \mathrm{tr}(h_\varphi^{1/2} h_\psi h_\varphi^{1/2})^{1/2} = \|h_\psi^{1/2} h_\varphi^{1/2}\|_1$ is the so-called *fidelity* and will be discussed in Example 5.18 of Sect. 5.3. On the other hand, as for the Rényi divergence for $\alpha = 1/2$ note that

$$Q_{1/2}(\psi\|\varphi) = P(\psi, \varphi), \qquad D_{1/2}(\psi\|\varphi) = -2\log\frac{P(\psi, \varphi)}{\psi(1)}, \tag{3.30}$$

where $P(\psi, \varphi) := \mathrm{tr}(h_\psi^{1/2} h_\varphi^{1/2}) = \langle h_\psi^{1/2}, h_\varphi^{1/2}\rangle$ is the so-called *transition probability*, introduced by Raggio [112] in the von Neumann algebra setting.

Properties of $\widetilde{D}_\alpha$ and $\widetilde{Q}_\alpha$ are summarized in the next theorem, which are more or less similar to those of D_α and Q_α in Theorem 3.2.

Theorem 3.16 ([20, 66, 67]) *Let $\psi, \varphi \in M_*^+$ with $\psi \neq 0$.*

(1) *The function $\alpha \mapsto \widetilde{D}_\alpha(\psi\|\varphi)$ is monotone increasing on $[1/2, 1) \cup (1, \infty)$.*
(2) Limit values:

$$\lim_{\alpha\nearrow 1} \widetilde{D}_\alpha(\psi\|\varphi) = \frac{D(\psi\|\varphi)}{\psi(1)},$$

where $D(\psi\|\varphi)$ is the relative entropy. *Moreover, if $\widetilde{D}_\alpha(\psi\|\varphi) < +\infty$ for some $\alpha > 1$, then $\lim_{\alpha\searrow 1} \widetilde{D}_\alpha(\psi\|\varphi) = D(\psi\|\varphi)/\psi(1)$. On the other hand,*

$$\lim_{\alpha\to\infty} \widetilde{D}_\alpha(\psi\|\varphi) = D_{\max}(\psi\|\varphi),$$

where

$$D_{\max}(\psi\|\varphi) := \log\inf\{t > 0 : \psi \le t\varphi\}$$

is the max-relative entropy *introduced in [31] (here $\inf\emptyset = +\infty$ as usual).*
(3) Joint lower semicontinuity: *The map $(\psi, \varphi) \in (M_*^+ \setminus \{0\}) \times M_*^+ \mapsto \widetilde{D}_\alpha(\psi\|\varphi)$ is jointly lower semicontinuous in the norm topology for every $\alpha \in (1, \infty)$ and jointly continuous in the norm topology for every $\alpha \in [1/2, 1)$.*
(4) Monotonicity: *For each $\alpha \in [1/2, \infty) \setminus \{1\}$ and for any unital normal positive map $\gamma : N \to M$ between von Neumann algebras,*

$$\widetilde{D}_\alpha(\psi\circ\gamma\|\varphi\circ\gamma) \le \widetilde{D}_\alpha(\psi\|\varphi).$$

(5) *The map $(\psi, \varphi) \in M_*^+ \times M_*^+ \mapsto \widetilde{Q}_\alpha(\psi\|\varphi)$ is jointly convex for $1 < \alpha < \infty$ and jointly concave for $1/2 \le \alpha < 1$. Hence, when $1/2 \le \alpha < 1$, $\widetilde{D}_\alpha(\psi\|\varphi)$ is jointly convex on $\{(\psi, \varphi) \in M_*^+ \times M_*^+ : \psi(1) = c\}$ for any fixed $c > 0$.*

(6) Relation between $\widetilde{D}_\alpha$ and D_α: *For every* $\alpha \in [1/2, \infty) \setminus \{1\}$,

$$\widetilde{D}_\alpha(\psi\|\varphi) \le D_\alpha(\psi\|\varphi). \tag{3.31}$$

In particular, $F(\psi, \varphi) \ge P(\psi, \varphi)$, *see* (3.29) *and* (3.30). *If* $\psi(1) = 1$, *then for every* $\alpha \in (1, \infty)$,

$$D_{2-\frac{1}{\alpha}}(\psi\|\varphi) \le \widetilde{D}_\alpha(\psi\|\varphi). \tag{3.32}$$

Moreover, $D_2(\psi\|\varphi) \le D_{\max}(\psi\|\varphi) \le D_\infty(\psi\|\varphi) := \lim_{\alpha\to\infty} D_\alpha(\psi\|\varphi)$.

(7) *Let* $\psi_i, \varphi_i \in M_*^+$ *for* $i = 1, 2$. *If* $1/2 \le \alpha < 1$, $\psi_1 \le \psi_2$ *and* $\varphi_1 \le \varphi_2$, *then* $\widetilde{Q}_\alpha(\psi_1\|\varphi_1) \le \widetilde{Q}_\alpha(\psi_2\|\varphi_2)$. *If* $1 < \alpha < \infty$, $\psi_1 \ge \psi_2$ *and* $\varphi_1 \le \varphi_2$, *then* $\widetilde{Q}_\alpha(\psi_1\|\varphi_1) \ge \widetilde{Q}_\alpha(\psi_2\|\varphi_2)$. *Hence, if* $\varphi_1 \le \varphi_2$, *then* $\widetilde{D}_\alpha(\psi\|\varphi_1) \ge \widetilde{D}_\alpha(\psi\|\varphi_2)$ *for all* $\alpha \in [1/2, \infty) \setminus \{1\}$.

(8) Strict positivity: *Let* $\alpha \in [1/2, \infty) \setminus \{1\}$ *and* $\psi, \varphi \ne 0$. *The inequality*

$$\widetilde{D}_\alpha(\psi\|\varphi) \ge \log \frac{\psi(1)}{\varphi(1)}$$

holds, and equality holds if and only if $\psi = (\psi(1)/\varphi(1))\varphi$. *If* $\psi(1) = \varphi(1)$, *then* $\widetilde{D}_\alpha(\psi\|\varphi) \ge 0$, *and* $\widetilde{D}_\alpha(\psi\|\varphi) = 0$ *if and only if* $\psi = \varphi$.

Proof

(1) See [20, Lemma 8] and [66, Proposition 3.7].

(2) See [20, Theorem 13, Lemma 9] and [66, Proposition 3.8].

(3) The joint lower semicontinuity of $\widetilde{D}_\alpha$ for $1 < \alpha < \infty$ was given in [66, Proposition 3.10]. For the joint continuity of $\widetilde{D}_\alpha$ (or $\widetilde{Q}_\alpha$) for $1/2 \le \alpha < 1$, by (3.16) it suffices to show that $(\psi, \varphi) \in M_*^+ \times M_*^+ \mapsto h_\varphi^{\frac{1-\alpha}{2\alpha}} h_\psi h_\varphi^{\frac{1-\alpha}{2\alpha}} \in L^\alpha(M)$ is continuous, whose proof is similar to that of Corollary 3.8 by using $\|a + b\|_p^p \le \|a\|_p^p + \|b\|_p^p$ for $a, b \in L^p(M)$ $(0 < p \le 1)$ in [37, Theorem 4.9 (iii)], Hölder's inequality in Theorem A.38 and Kosaki's generalized Powers–Størmer inequality in the appendix of [59].

(4) See [66, Theorem 3.14] and [67, Theorem 4.1] (see also Remark 3.20 (2) below). The monotonicity of $\widetilde{D}_\alpha$ under a unital normal CP map was shown in [20, Theorem 14] as well. (Note that we have used [66, Theorem 3.14] in the proof of Lemma 3.12 (3).)

(5) The joint convexity of $\widetilde{Q}_\alpha$ for $1 < \alpha < \infty$ was given in [66, Corollary 3.16]. For $1/2 \le \alpha < 1$ the monotonicity of $\widetilde{D}_\alpha$ in (4) means the "reverse" monotonicity of $\widetilde{Q}_\alpha$. Then the joint concavity of $\widetilde{Q}_\alpha$ for $1/2 \le \alpha < 1$ can be shown similarly to the proof of [66, Corollary 3.16] (or Lemma 3.12 (3)) by considering $\gamma : M \to M \oplus M$, $\gamma(x) := x \oplus x$, and the expression of $\widetilde{Q}_\alpha$ in Theorem 3.11 (2).

(6) See [20, Theorem 12] and [66, Corollary 3.6], and see [20, (95)] for $D_2 \le D_{\max}$. (More information will be given in Remark 3.18 (1) for the proofs of inequalities (3.31) and (3.32).)

(7) The case $1 < \alpha < \infty$ was given in [66, Proposition 3.9]. When $1/2 \le \alpha < 1$, Lemma B.7 of Appendix B implies that $\Big(h_{\varphi_1}^{\frac{1-\alpha}{2\alpha}} h_{\psi_1} h_{\varphi_1}^{\frac{1-\alpha}{2\alpha}}\Big)^\alpha \le \Big(h_{\varphi_1}^{\frac{1-\alpha}{2\alpha}} h_{\psi_2} h_{\varphi_1}^{\frac{1-\alpha}{2\alpha}}\Big)^\alpha$. Since $\frac{1-\alpha}{\alpha} \in (0,1]$, by Lemma B.7 again one has $h_{\varphi_1}^{\frac{1-\alpha}{\alpha}} \le h_{\varphi_2}^{\frac{1-\alpha}{\alpha}}$ and hence $\Big(h_{\psi_2}^{1/2} h_{\varphi_1}^{\frac{1-\alpha}{\alpha}} h_{\psi_2}^{1/2}\Big)^\alpha \le \Big(h_{\psi_2}^{1/2} h_{\varphi_2}^{\frac{1-\alpha}{\alpha}} h_{\psi_2}^{1/2}\Big)^\alpha$. Therefore,

$$\begin{aligned}\mathrm{tr}\Big(h_{\varphi_1}^{\frac{1-\alpha}{2\alpha}} h_{\psi_1} h_{\varphi_1}^{\frac{1-\alpha}{2\alpha}}\Big)^\alpha &\le \mathrm{tr}\Big(h_{\varphi_1}^{\frac{1-\alpha}{2\alpha}} h_{\psi_2} h_{\varphi_1}^{\frac{1-\alpha}{2\alpha}}\Big)^\alpha = \mathrm{tr}\Big(h_{\psi_2}^{1/2} h_{\varphi_1}^{\frac{1-\alpha}{\alpha}} h_{\psi_2}^{1/2}\Big)^\alpha \\ &\le \mathrm{tr}\Big(h_{\psi_2}^{1/2} h_{\varphi_2}^{\frac{1-\alpha}{\alpha}} h_{\psi_2}^{1/2}\Big)^\alpha = \mathrm{tr}\Big(h_{\varphi_2}^{\frac{1-\alpha}{2\alpha}} h_{\psi_2} h_{\varphi_2}^{\frac{1-\alpha}{2\alpha}}\Big)^\alpha,\end{aligned}$$

that is, $\widetilde{Q}_\alpha(\psi_1\|\varphi_1) \le \widetilde{Q}_\alpha(\psi_2\|\varphi_2)$.

(8) By (1) and (4) for $\gamma : \mathbb{C}1 \hookrightarrow M$ one has

$$\widetilde{D}_\alpha(\psi\|\varphi) \ge \widetilde{D}_{1/2}(\psi\|\varphi) \ge -2\log\frac{\psi(1)^{1/2}\varphi(1)^{1/2}}{\psi(1)} = \log\frac{\psi(1)}{\varphi(1)}.$$

If $\psi = (\psi(1)/\varphi(1))\varphi$, then $\widetilde{D}_\alpha(\psi\|\varphi) \le D_\alpha(\psi\|\varphi) = \log\psi(1)/\varphi(1)$ by (6) and Theorem 3.2 (9), so that $\widetilde{D}_\alpha(\psi\|\varphi) = \log\psi(1)/\varphi(1)$. Conversely, if this equality holds, then $\widetilde{D}_{1/2}(\psi\|\varphi) = \log\psi(1)/\varphi(1)$, and $\psi = (\psi(1)/\varphi(1))\varphi$ follows as in the proof of [54, Corollary 4.2 (1)].

Remark 3.17 The expression of $\widetilde{D}_\alpha$ in (3.27) with $\widetilde{Q}_\alpha(\psi\|\varphi) = \mathrm{tr}\Big(h_\varphi^{\frac{1-\alpha}{2\alpha}} h_\psi h_\varphi^{\frac{1-\alpha}{2\alpha}}\Big)^\alpha$ in (3.28) makes sense for all $\alpha \in (0,1)$. In the matrix case, as seen from discussions in [38] that the monotonicity of $\widetilde{D}_\alpha$ $(0 < \alpha < 1)$ under CP trace-preserving maps is equivalent to the joint concavity of $\widetilde{Q}_\alpha$. But we note by [52, Proposition 5.1] that $\widetilde{Q}_\alpha$ is not jointly concave (even not separately concave in the second variable) for $0 < \alpha < 1/2$. Thus, the monotonicity of $\widetilde{D}_\alpha$ under unital normal CP maps in the above (4) is restricted to $\alpha \in [1/2, \infty)$, while that of D_α in Theorem 3.2 (8) is restricted to $\alpha \in [0,2]$. See also [19, Theorem 7] for violation of the monotonicity property of $\widetilde{D}_\alpha$ when $\alpha < 1/2$.

Remarks 3.18

(1) Inequality (3.31) was proved in [20, 66] by using the complex interpolation technique, while the proof of (3.32) in [66] is by Hölder's inequality for Haagerup's L^p-norms. In the case $M = B(\mathcal{H})$ with $\dim\mathcal{H} < \infty$, the inequality $\widetilde{D}_\alpha \le D_\alpha$ for any $\alpha \in (0,\infty)\setminus\{1\}$ is an immediate consequence of the *ALT (Araki–Lieb–Thirring) inequality* [12] (also [5]). The ALT inequality was extended to Haagerup's L^p-norms by Kosaki [80, Theorem 4]. In view of (3.27) and (3.28) note that (3.31) for $1/2 \le \alpha < 1$ is equivalently written as

$$\big\|h_\psi^{1/2} h_\varphi^{\frac{1-\alpha}{2\alpha}}\big\|_{2\alpha} \ge \big\|\,|h_\psi^{\alpha/2} h_\varphi^{(1-\alpha)/2}|^{1/\alpha}\big\|_{2\alpha},$$

which is a special case of Kosaki's extension of the ALT inequality. But when $\alpha > 1$, we are not able to derive (3.31) from the ALT inequality in the von Neumann algebra setting.

(2) Let $\psi, \varphi \in M_*^+$ with $\psi \neq 0$. If ψ, φ commute in the sense that h_ψ, h_φ commute (see Lemma 4.20 and Proposition A.56 of Sect. A.7), then

$$\widetilde{D}_\alpha(\psi\|\varphi) = D_\alpha(\psi\|\varphi) \quad \text{for all } \alpha \in (0, \infty) \setminus \{1\}.$$

This is obvious for $\alpha \in (0, 1)$, since $\widetilde{Q}_\alpha(\psi\|\varphi) = \operatorname{tr} h_\psi^\alpha h_\varphi^{1-\alpha} = Q_\alpha(\psi\|\varphi)$ by (3.28) and (3.11). When $\alpha > 1$, since $\widetilde{D}_\alpha(\psi\|\varphi) \le D_\alpha(\psi\|\varphi)$ by (3.31), we may assume that $\widetilde{D}_\alpha(\psi\|\varphi) < +\infty$. Then there exists an $x \in (s(\varphi)L^\alpha(M)s(\varphi))_+$ such that $h_\psi = h_\varphi^{\frac{\alpha-1}{2\alpha}} x h_\varphi^{\frac{\alpha-1}{2\alpha}}$. Since h_ψ, h_φ commute, by taking the commuting spectral resolutions of h_ψ, h_φ one can show that $h_\psi^{\alpha/2} = \eta h_\varphi^{(\alpha-1)/2}$ with $\eta := x^{\alpha/2} \in s(\varphi)L^2(M)s(\varphi)$. Theorem 3.6 gives $Q_\alpha(\psi\|\varphi) = \|\eta\|_2^2 = \|x\|_\alpha^\alpha = \|h_\psi\|_{\alpha,\varphi}^\alpha = \widetilde{Q}(\psi\|\varphi)$.

At the end of the section we give a variational expression of $\widetilde{Q}_\alpha$ for $\alpha \in (0, 1)$, whose special case for $\alpha = 1/2$ will be used in Example 5.18 of Sect. 5.3.

Lemma 3.19 *Let $0 < \alpha < 1$ and $\psi, \varphi \in M_*^+$. Define $\widetilde{Q}_\alpha(\psi\|\varphi) := \operatorname{tr}\Big(h_\varphi^{\frac{1-\alpha}{2\alpha}} h_\psi h_\varphi^{\frac{1-\alpha}{2\alpha}}\Big)^\alpha$. Then*

$$\widetilde{Q}_\alpha(\psi\|\varphi) \le \inf_{x \in M_{++}} \Big[\alpha \operatorname{tr}(h_\psi x) + (1-\alpha)\operatorname{tr}\big(h_\varphi^{\frac{1-\alpha}{2\alpha}} x^{-1} h_\varphi^{\frac{1-\alpha}{2\alpha}}\big)^{\frac{\alpha}{1-\alpha}}\Big], \tag{3.33}$$

where M_{++} is the set of positive invertible operators in M. Furthermore, if $1/2 \le \alpha < 1$ or $\psi \le \lambda\varphi$ for some $\lambda > 0$, then the above inequality becomes an equality, i.e.,

$$\widetilde{Q}_\alpha(\psi\|\varphi) = \inf_{x \in M_{++}} \Big[\alpha \operatorname{tr}(h_\psi x) + (1-\alpha)\operatorname{tr}\big(h_\varphi^{\frac{1-\alpha}{2\alpha}} x^{-1} h_\varphi^{\frac{1-\alpha}{2\alpha}}\big)^{\frac{\alpha}{1-\alpha}}\Big]. \tag{3.34}$$

Proof Write $R_\alpha(\psi\|\varphi)$ for the RHS of (3.33). For every $x \in M_{++}$ we have

$$\begin{aligned}
\widetilde{Q}_\alpha(\psi\|\varphi) &= \big\|h_\psi^{1/2} h_\varphi^{\frac{1-\alpha}{2\alpha}}\big\|_{2\alpha}^{2\alpha} = \big\|h_\psi^{1/2} x^{1/2} x^{-1/2} h_\varphi^{\frac{1-\alpha}{2\alpha}}\big\|_{2\alpha}^{2\alpha} \\
&\le \Big(\|h_\psi^{1/2} x^{1/2}\|_2 \big\|x^{-1/2} h_\varphi^{\frac{1-\alpha}{2\alpha}}\big\|_{\frac{2\alpha}{1-\alpha}}\Big)^{2\alpha} \quad \text{(by Hölder's inequality)} \\
&= \big[\operatorname{tr}(h_\psi^{1/2} x h_\psi^{1/2})\big]^\alpha \Big[\operatorname{tr}\big(h_\varphi^{\frac{1-\alpha}{2\alpha}} x^{-1} h_\sigma^{\frac{1-\alpha}{2\alpha}}\big)^{\frac{\alpha}{1-\alpha}}\Big]^{1-\alpha} \\
&\le \alpha \operatorname{tr}(h_\psi x) + (1-\alpha)\operatorname{tr}\big(h_\varphi^{\frac{1-\alpha}{2\alpha}} x^{-1} h_\varphi^{\frac{1-\alpha}{2\alpha}}\big)^{\frac{\alpha}{1-\alpha}},
\end{aligned}$$

which implies that $\widetilde{Q}_\alpha(\psi\|\varphi) \le R_\alpha(\psi\|\varphi)$.

To prove the latter assertion, first assume that $\lambda^{-1}\varphi \le \psi \le \lambda\varphi$ for some $\lambda > 1$. We may assume that ψ, φ are faithful. Write $\beta := \frac{1-\alpha}{\alpha} \in (0, \infty)$, so $(1+\beta)(1-\alpha) = \beta$. Since $\lambda^{-1}h_\varphi^{1+\beta} \le h_\varphi^{\beta/2} h_\psi h_\varphi^{\beta/2} \le \lambda h_\varphi^{1+\beta}$, it follows from (A.24) of Lemma A.58 that there are $a_0, b_0 \in M$ such that

$$h_\varphi^{\beta/2} = a_0 (h_\varphi^{\beta/2} h_\psi h_\varphi^{\beta/2})^{\frac{1-\alpha}{2}}, \qquad (h_\varphi^{\beta/2} h_\psi h_\varphi^{\beta/2})^{\frac{1-\alpha}{2}} = b_0 h_\varphi^{\beta/2}.$$

It is immediate to verify that $a_0 b_0 = b_0 a_0 = 1$ so that $b_0 = a_0^{-1}$. Set $x_0 := a_0 a_0^*$ so that $x_0^{-1} = (a_0^{-1})^* a_0^{-1} = b_0^* b_0$. Since

$$h_\varphi^{\beta/2} h_\psi h_\varphi^{\beta/2} = (h_\varphi^{\beta/2} h_\psi h_\varphi^{\beta/2})^{\frac{1-\alpha}{2}} a_0^* h_\psi a_0 (h_\varphi^{\beta/2} h_\psi h_\varphi^{\beta/2})^{\frac{1-\alpha}{2}},$$

one has $a_0^* h_\psi a_0 = (h_\varphi^{\beta/2} h_\psi h_\varphi^{\beta/2})^\alpha$ so that

$$\mathrm{tr}(h_\psi x_0) = \mathrm{tr}(a_0^* h_\psi a_0) = \mathrm{tr}(h_\varphi^{\beta/2} h_\psi h_\varphi^{\beta/2})^\alpha. \tag{3.35}$$

Moreover, since

$$h_\varphi^{\beta/2} x_0^{-1} h_\varphi^{\beta/2} = h_\varphi^{\beta/2} b_0^* b_0 h_\varphi^{\beta/2} = (h_\varphi^{\beta/2} h_\psi h_\varphi^{\beta/2})^{1-\alpha},$$

one has

$$\mathrm{tr}(h_\varphi^{\beta/2} x_0^{-1} h_\varphi^{\beta/2})^{1/\beta} = \mathrm{tr}(h_\varphi^{\beta/2} h_\psi h_\varphi^{\beta/2})^\alpha. \tag{3.36}$$

From (3.35) and (3.36), $R_\alpha(\psi\|\varphi) \le \mathrm{tr}(h_\varphi^{\beta/2} h_\psi h_\varphi^{\beta/2})^\alpha = \widetilde{Q}_\alpha(\psi\|\varphi)$. Therefore, $Q_\alpha(\psi\|\varphi) = R_\alpha(\psi\|\varphi)$ holds.

Now, assume that $1/2 \le \alpha < 1$, and let $\psi, \varphi \in M_*^+$ be general. From the joint monotonicity of $(\psi, \varphi) \mapsto \widetilde{Q}_\alpha(\psi\|\varphi)$ in Theorem 3.16 (7) and the joint continuity in Theorem 3.16 (3) we have

$$\begin{aligned}
&\widetilde{Q}_\alpha(\psi\|\varphi) \\
&\quad = \inf_{\varepsilon>0} \widetilde{Q}_\alpha(\psi + \varepsilon\varphi \| \varphi + \varepsilon\psi) \\
&\quad = \inf_{\varepsilon>0} \inf_{x\in M_{++}} \Big[\alpha \mathrm{tr}((h_\psi + \varepsilon h_\varphi)x) \\
&\qquad\qquad + (1-\alpha)\mathrm{tr}\big((h_\varphi + \varepsilon h_\psi)^{\beta/2} x^{-1} (h_\varphi + \varepsilon h_\psi)^{\beta/2}\big)^{1/\beta}\Big] \\
&\quad = \inf_{x\in M_{++}} \inf_{\varepsilon>0} \Big[\alpha \mathrm{tr}((h_\psi + \varepsilon h_\varphi)x) + (1-\alpha)\|x^{-1/2}(h_\varphi + \varepsilon h_\psi)^\beta x^{-1/2}\|_{1/\beta}^{1/\beta}\Big],
\end{aligned}$$

where the second equality follows from the case proved first. It is obvious that $\mathrm{tr}((h_\psi + \varepsilon h_\varphi)x) \searrow \mathrm{tr}(h_\psi x)$ as $\varepsilon \searrow 0$. On the other hand, by Kosaki's generalized

Powers–Størmer inequality we have

$$\|(h_\varphi+\varepsilon h_\psi)^\beta - h_\varphi^\beta\|_{1/\beta} \le \|(h_\varphi+\varepsilon h_\psi) - h_\varphi\|_1^\beta = \|\varepsilon h_\psi\|_1^\beta \longrightarrow 0$$

so that $x^{-1/2}(h_\varphi+\varepsilon h_\psi)^\beta x^{-1/2} \to x^{-1/2}h_\varphi^\beta x^{-1/2}$ as $\varepsilon \searrow 0$ in the measure topology in $\widetilde{N}_+$, where $N := M \rtimes_{\sigma^\omega} \mathbb{R}$, see Sect. A.6. Moreover, since $\beta \in (0,1]$, note by Lemma B.7 that $x^{-1/2}(h_\varphi+\varepsilon h_\psi)^\beta x^{-1/2}$ is monotone decreasing as $\varepsilon \searrow 0$ in the sense of Proposition B.4 (equivalently, in the order in $\widetilde{N}_+$). Hence by [37, Lemmas 3.4 and 4.8] we have

$$\begin{aligned}\|x^{-1/2}(h_\varphi+\varepsilon h_\psi)^\beta x^{-1/2}\|_{1/\beta} &= \mu_1(x^{-1/2}(h_\varphi+\varepsilon h_\psi)^\beta x^{-1/2}) \\ &\searrow \mu_1(x^{-1/2}h_\varphi^\beta x^{-1/2}) = \|x^{-1/2}h_\varphi^\beta x^{-1/2}\|_{1/\beta}\end{aligned}$$

as $\varepsilon \searrow 0$. Therefore,

$$\begin{aligned}\widetilde{Q}_\alpha(\psi\|\varphi) &= \inf_{x\in M_{++}} \Big[\alpha\mathrm{tr}(h_\psi x) + (1-\alpha)\|x^{-1/2}h_\varphi^\beta x^{-1/2}\|_{1/\beta}^{1/\beta}\Big] \\ &= \inf_{x\in M_{++}} \Big[\alpha\mathrm{tr}(h_\psi x) + (1-\alpha)\mathrm{tr}(h_\varphi^{\beta/2}x^{-1}h_\varphi^{\beta/2})^{1/\beta}\Big].\end{aligned}$$

Next, assume that $\psi \le \lambda\varphi$ for some $\lambda > 0$. From the proofs of Theorem 3.16 (3) and (7) we note that the continuity property of Theorem 3.16 (3) and the monotonicity of $\psi \mapsto \widetilde{Q}_\alpha(\psi\|\varphi)$ hold for all $\alpha \in (0,1)$. Hence we have

$$\begin{aligned}\widetilde{Q}_\alpha(\psi\|\varphi) &= \inf_{\varepsilon>0} \widetilde{Q}_\alpha(\psi+\varepsilon\varphi\|\varphi) \\ &= \inf_{\varepsilon>0}\inf_{x\in M_{++}} \Big[\alpha\mathrm{tr}((h_\psi+\varepsilon h_\varphi)x) + (1-\alpha)\mathrm{tr}\big(h_\varphi^{\beta/2}x^{-1}h_\varphi^{\beta/2}\big)^{1/\beta}\Big] \\ &= \inf_{x\in M_{++}} \Big[\alpha\mathrm{tr}(h_\psi x) + (1-\alpha)\mathrm{tr}\big(h_\varphi^{\beta/2}x^{-1}h_\varphi^{\beta/2}\big)^{1/\beta}\Big],\end{aligned}$$

where the second equality holds by the first case.

Remarks 3.20

(1) The definition of $\widetilde{D}_\alpha$ for $\alpha > 1$ in Definition 3.9 ([66]) is based on Kosaki's *symmetric* L^p-spaces. In [67] Jenčová introduced a sightly different definition of $\widetilde{D}_\alpha$ for all $\alpha \in [1/2,\infty)\setminus\{1\}$ based on Kosaki's *right* L^p-spaces (see Sect. A.8), and proved that the new definition coincides with $\widetilde{D}_\alpha^{(\mathrm{BST})}$ in Definition 3.10 for all $\alpha \in [1/2,\infty)\setminus\{1\}$. Therefore, all the definitions of $\widetilde{D}_\alpha$ in [20, 66, 67] are equivalent.

(2) In [67] the variational expression in (3.34) was used to prove the monotonicity of $\widetilde{D}_\alpha$ $(1/2 \le \alpha < 1)$ under a unital normal general positive map. This and the same monotonicity in [66] of $\widetilde{D}_\alpha$ $(\alpha > 1)$ together give Theorem 3.16 (4). The variational expression

$$\widetilde{Q}_\alpha = \sup_{x \in M_+} \left[\alpha \mathrm{tr}(h_\psi x) - (\alpha - 1) \mathrm{tr}\big(h_\varphi^{\frac{\alpha-1}{2\alpha}} x^{-1} h_\varphi^{\frac{\alpha-1}{2\alpha}}\big)^{\frac{\alpha}{\alpha-1}} \right] \tag{3.37}$$

for $\alpha > 1$ was also given in [67, Proposition 3.4]. Hence (3.34) and (3.37) together are the von Neumann algebra version of [38, Lemma 4] in the finite-dimensional case.

Chapter 4
Maximal f-Divergences

4.1 Definition and Basic Properties

Let M be a von Neumann algebra with its standard form $(M, \mathcal{H}, J, \mathcal{P})$ as before. Throughout this chapter, we assume that f is an operator convex function on $(0, \infty)$. For $\psi, \varphi \in M_*^+$ we write $\psi \sim \varphi$ if $\delta\varphi \le \psi \le \delta^{-1}\varphi$ for some $\delta > 0$, and for convenience we set

$$(M_*^+ \times M_*^+)_\sim := \{(\psi, \varphi) \in M_*^+ \times M_*^+ : \psi \sim \varphi\},$$

$$(M_*^+ \times M_*^+)_\le := \{(\psi, \varphi) \in M_*^+ \times M_*^+ : \psi \le \alpha\varphi \text{ for some } \alpha > 0\},$$

which are convex sets. We first define the maximal f-divergence for $(\psi, \varphi) \in (M_*^+ \times M_*^+)_\sim$ and then extend it to general $\psi, \varphi \in M_*^+$.

Definition 4.1 For $(\psi, \varphi) \in (M_*^+ \times M_*^+)_\le$ with respective vector representatives $\Psi, \Phi \in \mathcal{P}$, by Lemma A.24 of Sect. A.3 we have a unique $A \in s(\varphi)Ms(\varphi)$ such that $\Psi = A\Phi$. We write $T_{\psi/\varphi}$ for $A^*A \in (s(\varphi)Ms(\varphi))_+$. Now assume that $(\psi, \varphi) \in (M_*^+ \times M_*^+)_\sim$, so $T_{\psi/\varphi}$ is a positive invertible operator in $s(\varphi)Ms(\varphi)$. Then we have a self-adjoint operator $f(T_{\psi/\varphi})$ in $s(\varphi)Ms(\varphi)$ via functional calculus and define the *maximal f-divergence* of ψ with respect to φ by

$$\widehat{S}_f(\psi\|\varphi) := \varphi(f(T_{\psi/\varphi})) \in \mathbb{R}. \tag{4.1}$$

When $M = B(\mathcal{H})$ on a finite-dimensional Hilbert space $\mathcal{H}$ and $\psi, \varphi \in B(\mathcal{H})_*^+$ are faithful with the density operators D_ψ, D_φ, the above (4.1) is given as

$$\widehat{S}_f(\psi\|\varphi) = \operatorname{Tr} D_\varphi^{1/2} f(D_\varphi^{-1/2} D_\psi D_\varphi^{-1/2}) D_\varphi^{1/2}. \tag{4.2}$$

This notion of quantum divergence was first introduced by Petz and Ruskai [110] and later developed in more detail in [58, 92]. The above expression

F. Hiai, *Quantum f-Divergences in von Neumann Algebras*,
Mathematical Physics Studies, https://doi.org/10.1007/978-981-33-4199-9_4

$D_\varphi^{1/2} f(D_\varphi^{-1/2} D_\psi D_\varphi^{-1/2}) D_\varphi^{1/2}$ is the *operator perspective* [35, 36] of D_ψ, D_φ by the function f, which will appear in Sect. 8.1 again.

To extend $\widehat{S}_f(\psi\|\varphi)$ to general $\psi, \varphi \in M_*^+$, we need the following:

Lemma 4.2 ([55]) *Let $\psi, \varphi \in M_*^+$.*

(1) *For every $\omega \in M_*^+$ with $\omega \sim \psi + \varphi$, the limit*

$$\lim_{\varepsilon \searrow 0} \widehat{S}_f(\psi + \varepsilon\omega \| \varphi + \varepsilon\omega) \in (-\infty, +\infty]$$

exists, and moreover the limit is independent of the choice of ω as above.

(2) *If $(\psi, \varphi) \in (M_*^+ \times M_*^+)_\sim$, then*

$$\widehat{S}_f(\psi\|\varphi) = \lim_{\varepsilon \searrow 0} \widehat{S}_f(\psi + \varepsilon\varphi \| (1+\varepsilon)\varphi).$$

The lemma says that the RHS of (4.3) is well defined independently of the choice of ω and extends Definition 4.1 for the case $\psi \sim \varphi$.

Definition 4.3 For every $\psi, \varphi \in M_*^+$ define the *maximal f-divergence* $\widehat{S}_f(\psi\|\varphi)$ by

$$\widehat{S}_f(\psi\|\varphi) := \lim_{\varepsilon \searrow 0} \widehat{S}_f(\psi + \varepsilon\omega \| \varphi + \varepsilon\omega) \in (-\infty, +\infty] \tag{4.3}$$

for any $\omega \in M_*^+$ with $\omega \sim \psi + \varphi$, where $\widehat{S}_f(\psi + \varepsilon\omega \| \varphi + \varepsilon\omega)$ is defined in Definition 4.1.

The most basic properties of $\widehat{S}_f(\psi\|\varphi)$ are the following:

Theorem 4.4 ([55])

(i) Monotonicity: *Let N be another von Neumann algebra and $\gamma : N \to M$ be a unital normal positive map. Then for every $\psi, \varphi \in M_*^+$,*

$$\widehat{S}_f(\psi \circ \gamma \| \varphi \circ \gamma) \le \widehat{S}_f(\psi\|\varphi).$$

(ii) Joint convexity: *For every $\psi_i, \varphi_i \in M_*^+$ and $\lambda_i \ge 0$ $(1 \le i \le n)$,*

$$\widehat{S}_f\left(\sum_{i=1}^n \lambda_i \psi_i \,\middle\|\, \sum_{i=1}^n \lambda_i \varphi_i\right) \le \sum_{i=1}^n \lambda_i \widehat{S}_f(\psi_i \| \varphi_i).$$

Our proofs of Lemma 4.2 and Theorem 4.4 in [55] go as follows: The properties (i) and (ii) of Theorem 4.4 are first shown restricted to $(\psi, \varphi) \in (M_*^+ \times M_*^+)_\sim$, from which we can prove the convergence results in Lemma 4.2. Then from definition (4.3) the properties of Theorem 4.4 can extend to general $\psi, \varphi \in M_*^+$.

Another significant property of $\widehat{S}_f(\psi\|\varphi)$ is the joint lower semicontinuity, which will be given later after developing a general integral formula in the next section. Some other basic properties are summarized in the next proposition.

Proposition 4.5 ([55]) *Let $\psi, \varphi \in M_*^+$.*

(1) *Let $\widetilde{f}$ be the transpose of f (see Proposition 2.3 (5)). Then*

$$\widehat{S}_{\widetilde{f}}(\psi\|\varphi) = \widehat{S}_f(\varphi\|\psi).$$

(2) *Let $\psi_i, \varphi_i \in M_*^+$ $(i = 1, 2)$. If $s(\psi_1) \vee s(\varphi_1) \perp s(\psi_2) \vee s(\varphi_2)$, then*

$$\widehat{S}_f(\psi_1 + \psi_2\|\varphi_1 + \varphi_2) = \widehat{S}_f(\psi_1\|\varphi_1) + \widehat{S}_f(\psi_2\|\varphi_2).$$

(3) *When $(\psi, \varphi) \in (M_*^+ \times M_*^+)_\le$,*

$$\widehat{S}_f(\psi\|\varphi) = \lim_{\varepsilon\searrow 0} \widehat{S}_f(\psi + \varepsilon\varphi\|\varphi).$$

When $(\varphi, \psi) \in (M_*^+ \times M_*^+)_\le$,

$$\widehat{S}_f(\psi\|\varphi) = \lim_{\varepsilon\searrow 0} \widehat{S}_f(\psi\|\varphi + \varepsilon\psi).$$

(4) *If $s(\psi) \not\le s(\varphi)$ and $f'(\infty) = +\infty$, then $\widehat{S}_f(\psi\|\varphi) = +\infty$. If $s(\varphi) \not\le s(\psi)$ and $f(0^+) = +\infty$, then $\widehat{S}_f(\psi\|\varphi) = +\infty$.*

(5) *If $f(0^+) < +\infty$, then*

$$\widehat{S}_f(\psi\|\varphi) = \lim_{\varepsilon\searrow 0} \widehat{S}_f(\psi\|\varphi + \varepsilon\psi).$$

If $f'(\infty) < +\infty$, then

$$\widehat{S}_f(\psi\|\varphi) = \lim_{\varepsilon\searrow 0} \widehat{S}_f(\psi + \varepsilon\varphi\|\varphi).$$

(6) *If $f(0^+) < +\infty$, then expression* (4.1) *holds for every $(\psi, \varphi) \in (M_*^+ \times M_*^+)_\le$, where $f(T_{\psi/\varphi})$ is defined for f on $[0, \infty)$ with $f(0) = f(0^+)$.*

Examples 4.6

(1) Assume that $M = L^\infty(X, \mu)$ is an abelian von Neumann algebra on a σ-finite measure space $(X, \mathcal{X}, \mu)$. Let $\psi, \varphi \in M_*^+$, identified with functions in $L^1(X, \mu)_+$ so that $\psi(\phi) = \int_X \phi\psi\, d\mu$, $\varphi(\phi) = \int_X \phi\varphi\, d\mu$ for $\phi \in L^\infty(X, \mu)$. With $\omega = \psi + \varphi$ we have

$$\widehat{S}_f(\psi + \varepsilon\omega\|\varphi + \varepsilon\omega) = S_f(\psi + \varepsilon\omega\|\varphi + \varepsilon\omega),$$

see Example 2.2 (1). Take the limit of the above as $\varepsilon \searrow 0$ to see that $\widehat{S}_f(\psi\|\varphi)$ coincides with the classical f-divergence $S_f(\psi\|\varphi) = \int_X \varphi f(\psi/\varphi)\,d\mu$.

(2) Assume that $M = B(\mathcal{H})$ on a finite-dimensional Hilbert space $\mathcal{H}$. Let $\psi, \varphi \in B(\mathcal{H})_*^+$ with the density operators D_ψ, D_φ. When ψ, φ are faithful, (4.1) becomes (4.2). For general $\psi, \varphi \in B(\mathcal{H})_*^+$ let e be the support projection of $\psi + \varphi$. By Proposition 4.5 (2) we have

$$\begin{aligned}\lim_{\varepsilon\searrow 0} \widehat{S}_f(\psi + \varepsilon I\|\varphi + \varepsilon I) &= \lim_{\varepsilon\searrow 0} \big\{\widehat{S}_f(\psi + \varepsilon e\|\varphi + \varepsilon e) + \widehat{S}_f(\varepsilon(I - e)\|\varepsilon(I - e))\big\} \\ &= \lim_{\varepsilon\searrow 0} \big\{\widehat{S}_f(\psi + \varepsilon e\|\varphi + \varepsilon e) + \varepsilon \mathrm{Tr}(I - e) f(1)\big\} \\ &= \lim_{\varepsilon\searrow 0} \widehat{S}_f(\psi + \varepsilon e\|\varphi + \varepsilon e),\end{aligned}$$

so that $\widehat{S}_f(\psi\|\varphi)$ coincides with that defined in [58, Definition 3.21].

Examples 4.7

(1) Consider a linear function $f(t) = a + bt$ with $a, b \in \mathbb{R}$. Let $(\psi, \varphi) \in (M_*^+ \times M_*^+)_\sim$ with $e = s(\psi) = s(\varphi)$. Let $A \in eMe$ be as in Lemma A.24 so that $\Psi = A\Phi$; then

$$\widehat{S}_{a+bt}(\psi\|\varphi) = \varphi(ae + bA^*A) = a\varphi(e) + b\|A\Phi\|^2 = a\varphi(1) + b\psi(1).$$

Hence by Definition 4.3 and (2.4) we have for every $\psi, \varphi \in M_*^+$,

$$\widehat{S}_{a+bt}(\psi\|\varphi) = S_{a+bt}(\psi\|\varphi) = a\varphi(1) + b\psi(1). \tag{4.4}$$

(2) Consider $f(t) = t^2$ and show that for every $\psi, \varphi \in M_*^+$,

$$\widehat{S}_{t^2}(\psi\|\varphi) = S_{t^2}(\psi\|\varphi). \tag{4.5}$$

Let $(\psi, \varphi) \in (M_*^+ \times M_*^+)_\sim$ with corresponding $\Psi, \Phi \in \mathcal{P}$. With A in Lemma A.24 we have

$$\widehat{S}_{t^2}(\psi\|\varphi) = \langle\Phi, (A^*A)^2\Phi\rangle = \langle\Psi, AA^*\Psi\rangle = \|A^*\Psi\|_2^2.$$

On the other hand,

$$S_{t^2}(\psi\|\varphi) = \|\Delta_{\psi,\varphi}\Phi\|_2^2 = \|\Delta_{\psi,\varphi}^{1/2}\Psi\|_2^2 = \|\Delta_{\psi,\varphi}^{1/2}A\Phi\|_2^2 = \|A^*\Psi\|_2^2,$$

so (4.5) follows in this case. Therefore, in view of (4.3) and (2.16), this holds for all $\psi, \varphi \in M_*^+$. By (4.4) and (4.5) together, $\widehat{S}_f = S_f$ if f is a quadratic polynomial.

4.2 Further Properties of Maximal f-Divergences

Let $\psi, \varphi \in M_*^+$ and let $\Psi, \Omega \in \mathcal{P}$ be the vector representatives of ψ and $\omega := \psi + \varphi$. We have a unique $A \in s(\omega)Ms(\omega)$ such that $\Psi = A\Omega$. Let $T_{\psi/\omega} := A^*A$ in $s(\omega)Ms(\omega)$ as in Definition 4.1. Since $0 \le T_{\psi/\omega} \le 1$, we write the spectral decomposition of $T_{\psi/\omega}$ as

$$T_{\psi/\omega} = \int_0^1 t\, dE_{\psi/\omega}(t) \tag{4.6}$$

with the spectral resolution $\{E_{\psi/\omega}(t) : 0 \le t \le 1\}$ in $s(\omega)Ms(\omega)$. In the next theorem we present a general integral formula of $\widehat{S}_f(\psi\|\varphi)$, which may serve as the second definition of maximal f-divergences.

Theorem 4.8 ([55]) *Let $\psi, \varphi \in M_*^+$ and $\omega := \psi + \varphi$ with the vector representative Ω. Then for every operator convex function f on $(0, \infty)$,*

$$\widehat{S}_f(\psi\|\varphi) = \int_0^1 (1-t) f\left(\frac{t}{1-t}\right) d\|E_{\psi/\omega}(t)\Omega\|_2^2, \tag{4.7}$$

where $(1-t)f\left(\frac{t}{1-t}\right)$ is understood as $f(0^+)$ for $t = 0$ and $f'(\infty)$ for $t = 1$.

Corollary 4.9 *If $f(0^+) < +\infty$ and $f'(+\infty) < +\infty$, then $\widehat{S}_f(\psi\|\varphi)$ is finite for every $\psi, \varphi \in M_*^+$.*

Proof If $f(0^+) < +\infty$ and $f'(+\infty) < +\infty$, then the function $(1-t)f\left(\frac{t}{1-t}\right)$ is bounded on $[0, 1]$. Hence the result is obvious from expression (4.7).

The proof of Theorem 4.8 in [55] is based on the integral representation (2.5) of f. Here it is worth noting that the integrand of (4.7) enjoys a special theoretical meaning as follows. Let f be a real function on $(0, \infty)$ and g be given as $g(t) := (1-t)f\left(\frac{t}{1-t}\right)$, $t \in (0, 1)$. Then f is operator convex on $(0, \infty)$ if and only if g is operator convex on $(0, 1)$. Moreover, in this case, $g(0^+) = f(0^+)$ and $g(1^-) = f'(\infty)$. We include the proof of this fact in Appendix C.

Remark 4.10 Similarly to the definition of the standard f-divergence $S_f(\psi\|\varphi)$ in (2.3), one can write (4.7) as the sum of three terms

$$\begin{aligned}\widehat{S}_f(\psi\|\varphi) = &\int_{(0,1)} (1-t) f\left(\frac{t}{1-t}\right) d\|E_{\psi/\omega}(t)\Omega\|_2^2 \\ &+ f(0^+)\langle\Omega, E_{\psi/\omega}(\{0\})\Omega\rangle + f'(\infty)\langle\Omega, E_{\psi/\omega}(\{1\})\Omega\rangle,\end{aligned} \tag{4.8}$$

where $E_{\psi/\omega}(\{0\})$ and $E_{\psi/\omega}(\{1\})$ are the spectral projections of $T_{\psi/\omega}$ for $\{0\}$ and $\{1\}$. One might expect that the above boundary terms including $f(0^+)$ and $f'(\infty)$ are equal to the corresponding terms $f(0^+)\varphi(1 - s(\psi))$ and $f'(\infty)\psi(1 - s(\varphi))$, respectively, in (2.3). But it is not true, as shown in [55, Example 4.6].

In the following we state the joint lower semicontinuity in the norm and the martingale convergence for $\widehat{S}_f(\psi\|\varphi)$.

Theorem 4.11 ([55]) *The function* $(\psi,\varphi)\in M_*^+\times M_*^+\mapsto\widehat{S}_f(\psi\|\varphi)$ *is jointly lower semicontinuous in the norm topology.*

Theorem 4.12 ([55]) *Let* $\{M_\alpha\}$ *be an increasing net of unital von Neumann subalgebras of* M *such that* $\left(\bigcup_\alpha M_\alpha\right)''=M$. *Then for every* $\psi,\varphi\in M_*^+$,

$$\widehat{S}_f(\psi|_{M_\alpha}\|\varphi|_{M_\alpha})\nearrow\widehat{S}_f(\psi\|\varphi).$$

The proofs of these properties are based on the integral formula (4.7). For each $n\in\mathbb{N}$ consider the approximation f_n of f in (2.6) with integral on the cut-off interval $[1/n,n]$. From (2.9) we have as $n\to\infty$,

$$(1-t)f_n\left(\frac{t}{1-t}\right)\nearrow(1-t)f\left(\frac{t}{1-t}\right),\qquad t\in[0,1].$$

Hence from the monotone convergence theorem applied to the integral formula in (4.7) for f_n and f we have for every $\psi,\varphi\in M_*^+$,

$$\widehat{S}_f(\psi\|\varphi)=\lim_{n\to\infty}\widehat{S}_{f_n}(\psi\|\varphi)\quad\text{increasingly},$$

so that $\widehat{S}_f(\psi\|\varphi)=\sup_{n\ge1}\widehat{S}_{f_n}(\psi\|\varphi)$. Moreover, since

$$\lim_{t\searrow0}f_n'(t)=\lim_{t\searrow0}\frac{f_n(t)-f_n(0^+)}{t}>-\infty,$$

to prove Theorems 4.11 and 4.12, we may assume that $f(0^+)<+\infty$, $f'(\infty)<+\infty$ and $\lim_{t\searrow0}f'(t)>-\infty$. See [55, Sec. V] for the details of the proofs of the theorems.

Problem 4.13 Theorem 2.7 (i) says that the standard f-divergence $S_f(\psi\|\varphi)$ is jointly lower semicontinuous in the $\sigma(M_*,M)$-topology. It follows from Theorem 4.12 that this property (stronger than Theorem 4.11) holds for $\widehat{S}_f(\psi\|\varphi)$ as well when M is injective, or equivalently, approximately finite dimensional (AFD), that is, there is an increasing net $\{M_\alpha\}$ of finite-dimensional unital subalgebras of M such that $M=\left(\bigcup_\alpha M_\alpha\right)''$, see Sect. A.1. In fact, in this case, $\widehat{S}_f(\psi\|\varphi)=\sup_\alpha\widehat{S}_f(\psi|_{M_\alpha}\|\varphi|_{M_\alpha})$ by Theorem 4.12 and $(\psi,\varphi)\mapsto\widehat{S}_f(\psi|_{M_\alpha}\|\varphi|_{M_\alpha})$ is lower semicontinuous in the $\sigma(M_*,M)$-topology. However, it is unknown whether or not $\widehat{S}_f(\psi\|\varphi)$ is jointly lower semicontinuous in the $\sigma(M_*,M)$-topology for general M.

Another martingale convergence like Proposition 2.10 for S_f holds for $\widehat{S}_f$ as well.

Proposition 4.14 *Let $\{e_\alpha\}$ be an increasing net of projections in M such that $e_\alpha \nearrow 1$. Then for every $\psi, \varphi \in M_*^+$,*

$$\lim_\alpha \widehat{S}_f(e_\alpha \psi e_\alpha \| e_\alpha \varphi e_\alpha) = \widehat{S}_f(\psi \| \varphi),$$

where $e_\alpha \varphi e_\alpha$ is the restriction of φ to the reduced von Neumann algebra $e_\alpha M e_\alpha$.

4.3 Minimal Reverse Test

For every $\psi, \varphi \in M_*^+$ let $\omega := \psi + \varphi$ with the vector representative Ω of ω as well as Ψ of ψ. We have a positive operator $T_{\psi/\omega} = A^*A = \int_0^1 t\, dE_{\psi/\omega}(t)$ in $s(\omega)Ms(\omega)$ as in (4.6), where $A \in s(\omega)Ms(\omega)$ satisfies $\Psi = A\Omega$. Define a finite Borel measure ν on $[0, 1]$ by

$$\nu := \omega(E_{\psi/\omega}(\cdot)) = \|E_{\psi/\omega}(\cdot)\Omega\|^2, \tag{4.9}$$

and consider an abelian von Neumann algebra $L^\infty([0, 1], \nu) = L^1([0, 1], \nu)^*$.

Lemma 4.15 ([55]) *Define $\lambda_0 : M \to L^\infty([0, 1], \nu)$ by*

$$\lambda_0(x) = \frac{d\langle Jx\Omega, E_{\psi/\omega}(\cdot)\Omega\rangle}{d\nu} \quad \textit{(the Radon–Nikodym derivative)}, \qquad x \in M.$$

Then λ_0 is a unital normal positive map and its predual map $\lambda_{0*} : L^1([0, 1], \nu) \to M_*$ is given by

$$\lambda_{0*}(\phi)(x) = \left\langle Jx\Omega, \left(\int_0^1 \phi\, dE_{\psi/\omega}\right)\Omega\right\rangle$$

for every $\phi \in L^\infty([0, 1], \nu)$ ($\subset L^1([0, 1], \nu)$) and $x \in M$.

The above λ_0 satisfies

$$\lambda_{0*}(\phi)(1) = \omega\left(\int_0^1 \phi\, dE_{\psi/\omega}\right) = \int_0^1 \phi\, d\nu$$

for all $\phi \in L^\infty([0, 1], \nu)$ (hence all $\phi \in L^1([0, 1], \nu)$). Moreover, for every $x \in M$,

$$\begin{aligned} \lambda_{0*}(1)(x) &= \langle Jx\Omega, \Omega\rangle = \langle \Omega, x\Omega\rangle = \omega(x), \\ \lambda_{0*}(t)(x) &= \langle Jx\Omega, T_{\psi,\omega}\Omega\rangle = \langle JxJ\Omega, A^*A\Omega\rangle \\ &= \langle JxJA\Omega, A\Omega\rangle = \langle Jx\Psi, \Psi\rangle = \langle \Psi, x\Psi\rangle = \psi(x), \end{aligned}$$

where t denotes the identity function $t \mapsto t$ on $[0, 1]$. Hence $\lambda_{0*}(t) = \psi$ and $\lambda_{0*}(1-t) = \omega - \psi = \varphi$.

Now, following Matsumoto's idea [92], we introduce a key notion below.

Definition 4.16 Let $(X, \mathcal{X}, \mu)$ be a σ-finite measure space and $\lambda : M \to L^\infty(X, \mu)$ be a positive linear map which is unital and normal. Then the predual map $\lambda_* : L^1(X, \mu) \to M_*$ is trace-preserving in the sense that $\int_X \phi\, d\mu = \lambda_*(\phi)(1)$ $(= \|\lambda_*(\phi)\|)$ for every $\phi \in L^1(X, \mu)_+$. We call a triplet (λ, p, q) of such a map λ and $p, q \in L^1(X, \mu)_+$ a *reverse test* for ψ, φ if $\lambda_*(p) = \psi$ and $\lambda_*(q) = \varphi$, i.e., $p \circ \lambda = \psi$ and $q \circ \lambda = \varphi$ under identification $L^1(X, \mu)_+ = L^\infty(X, \mu)_*^+$.

The next variational formula of $\widehat{S}_f(\psi\|\varphi)$ may serve as the third definition of maximal f-divergences.

Theorem 4.17 ([55]) *For every* $\psi, \varphi \in M_*^+$,

$$\widehat{S}_f(\psi\|\varphi) = \min\{S_f(p\|q) : (\lambda, p, q) \text{ a reverse test for } \psi, \varphi\}. \tag{4.10}$$

Moreover, $\lambda : M \to L^\infty(X, \mu)$ *in* (4.10) *can be restricted to those with a standard Borel probability space* $(X, \mathcal{X}, \mu)$ *or more specifically to those with a Borel probability space on* $[0, 1]$.

In fact, $(\lambda_0, t, 1-t)$ given in Lemma 4.15 is a reverse test, for which the equality $\widehat{S}_f(\psi\|\varphi) = S_f(t\|1-t)$ holds since from (4.7) and (4.9) we have

$$\widehat{S}_f(\psi\|\varphi) = \int_0^1 (1-t) f\left(\frac{t}{1-t}\right) d\nu(t) = S_f(t\|1-t).$$

Therefore, the reverse test $(\lambda_0, t, 1-t)$ is a minimizer for expression (4.10), which is considered as the von Neumann algebra version of Matsumoto's *minimal* or *optimal reverse test* [92] for ψ, φ. Apply the monotonicity property of S_f (Theorem 2.7 (iv)) to this λ_0 (that is a unital and normal CP map) to have

$$S_f(\psi\|\varphi) \le S_f(t\|1-t) = \widehat{S}_f(\psi\|\varphi),$$

so we have the following:

Theorem 4.18 ([55]) *For every* $\psi, \varphi \in M_*^+$,

$$S_f(\psi\|\varphi) \le \widehat{S}_f(\psi\|\varphi).$$

The following corollary is worth giving, which is immediate from the above theorem and Definition 4.1.[1]

[1] A simple direct proof of $S_f(\psi\|\varphi) < +\infty$ for $(\psi, \varphi) \in (M_*^+ \times M_*^+)_\sim$ is unknown to us.

Corollary 4.19 *For every* $(\psi, \varphi) \in (M_*^+ \times M_*^+)_\sim$,

$$S_f(\psi\|\varphi) \le \widehat{S}_f(\psi\|\varphi) < +\infty.$$

When $\mathcal{H}$ is finite-dimensional, it is easy to verify that if $\psi, \varphi \in B(\mathcal{H})_*^+$ commute in the sense that the density operators D_ψ, D_φ commute, then $S_f(\psi\|\varphi) = \widehat{S}_f(\psi\|\varphi)$ for every operator convex (even simply convex) function on $(0, \infty)$. We extend this to the general von Neumann algebra setting. To do this, we first state the next lemma whose proof was given in [55, Appendix B].

Lemma 4.20 *For every* $\psi, \varphi \in M_*^+$ *the following conditions are equivalent:*

(i) $\psi \circ \sigma_t^{\psi+\varphi} = \psi$ *on* $s(\psi+\varphi)Ms(\psi+\varphi)$ *for all* $t \in \mathbb{R}$ *(note that the modular automorphism group* $\sigma_t^{\psi+\varphi}$ *is defined on* $s(\psi+\varphi)Ms(\psi+\varphi)$*);*
(ii) $\varphi \circ \sigma_t^{\psi+\varphi} = \varphi$ *on* $s(\psi+\varphi)Ms(\psi+\varphi)$ *for all* $t \in \mathbb{R}$*;*
(iii) $h_\psi h_\varphi = h_\varphi h_\psi$*, where* h_ψ, h_φ *are the corresponding elements of Haagerup's* $L^1(M)$ *(see Sect. A.6).*

When $s(\psi) \le s(\varphi)$*, the above are also equivalent to* $\psi \circ \sigma_t^\varphi = \psi$ *for all* $t \in \mathbb{R}$.

More equivalent conditions on the commutativity of ψ, φ are provided in Proposition A.56 of Sect. A.7.

Proposition 4.21 ([55]) *If* $\psi, \varphi \in M_*^+$ *commute in the sense that the equivalent conditions of Lemma 4.20 hold, then*

$$S_f(\psi\|\varphi) = \widehat{S}_f(\psi\|\varphi)$$

for any operator convex function f *on* $(0, \infty)$.

Example 4.22 In [17] Belavkin and Staszewski introduced a type of relative entropy for states on a C^*-algebra. In the von Neumann algebras case, their relative entropy is realized as $\widehat{S}_f$ for $f(t) = t\log t$ (see [55, Example 3.5] for a detailed explanation). So *Belavkin and Staszewski's relative entropy* $D_{\mathrm{BS}}(\psi\|\varphi)$ for every $\psi, \varphi \in M_*^+$ is defined to be

$$D_{\mathrm{BS}}(\psi\|\varphi) := \widehat{S}_{t\log t}(\psi\|\varphi).$$

Since $D(\psi\|\varphi) = S_{t\log t}(\psi\|\varphi)$, by Theorem 4.18 and Proposition 4.21 we have for every $\psi, \varphi \in M_*^+$,

$$D(\psi\|\varphi) \le D_{\mathrm{BS}}(\psi\|\varphi),$$

and $D(\psi\|\varphi) = D_{\mathrm{BS}}(\psi\|\varphi)$ if ψ, φ commute (in the sense of Lemma 4.20).

Remark 4.23 In Theorem 7.11 of Sect. 7.2 we will prove that the converse direction of Proposition 4.21 is also true, in such a way that $\psi, \varphi \in M_*^+$ commute if $S_f(\psi\|\varphi) = \widehat{S}_f(\psi\|\varphi) < +\infty$ under $s(\psi) \le s(\varphi)$ and a certain mild assumption on f. (This was shown in [58, Theorem 4.3] in the finite-dimensional case.) In particular, for $\psi, \varphi \in M_*^+$ with $D(\psi\|\varphi) < +\infty$, it will follow that ψ, φ commute if and only if $D(\psi\|\varphi) = D_{\mathrm{BS}}(\psi\|\varphi)$.

Chapter 5
Measured f-Divergences

5.1 Definition

Let f be a convex function on $(0, \infty)$, not necessarily operator convex unless we specify that. We use the convention in (2.2). Let M be a general von Neumann algebra. A *measurement* $\mathcal{M}$ in M is given by $\mathcal{M} = (A_j)_{1 \le j \le n}$ for some $n \in \mathbb{N}$, where $A_j \in M_+$ for $1 \le j \le n$ and $\sum_{j=1}^{n} A_j = 1$. The measurement $\mathcal{M}$ is identified with a unital positive map $\alpha : \mathbb{C}^n \to M$, determined by $\alpha(\delta_j) := A_j$, $1 \le j \le n$. For $\psi, \varphi \in M_*^+$ we write $\mathcal{M}(\psi) := \psi \circ \alpha$, i.e., $\mathcal{M}(\psi) := (\psi(A_j))_{1 \le j \le n}$, and similarly for $\mathcal{M}(\varphi)$. Then the classical f-divergence $S_f(\mathcal{M}(\psi) \| \mathcal{M}(\varphi))$ is defined by

$$S_f(\mathcal{M}(\psi) \| \mathcal{M}(\varphi)) := \sum_{j=1}^{n} \varphi(A_j) f\left(\frac{\psi(A_j)}{\varphi(A_j)}\right)$$

under the convention (2.2). A measurement $\mathcal{E} = (E_j)_{1 \le j \le n}$ in M is said to be *projective* if E_j's are orthogonal projections with $\sum_{j=1}^{n} E_j = 1$.

Definition 5.1 For $\psi, \varphi \in M_*^+$ we define the *measured f-divergence* and its *projective* variant as follows:

$$S_f^{\mathrm{meas}}(\psi \| \varphi) := \sup\{S_f(\mathcal{M}(\psi) \| \mathcal{M}(\varphi)) : \mathcal{M} \text{ a measurement in } M\}, \tag{5.1}$$

$$S_f^{\mathrm{pr}}(\psi \| \varphi) := \sup\{S_f(\mathcal{E}(\psi) \| \mathcal{E}(\varphi)) : \mathcal{E} \text{ a projective measurement in } M\}. \tag{5.2}$$

Obviously, $S_f^{\mathrm{pr}}(\psi \| \varphi) \le S_f^{\mathrm{meas}}(\psi \| \varphi)$. Moreover, $S_f^{\mathrm{meas}}(\psi \| \varphi) \le S_f(\psi \| \varphi)$ whenever f is operator convex on $(0, \infty)$, due to the monotonicity of S_f, see Theorem 2.7 (iv).

F. Hiai, *Quantum f-Divergences in von Neumann Algebras*,
Mathematical Physics Studies, https://doi.org/10.1007/978-981-33-4199-9_5

Proposition 5.2 *For every $\psi, \varphi \in M_*^+$,*

$$S_f^{\mathrm{meas}}(\psi\|\varphi) = \sup\{S_f(\psi\circ\alpha\|\varphi\circ\alpha) : \alpha : \mathcal{A} \to M \text{ a unital normal positive map}, \mathcal{A} \text{ a commutative von Neumann algebra}\}. \tag{5.3}$$

Proof Since $S_f^{\mathrm{meas}}(\psi\|\varphi) \le$ the RHS of (5.3) is obvious, we may prove that

$$S_f^{\mathrm{meas}}(\psi\|\varphi) \ge S_f(\psi\circ\alpha\|\varphi\circ\alpha) \tag{5.4}$$

for every $\alpha : \mathcal{A} \to M$ as in (5.3). Let $e_0 := s(\psi\circ\alpha) \vee s(\varphi\circ\alpha)$ and define $\beta : e_0\mathcal{A}e_0 \to \mathcal{A}$ by $\beta(x) := x + \psi(x)(1 - e_0)$ for $x \in e_0\mathcal{A}e_0$. Then β is a unital normal positive map and it is obvious that $S_f(\psi\circ\alpha\|\varphi\circ\alpha) = S_f(\psi\circ\alpha\circ\beta\|\varphi\circ\alpha\circ\beta)$. Replacing α with $\alpha\circ\beta : e_0\mathcal{A}e_0 \to M$, we may assume that $\mathcal{A} = L^\infty(X, \mathcal{X}, \mu)$, where $(X, \mathcal{X}, \mu)$ is a finite measure space. Identifying $\psi\circ\alpha$, $\varphi\circ\alpha$ as functions $\hat\psi, \hat\varphi \in L^1(X, \mathcal{X}, \mu)_+$ we have

$$\begin{aligned} &S_f(\psi\circ\alpha\|\varphi\circ\alpha) \\ &= \int_{\{\hat\psi>0\}\cap\{\hat\varphi>0\}} \hat\varphi f(\hat\psi/\hat\varphi)\,d\mu + f(0^+)\int_{\{\hat\psi=0\}} \hat\varphi\,d\mu + f'(+\infty)\int_{\{\hat\varphi=0\}} \hat\psi\,d\mu. \end{aligned}$$

Hence, to prove (5.4), it suffices to prove that

$$\int_{X_0} f\left(\frac{dp}{dq}\right)dq \le \sup\left\{\sum_{j=1}^n q(X_j) f\left(\frac{p(X_j)}{q(X_j)}\right) : \{X_1, \dots, X_n\} \text{ a finite measurable partition of } X_0\right\}, \tag{5.5}$$

where $X_0 := \{\hat\psi > 0\} \cap \{\hat\varphi > 0\}$, $dp := \hat\psi\,d\mu$, $dq := \hat\varphi\,d\mu$, and $dp/dq = \hat\psi/\hat\varphi$, the Radon–Nikodym derivative.

To show (5.5), we may and do assume that $f \ge 0$. Indeed, there exist $a, b \in \mathbb{R}$ and a convex function $f_0 \ge 0$ on $(0, \infty)$ such that $f(t) = f_0(t) + at + b$, $t \in (0, \infty)$. Then

$$\int_{X_0} f\left(\frac{dp}{dq}\right)dq = \int_{X_0} f_0\left(\frac{dp}{dq}\right)dq + ap(X_0) + bq(X_0),$$

$$\sum_{j=1}^n q(X_j) f\left(\frac{p(X_j)}{q(X_j)}\right) = \sum_{j=1}^n q(X_j) f_0\left(\frac{p(X_j)}{q(X_j)}\right) + ap(X_0) + bq(X_0),$$

so that it suffices to show (5.5) for f_0. For every $n \in \mathbb{N}$ consider a finite measurable partition $\{X_1, \ldots, X_{n2^n+1}\}$ given by

$$X_j := \left\{x \in X : \frac{j-1}{2^n} \le \frac{dp}{dq}(x) < \frac{j}{2^n}\right\}, \qquad 1 \le j \le n2^n,$$

$$X_{n2^n+1} := \left\{x \in X : \frac{dp}{dq}(x) \ge n2^n\right\},$$

and define a simple function $\kappa_n := \sum_{j=1}^{n2^n+1} \lambda_j 1_{X_j}$, where $\lambda_j := \inf\{f(t) : \frac{j-1}{2^n} \le t \le \frac{j}{2^n}\}$ (with $f(0) = f(0^+)$), $1 \le j \le n2^n$, and $\lambda_{n2^n+1} := \inf\{f(t) : t \ge n2^n\}$. Then, since f is convex and so continuous on $(0, \infty)$, it is immediate to see that $\kappa_n(x) \to f\big(\frac{dp}{dq}(x)\big)$ for q-a.e. $x \in X$. By Fatou's lemma we have

$$\begin{aligned}\int_{X_0} f\left(\frac{dp}{dq}\right) dq &\le \liminf_{n\to\infty} \int_{X_0} \kappa_n \, dq = \liminf_{n\to\infty} \sum_{j=1}^{n2^n+1} \lambda_j q(X_j) \\ &\le \liminf_{n\to\infty} \sum_{j=1}^{n2^n+1} f\left(\frac{p(X_j)}{q(X_j)}\right) q(X_j) \\ &\le \text{the RHS of (5.5)},\end{aligned}$$

where the second inequality above follows since

$$\frac{p(X_j)}{q(X_j)} = \frac{1}{q(X_j)} \int_{X_j} \frac{dp}{dq} \, dq \in \begin{cases} \left[\frac{j-1}{2^n}, \frac{j}{2^n}\right], & 1 \le j \le n2^n, \\ [n2^n, +\infty), & j = n2^n + 1. \end{cases}$$

Hence (5.4) is obtained.

Let H be a self-adjoint operator affiliated with M and

$$H = \int_{-\infty}^{\infty} \lambda \, dE_H(\lambda)$$

be the spectral decomposition of H with the spectral resolution $\{E_H(\lambda) : \lambda \in \mathbb{R}\}$ in M, see the beginning of Sect. A.5. Write $\mathcal{E}_H(\psi)$ for the finite positive measures $d\psi(E_H(\lambda))$ on the Borel space $(\mathbb{R}, \mathcal{B}(\mathbb{R}))$. Then similarly to the proof of Proposition 5.2 we have the following:

Proposition 5.3 *For every $\psi, \varphi \in M_*^+$,*

$$S_f^{\mathrm{pr}}(\psi\|\varphi) = \sup\{S_f(\mathcal{E}_H(\psi)\|\mathcal{E}_H(\varphi)) : H \text{ a self-adjoint operator affiliated with } M\}. \tag{5.6}$$

In this way, S_f^{pr} is the supremum of the classical f-divergences over the von Neumann (or projective) measurements, while S_f^{meas} is that over all measurements of quantum-classical channels.

The next proposition summarizes basic properties of S_f^{meas} and S_f^{pr}.

Proposition 5.4

(1) Joint lower semicontinuity: *$S_f^{\mathrm{meas}}(\psi\|\varphi)$ and $S_f^{\mathrm{pr}}(\psi\|\varphi)$ are jointly lower semicontinuous in $(\psi,\varphi)\in M_*^+\times M_*^+$ in the $\sigma(M_*,M)$-topology.*

(2) Joint convexity: *$S_f^{\mathrm{meas}}(\psi\|\varphi)$ and $S_f^{\mathrm{pr}}(\psi\|\varphi)$ are jointly convex and jointly subadditive on $M_*^+\times M_*^+$.*

(3) Monotonicity: *Let $\gamma : N\to M$ be a unital normal positive map between von Neumann algebras. For every $\psi,\varphi\in M_*^+$,*

$$S_f^{\mathrm{meas}}(\psi\circ\gamma\|\varphi\circ\gamma)\le S_f^{\mathrm{meas}}(\psi\|\varphi).$$

(4) *For every $\psi,\varphi,\omega\in M_*^+$,*

$$S_f^{\mathrm{meas}}(\psi\|\varphi)=\lim_{\varepsilon\searrow 0}S_f^{\mathrm{meas}}(\psi+\varepsilon\omega\|\varphi+\varepsilon\omega),$$

$$S_f^{\mathrm{pr}}(\psi\|\varphi)=\lim_{\varepsilon\searrow 0}S_f^{\mathrm{pr}}(\psi+\varepsilon\omega\|\varphi+\varepsilon\omega).$$

Proof Note that the classical f-divergence $S_f(a\|b)=\sum_{j=1}^n b_jf(a_j/b_j)$ for $a=(a_j)_{j=1}^n, b=(b_j)_{j=1}^n\in[0,\infty)^n$ is jointly lower semicontinuous and jointly convex. Hence (1) and (2) immediately follow from Definition 5.1.

(3) Let $\gamma : N\to M$ be as stated. For any measurement $\mathcal{M}=(B_j)_{j=1}^n$ in N, since $\gamma(\mathcal{M}):=(\gamma(B_j))_{j=1}^n$ is a measurement in M, we have

$$S_f(\mathcal{M}(\psi\circ\gamma)\|\mathcal{M}(\varphi\circ\gamma))=S_f(\gamma(\mathcal{M})(\psi)\|\gamma(\mathcal{M})(\varphi))\le S_f^{\mathrm{meas}}(\psi\|\varphi),$$

which implies that $S_f^{\mathrm{meas}}(\psi\circ\gamma\|\varphi\circ\gamma)\le S_f^{\mathrm{meas}}(\psi\|\varphi)$.

(4) From (1) it follows that

$$S_f^{\mathrm{meas}}(\psi\|\varphi)\le\liminf_{\varepsilon\searrow 0}S_f^{\mathrm{meas}}(\psi+\varepsilon\omega\|\varphi+\varepsilon\omega).$$

On the other hand, by (2) we have

$$\begin{aligned}S_f^{\mathrm{meas}}(\psi+\varepsilon\omega\|\varphi+\varepsilon\omega)&\le S_f^{\mathrm{meas}}(\psi\|\varphi)+S_f^{\mathrm{meas}}(\varepsilon\omega\|\varepsilon\omega)\\&=S_f^{\mathrm{meas}}(\psi\|\varphi)+\varepsilon\omega(1)f(1).\end{aligned}$$

Hence the assertion for S_f^{meas} follows. The proof for S_f^{pr} is similar.

This is a good place for us to explain an abstract approach to quantum f-divergences. Assume that f is operator convex on $(0, \infty)$. We say that a function $S_f^q : M_*^+ \times M_*^+ \to (-\infty, +\infty]$, where M varies over all von Neumann algebras, is a *monotone quantum f-divergence* if the following two properties are satisfied:

(a) $S_f^q(\psi \circ \gamma \| \varphi \circ \gamma) \le S_f^q(\psi \| \varphi)$ for any unital normal CP map $\gamma : N \to M$ between von Neumann algebras and for every $\psi, \varphi \in M_*^+$.
(b) When M is an abelian von Neumann algebra with $M = L^\infty(X, \mu)$ on a σ-finite measure space $(X, \mathcal{X}, \mu)$, $S_f^q(\psi \| \varphi)$ coincides with the classical f-divergence of $\psi, \varphi \in L^1(X, \mu)_+$ as in Example 2.2 (1).

If S_f^q is a monotone quantum f-divergence, then it is readily seen from Definition 5.1 and Theorem 4.17 that for every $\psi, \varphi \in M_*^+$,

$$S_f^{\text{meas}}(\psi \| \varphi) \le S_f^q(\psi \| \varphi) \le \widehat{S}_f(\psi \| \varphi),$$

which justifies the name *maximal* f-divergence for $\widehat{S}_f$ and suggests that we call S_f^{meas} the *minimal* f-divergence. Since the *standard* f-divergence S_f is typical among monotone quantum f-divergences, we have the following:

Proposition 5.5 *For every $\psi, \varphi \in M_*^+$,*

$$S_f^{\text{meas}}(\psi \| \varphi) \le S_f(\psi \| \varphi) \le \widehat{S}_f(\psi \| \varphi). \tag{5.7}$$

5.2 Variational Expressions

We consider the following classes of functions on $(0, \infty)$:

(a) $\mathcal{F}_{\text{cv}}^{\nearrow}(0, \infty)$ is the set of non-decreasing convex real functions f on $(0, \infty)$ such that $f'(\infty) = +\infty$,
(b) $\mathcal{F}_{\text{cc}}^{\nearrow}(0, \infty)$ is the set of non-decreasing concave real functions g on $(0, \infty)$ such that $g'(\infty) = 0$.

The following variational formulas were shown in [53, Appendix A], which provides a theory of conjugate functions (or the Legendre transform) on the half-line $(0, \infty)$.

Lemma 5.6

(1) *For every $f \in \mathcal{F}_{\text{cv}}^{\nearrow}(0, \infty)$ define*

$$f^\sharp(t) := \sup_{s>0}\{st - f(s)\}, \qquad t \in (0, \infty).$$

Then $f^\sharp \in \mathcal{F}_{\mathrm{cv}}^{\nearrow}(0,\infty)$ and $f \mapsto f^\sharp$ is an involutive bijection on $\mathcal{F}_{\mathrm{cv}}^{\nearrow}(0,\infty)$, i.e., a bijection such that $f^{\sharp\sharp} = f$ for all $f \in \mathcal{F}_{\mathrm{cv}}^{\nearrow}(0,\infty)$. Therefore,

$$f(s) = \sup_{t>0}\{st - f^\sharp(t)\}, \qquad s \in (0,\infty),$$

and moreover $f^\sharp(0^+) = -f(0^+)$ $(\in \mathbb{R})$.

(2) *For every $g \in \mathcal{F}_{\mathrm{cc}}^{\nearrow}(0,\infty)$ define*

$$g^\flat(t) := \inf_{s>0}\{st - g(s)\}, \qquad t \in (0,\infty).$$

Then $g^\flat \in \mathcal{F}_{\mathrm{cc}}^{\nearrow}(0,\infty)$ and $g \mapsto g^\flat$ is an involutive bijection on $\mathcal{F}_{\mathrm{cc}}^{\nearrow}(0,\infty)$, i.e., a bijection such that $g^{\flat\flat} = g$ for all $g \in \mathcal{F}_{\mathrm{cc}}^{\nearrow}(0,\infty)$. Therefore,

$$g(s) = \inf_{t>0}\{st - g^\flat(t)\}, \qquad s \in (0,\infty),$$

and moreover $g^\flat(0^+) = -g(\infty)$ and $g(0^+) = -g^\flat(\infty)$.

For a, b with $-\infty \le a < b \le \infty$, we write $M_{(a,b)}$ for the set of self-adjoint operators $A \in M$ whose spectrum is included in (a,b). In particular, $M_{(0,\infty)}$ is the the set M_{++} of positive invertible operators in M, and $M_{(-\infty,\infty)}$ is the set M_{sa} of self-adjoint operators in M. For $-\infty < a < b < \infty$ we also write $M_{[a,b]}$ for the set of self-adjoint operators $A \in M$ whose spectrum is in $[a,b]$, i.e., $a1 \le A \le b1$.

From Lemma 5.6 we can show variational expressions of S_f^{pr} in the following theorem.

Theorem 5.7 *Let $\psi, \varphi \in M_*^+$.*

(1) *Assume that $f \in \mathcal{F}_{\mathrm{cv}}^{\nearrow}(0,\infty)$. Let $a := f'(0^+) \in [0,\infty)$ and $b := -f(0^+) \in (-\infty,\infty)$. Then the inverse $(f^\sharp)^{-1} : (b,\infty) \to (a,\infty)$ exists and*

$$S_f^{\mathrm{pr}}(\psi\|\varphi) = \sup_{A\in M_{++}}\{\psi(A) - \varphi(f^\sharp(A))\} \tag{5.8}$$

$$= \sup_{A\in M_{(a,\infty)}}\{\psi(A) - \varphi(f^\sharp(A))\} \tag{5.9}$$

$$= \sup_{A\in M_{(b,\infty)}}\{\psi((f^\sharp)^{-1}(A)) - \varphi(A)\}. \tag{5.10}$$

(2) *Assume that $g \in \mathcal{F}_{\mathrm{cc}}^{\nearrow}(0,\infty)$ and g is non-constant. Let $f := -g$, and $a := g'(0^+)$, $b := -g(0^+)$, $c := -g(\infty)$. Then $a \in (0,\infty]$, $-\infty \le c < b \le \infty$, the inverse $(g^\flat)^{-1} : (c,b) \to (0,a)$ exists, and*

$$S_f^{\mathrm{pr}}(\psi\|\varphi) = \sup_{A\in M_{++}}\{-\psi(A) + \varphi(g^\flat(A))\} \tag{5.11}$$

$$= \sup_{A \in M_{(0,a)}} \{-\psi(A) + \varphi(g^{\flat}(A))\} \tag{5.12}$$

$$= \sup_{A \in M_{(c,b)}} \{-\psi((g^{\flat})^{-1}(A)) + \varphi(A)\}. \tag{5.13}$$

Proof

(1) Let $\mathcal{E} = (E_j)_{1 \le j \le n}$ be a projective measurement in M. For each j we prove that

$$\varphi(E_j) f\left(\frac{\psi(E_j)}{\varphi(E_j)}\right) = \sup_{t>0}\{t\psi(E_j) - f^{\sharp}(t)\varphi(E_j)\}. \tag{5.14}$$

When $\psi(E_j) > 0$ and $\varphi(E_j) > 0$, by Lemma 5.6 (1) the LHS of (5.14) is

$$\varphi(E_j) \sup_{t>0}\left\{t\frac{\psi(E_j)}{\varphi(E_j)} - f^{\sharp}(t)\right\} = \sup_{t>0}\{t\psi(E_j) - f^{\sharp}(t)\varphi(E_j)\}.$$

When $\psi(E_j) = 0$ and $\varphi(E_j) > 0$, the LHS of (5.14) is

$$f(0^+)\varphi(E_j) = -f^{\sharp}(0^+)\varphi(E_j) = \sup_{t>0}\{-f^{\sharp}(t)\varphi(E_j)\}$$

by the convention (2.2) and $f(0^+) = -f^{\sharp}(0^+)$ stated in Lemma 5.6 (1). When $\psi(E_j) > 0$ and $\varphi(E_j) = 0$, the LHS of (5.14) is

$$f'(\infty)\psi(E_j) = +\infty = \sup_{t>0}\{t\psi(E_j)\}.$$

When $\psi(E_j) = \varphi(E_j) = 0$, both sides of (5.14) are 0. Hence (5.14) holds in all the cases.

Summing up (5.14) for all j we have

$$\begin{aligned}\sum_{j=1}^{n} \varphi(E_j) f\left(\frac{\psi(E_j)}{\varphi(E_j)}\right) &= \sup_{(t_j):t_j>0} \sum_{j=1}^{n}\{t_j\psi(E_j) - f^{\sharp}(t_j)\varphi(E_j)\} \\ &= \sup_{(t_j):t_j>0}\left\{\psi\left(\sum_j t_jE_j\right) - \varphi\left(\sum_j f^{\sharp}(t_j)E_j\right)\right\}.\end{aligned} \tag{5.15}$$

Therefore, we have

$$\begin{aligned}S_f^{\mathrm{pr}}(\psi\|\varphi) &= \sup_{(E_j)}\sup_{(t_j)}\left\{\psi\left(\sum_j t_jE_j\right) - \varphi\left(f^{\sharp}\left(\sum_j t_jE_j\right)\right)\right\} \\ &= \sup_{A \in M_{++}} \{\psi(A) - \varphi(f^{\sharp}(A))\},\end{aligned}$$

since $\psi(A) - \varphi(f^\sharp(A))$ for any $A \in M_{++}$ is approximated by $\psi(A_0) - \varphi(f^\sharp(A_0))$ for discrete $A_0 := \sum_{j=1}^n t_j E_j$. Hence (5.8) follows.

By Lemma 5.6 (1) note that $f^\sharp \in \mathcal{F}_{cv}^{\nearrow}(0,\infty)$ and $f^\sharp(0^+) = -f(0^+) = b$ so that $f^\sharp(t) = -f(0^+)$ for $0 < t \le a$. Hence we can restrict t_j to $t_j > a$ in the supremum in (5.15), so the supremum in (5.8) is the same as that taken over $A \in M_{(a,\infty)}$. Furthermore, we easily see that $f^\sharp$ is strictly increasing from (a,∞) onto (b,∞). Hence the inverse $(f^\sharp)^{-1} : (b,\infty) \to (a,\infty)$ exists. Since $X \mapsto (f^\sharp)^{-1}(X)$ is a bijective map from $M_{(b,\infty)}$ onto $M_{(a,\infty)}$, we have (5.10) as well.

(2) Let $\mathcal{E} = (E_j)_{1\le j\le n}$ be a projective measurement in M. For each j we prove that

$$\varphi(E_j) f\left(\frac{\psi(E_j)}{\varphi(E_j)}\right) = \sup_{t>0}\{-t\psi(E_j) + g^\flat(t)\varphi(E_j)\}. \tag{5.16}$$

When $\psi(E_j) > 0$ and $\varphi(E_j) > 0$, by Lemma 5.6 (2) the LHS of (5.16) is

$$\begin{aligned}-\varphi(E_j) g\left(\frac{\psi(E_j)}{\varphi(E_j)}\right) &= -\varphi(E_j)\inf_{t>0}\left\{\frac{\psi(E_j)}{\varphi(E_j)} t - g^\flat(t)\right\} \\ &= \sup_{t>0}\{-t\psi(E_j) + g^\flat(t)\varphi(E_j)\}.\end{aligned}$$

When $\psi(E_j) = 0$ and $\varphi(E_j) > 0$, the LHS of (5.16) is

$$-g(0^+)\varphi(E_j) = g^\flat(\infty)\varphi(E_j) = \sup_{t>0}\{g^\flat(t)\varphi(E_j)\}$$

from (2.2) and $g(0^+) = -g^\flat(\infty)$ stated in Lemma 5.6 (2). When $\psi(E_j) > 0$ and $\varphi(E_j) = 0$, the LHS of (5.16) is

$$-g'(\infty)\psi(E_j) = 0 = \sup_{t>0}\{-t\psi(E_j)\}.$$

When $\psi(E_j) = \varphi(E_j) = 0$, both sides of (5.16) are 0. Hence (5.16) has been shown. Since g is non-constant, $a \in (0,\infty]$ and $-\infty \le c < b \le \infty$ are obvious. Moreover, when $a < \infty$, since $-b = g(0^+) > -\infty$ and $g^\flat(t) = -b$ for $t \ge a$, we find that the supremum in (5.16) is the same as that taken over $t \in (0,a)$. Thus, (5.11) and (5.12) follow as in the proof of (1). Furthermore, we easily see that $g^\flat$ is strictly increasing from $(0,a)$ onto (c,b). Hence the inverse $(g^\flat)^{-1} : (c,b) \to (0,a)$ exists. Since $X \mapsto (g^\flat)^{-1}(X)$ is a bijective map from $M_{(c,b)}$ onto $M_{(0,a)}$, we have (5.13) as well.

Based on Theorem 5.7 we can show the equality $S_f^{\mathrm{pr}} = S_f^{\mathrm{meas}}$ in the following cases.

Theorem 5.8 *Let f be a convex function on $(0,\infty)$, and $\widetilde{f}$ be the transpose of f, i.e., $\widetilde{f}(t) := tf(t^{-1})$, $t > 0$. Let $g := -f$ and $\widetilde{g} := -\widetilde{f}$ (hence $\widetilde{g}$ is the transpose of g). Assume that one of the following conditions (1)–(4) is satisfied:*

(1) *$f \in \mathcal{F}_{\mathrm{cv}}^{\nearrow}(0,\infty)$ and either $f^\sharp$ is operator convex on $(f'(0^+),\infty)$ or $(f^\sharp)^{-1}$ is operator concave on $(-f(0^+),\infty)$,*
(2) *$\widetilde{f} \in \mathcal{F}_{\mathrm{cv}}^{\nearrow}(0,\infty)$ and either $\widetilde{f}^\sharp$ is operator convex on $(\widetilde{f}'(0^+),\infty)$ or $(\widetilde{f}^\sharp)^{-1}$ is operator concave on $(-\widetilde{f}(0^+),\infty)$,*
(3) *$g \in \mathcal{F}_{\mathrm{cc}}^{\nearrow}(0,\infty)$ and either $g^\flat$ is operator concave on $(0, g'(0^+))$ or $(g^\flat)^{-1}$ is operator convex on $(-g(\infty), -g(0^+))$,*
(4) *$\widetilde{g} \in \mathcal{F}_{\mathrm{cc}}^{\nearrow}(0,\infty)$ and either $\widetilde{g}^\flat$ is operator concave on $(0,\widetilde{g}'(0^+))$ or $(\widetilde{g}^\flat)^{-1}$ is operator convex on $(-\widetilde{g}(\infty), -\widetilde{g}(0^+))$.*

Then for every $\psi,\varphi \in M_*^+$,

$$S_f^{\mathrm{pr}}(\psi\|\varphi) = S_f^{\mathrm{meas}}(\psi\|\varphi).$$

Proof Since $S_f^{\mathrm{pr}}(\psi\|\varphi) \le S_f^{\mathrm{meas}}(\psi\|\varphi)$ holds trivially, we may prove the converse inequality.

Case (1). Assume that $f \in \mathcal{F}_{\mathrm{cv}}^{\nearrow}(0,\infty)$ and $f^\sharp$ is operator convex on $(f'(0^+),\infty)$. For any measurement $(A_j)_{1\le j\le n}$ in M, as in the proof of Theorem 5.7 (1) with $\psi(A_j)$, $\varphi(A_j)$ in place of $\psi(E_j)$, $\varphi(E_j)$, we have, with $a := f'(0^+)$,

$$\begin{aligned}\sum_j \varphi(A_j) f\biggl(\frac{\psi(A_j)}{\varphi(A_j)}\biggr) &= \sup_{(t_j):t_j>a}\biggl\{\psi\biggl(\sum_j t_jA_j\biggr) - \varphi\biggl(\sum_j f^\sharp(t_j)A_j\biggr)\biggr\}\\ &\le \sup_{(t_j):t_j>a}\biggl\{\psi\biggl(\sum_j t_jA_j\biggr) - \varphi\biggl(f^\sharp\biggl(\sum_j t_jA_j\biggr)\biggr)\biggr\}\\ &= \sup_{A\in M_{(a,\infty)}}\{\psi(A) - \varphi(f^\sharp(A))\} = S_f^{\mathrm{pr}}(\psi\|\varphi).\end{aligned}$$

In the above, the last equality holds by Theorem 5.7 (1), and the inequality is seen from the operator convexity of $f^\sharp$ on (a,∞) as follows:

$$f^\sharp\biggl(\sum_j t_jA_j\biggr) = f^\sharp\biggl(\sum_j A_j^{1/2}t_jA_j^{1/2}\biggr) \le \sum_j A_j^{1/2}f^\sharp(t_j)A_j^{1/2} = \sum_j f^\sharp(t_j)A_j$$

by [25, Theorem 2.1] (also [51, Theorem 2.5.7]). Therefore, we have $S_f^{\mathrm{meas}}(\psi\|\varphi) \le S_f^{\mathrm{pr}}(\psi\|\varphi)$.

Next, assume that $(f^\sharp)^{-1}$ is operator concave on $(-f(0^+),\infty)$. We have, with $b := -f(0^+)$,

$$\begin{aligned}
\sum_j \varphi(A_j) f\left(\frac{\psi(A_j)}{\varphi(A_j)}\right) &= \sup_{(t_j):t_j>a} \left\{\psi\left(\sum_j t_j A_j\right) - \varphi\left(\sum_j f^\sharp(t_j)A_j\right)\right\} \\
&= \sup_{(s_j):s_j>b} \left\{\psi\left(\sum_j (f^\sharp)^{-1}(s_j)A_j\right) - \varphi\left(\sum_j s_j A_j\right)\right\} \\
&\le \sup_{(s_j):s_j>b} \left\{\psi\left((f^\sharp)^{-1}\left(\sum_j s_j A_j\right)\right) - \varphi\left(\sum_j s_j A_j\right)\right\} \\
&= \sup_{A\in M_{(b,\infty)}} \left\{\psi(f^\sharp)^{-1}(A) - \varphi A\right\} = S_f^{\mathrm{pr}}(\psi\|\varphi).
\end{aligned}$$

In the above, the last equality holds by Theorem 5.7 (1), and the inequality follows from the operator concavity of $(f^\sharp)^{-1}$ on (b,∞).

Case (2). We readily see that

$$S_f^{\mathrm{pr}}(\psi\|\varphi) = S_{\widetilde{f}}^{\mathrm{pr}}(\varphi\|\psi), \qquad S_f^{\mathrm{meas}}(\psi\|\varphi) = S_{\widetilde{f}}^{\mathrm{meas}}(\varphi\|\psi).$$

Hence the result is immediate from case (1).

Case (3). Assume that $g \in \mathcal{F}_{\mathrm{cc}}^{\nearrow}(0,\infty)$. Since the assertion is obvious when g is a constant function, we may assume that g is non-constant; then $g'(0^+) > 0$ and $g(0^+) < g(\infty)$. Assume that $g^\flat$ is operator concave on $(0, g'(0^+))$. For any measurement $(A_j)_{1\le j\le n}$ in M, as in the proof of Theorem 5.7 (2), we have, with $a := g'(0^+)$,

$$\begin{aligned}
\sum_j \varphi(A_j) f\left(\frac{\psi(A_j)}{\varphi(A_j)}\right) &= \sup_{(t_j):0<t_j<a} \left\{-\psi\left(\sum_j t_j A_j\right) + \varphi\left(\sum_j g^\flat(t_j)A_j\right)\right\} \\
&\le \sup_{(t_j):0<t_j<a} \left\{-\psi\left(\sum_j t_j A_j\right) + \varphi\left(g^\flat\left(\sum_j t_j A_j\right)\right)\right\} \\
&= \sup_{A\in M_{(0,a)}} \left\{-\psi(A) + \varphi(g^\flat(A))\right\} = S_f^{\mathrm{pr}}(\psi\|\varphi).
\end{aligned}$$

In the above, the last equality is due to Theorem 5.7 (2), and the inequality follows from the operator concavity of $g^\flat$ on $(0,a)$. When $(g^\flat)^{-1}$ is operator convex on $(-g(\infty), -g(0^+))$, the proof is similar to the second part of the proof of case (1).

Case (4). Immediate from case (3) as in case (2).

Example 5.9 (Rényi divergence) Consider

$$f_\alpha(t) := t^\alpha \in \mathcal{F}_{\mathrm{cv}}^{\nearrow}(0,\infty) \quad \text{for } \alpha \in (1,\infty),$$

$$g_\alpha(t) := t^\alpha \in \mathcal{F}_{\mathrm{cc}}^{\nearrow}(0,\infty), \quad f_\alpha := -g_\alpha \quad \text{for } \alpha \in (0,1).$$

(1) Assume that $\alpha \in (1,\infty)$. We have $f_\alpha(0^+) = f'_\alpha(0^+) = 0$ and

$$f_\alpha^\sharp(t) = (\alpha-1)\left(\frac{t}{\alpha}\right)^{\frac{\alpha}{\alpha-1}}, \quad (f_\alpha^\sharp)^{-1}(t) = \alpha\left(\frac{t}{\alpha-1}\right)^{\frac{\alpha-1}{\alpha}}, \qquad t \in (0,\infty).$$

Hence $(f_\alpha^\sharp)^{-1}$ is operator concave on $(0,\infty)$. By Theorem 5.7 (1) and case (1) of Theorem 5.8, for every $\psi, \varphi \in M_*^+$ we have

$$\begin{aligned} S_{f_\alpha}^{\mathrm{pr}}(\psi\|\varphi) = S_{f_\alpha}^{\mathrm{meas}}(\psi\|\varphi) &= \sup_{A\in M_{++}} \left\{\psi(A) - (\alpha-1)\varphi\left(\left(\frac{A}{\alpha}\right)^{\frac{\alpha}{\alpha-1}}\right)\right\} \\ &= \sup_{A\in M_{++}} \left\{\alpha\psi\left(\left(\frac{A}{\alpha-1}\right)^{\frac{\alpha-1}{\alpha}}\right) - \varphi(A)\right\} \\ &= \sup_{A\in M_{++}} \left\{\alpha\psi\left(A^{\frac{\alpha-1}{\alpha}}\right) - (\alpha-1)\varphi(A)\right\}. \end{aligned} \tag{5.17}$$

(2) Assume that $\alpha \in (0,1)$. We have $g_\alpha(0^+) = 0$, $g_\alpha(\infty) = \infty$, $g'_\alpha(0^+) = \infty$, and

$$g_\alpha^\flat(t) = (\alpha-1)\left(\frac{t}{\alpha}\right)^{\frac{\alpha}{\alpha-1}}, \quad (g_\alpha^\flat)^{-1}(-t) = \alpha\left(\frac{t}{1-\alpha}\right)^{\frac{\alpha-1}{\alpha}}, \qquad t \in (0,\infty).$$

Note that $\widetilde{g}_\alpha(t) = t^{1-\alpha} \in \mathcal{F}_{\mathrm{cc}}^{\nearrow}(0,\infty)$ with $\widetilde{g}_\alpha(0^+) = 0$, $\widetilde{g}_\alpha(\infty) = \infty$ and $\widetilde{g}'_\alpha(0^+) = \infty$. When $\alpha \in [1/2,1)$, $\widetilde{g}_\alpha^\flat(t) = -\alpha\left(\frac{t}{1-\alpha}\right)^{\frac{\alpha-1}{\alpha}}$ is operator concave on $(0,\infty)$. On the other hand, when $\alpha \in (0,1/2]$, $(\widetilde{g}_\alpha^\flat)^{-1}(-t) = (1-\alpha)\left(\frac{t}{\alpha}\right)^{\frac{\alpha}{\alpha-1}}$ is operator convex on $(0,\infty)$. Hence by Theorem 5.7 (2) and case (4) of Theorem 5.8, we have

$$\begin{aligned} S_{f_\alpha}^{\mathrm{pr}}(\psi\|\varphi) = S_{f_\alpha}^{\mathrm{meas}}(\psi\|\varphi) &= \sup_{A\in M_{++}} \left\{-\psi(A) - (1-\alpha)\varphi\left(\left(\frac{A}{\alpha}\right)^{\frac{\alpha}{\alpha-1}}\right)\right\} \\ &= \sup_{A\in M_{++}} \left\{-\alpha\psi\left(\left(\frac{A}{1-\alpha}\right)^{\frac{\alpha-1}{\alpha}}\right) - \varphi(A)\right\} \\ &= -\inf_{A\in M_{++}} \left\{\alpha\psi\left(A^{\frac{\alpha-1}{\alpha}}\right) + (1-\alpha)\varphi(A)\right\}. \end{aligned} \tag{5.18}$$

The variational formulas in (5.17) and (5.18) for $S_{f_\alpha}^{\mathrm{pr}}$ were given in [19, Lemma 3, (20)] in the finite-dimensional setting.

(3) For each $\alpha \in (0,\infty)\setminus\{1\}$ and $\psi,\varphi \in M_*^+$ with $\psi \neq 0$, apart from the α-Rényi divergence $D_\alpha(\psi\|\varphi)$ (Sect. 3.1) and the sandwiched α-Rényi divergence $\widetilde D_\alpha(\psi\|\varphi)$ (Sect. 3.3), we define the *measured α-Rényi divergence* by

$$D_\alpha^{\mathrm{meas}}(\psi\|\varphi) := \sup\{D_\alpha(\mathcal{M}(\psi)\|\mathcal{M}(\varphi)) : \mathcal{M} \text{ a measurement in } M\}.$$

Alternatively,

$$D_\alpha^{\mathrm{meas}}(\psi\|\varphi) = \frac{1}{\alpha-1}\log\frac{Q_\alpha^{\mathrm{meas}}(\psi\|\varphi)}{\psi(1)}$$

with

$$Q_\alpha^{\mathrm{meas}}(\psi\|\varphi) := \begin{cases} S_{f_\alpha}^{\mathrm{meas}}(\psi\|\varphi) & \text{if } \alpha \in (1,\infty), \\ -S_{f_\alpha}^{\mathrm{meas}}(\psi\|\varphi) & \text{if } \alpha \in (0,1). \end{cases}$$

The projective version $D_\alpha^{\mathrm{pr}}(\psi\|\varphi)$ is similarly defined (see Definition 5.1). But from (5.17) and (5.18) above we find that for all $\alpha \in (0,\infty)\setminus\{1\}$ and all $\psi,\varphi \in M_*^+$,

$$D_\alpha^{\mathrm{pr}}(\psi\|\varphi) = D_\alpha^{\mathrm{meas}}(\psi\|\varphi),$$

which was given in [19, Theorem 4] in the finite-dimensional setting. By Remark 3.18 (2) note that the (projectively) measured versions of $\widetilde D_\alpha$ are the same as D_α^{meas} and D_α^{pr}.

By Proposition 5.4 (1), $D_\alpha^{\mathrm{pr}} = D_\alpha^{\mathrm{meas}}$ is jointly lower semicontinuous in the $\sigma(M_*, M)$-topology for all $\alpha \in (0,\infty)\setminus\{1\}$, while the same lower semicontinuity is unknown for D_α and $\widetilde D_\alpha$, see Theorems 3.2 (4) and 3.16 (3). When $0 < \alpha \le 1/2$, $D_\alpha^{\mathrm{meas}} \le D_\alpha$ follows from Theorem 3.2 (8). When $\alpha \in [1/2,\infty)\setminus\{1\}$, $D_\alpha^{\mathrm{meas}} \le \widetilde D_\alpha$ follows from Theorem 3.16 (4). Combining this with the above $D_\alpha^{\mathrm{pr}} = D_\alpha^{\mathrm{meas}}$ and Theorem 3.16 (6) we have

$$D_\alpha^{\mathrm{pr}}(\psi\|\varphi) = D_\alpha^{\mathrm{meas}}(\psi\|\varphi) \le \widetilde D_\alpha(\psi\|\varphi) \le D_\alpha(\psi\|\varphi) \tag{5.19}$$

for all $\psi,\varphi \in M_*^+$ and all $\alpha \in [1/2,\infty)\setminus\{1\}$.

Example 5.10 (Relative Entropy) Consider

$$g_0(t) := \log t \in \mathcal{F}_{\mathrm{cc}}^{\nearrow}(0,\infty), \qquad f_0 := -g_0.$$

Recall the relative entropy $D(\psi\|\varphi) = S_{f_0}(\varphi\|\psi) = S_{\widetilde{f_0}}(\psi\|\varphi)$ for $\psi,\varphi \in M_*^+$. We define the (projectively) *measured relative entropies* $D^{\mathrm{meas}}(\psi\|\varphi)$ and $D^{\mathrm{pr}}(\psi\|\varphi)$ by

Definition 5.1. We have $g_0(0^+) = -\infty$, $g_0(+\infty) = \infty$, $g_0'(0^+) = \infty$, and

$$g_0^\flat(t) = 1 + \log t, \quad t \in (0, \infty); \qquad (g_0^\flat)^{-1}(t) = e^{t-1}, \quad t \in (-\infty, \infty).$$

Since $g_0^\flat$ is operator concave on $(0, \infty)$, by Theorem 5.7 (2) and case (3) of Theorem 5.8 we have

$$\begin{aligned} D^{\mathrm{pr}}(\psi\|\varphi) = D^{\mathrm{meas}}(\psi\|\varphi) &= \sup_{A \in M_{++}} \{-\varphi(A) + \psi(1 + \log A)\} \\ &= \sup_{H \in M_{\mathrm{sa}}} \{\psi(H) + \psi(1) - \varphi(e^H)\}. \end{aligned} \tag{5.20}$$

The above first variational formula and $D^{\mathrm{pr}} = D^{\mathrm{meas}}$ were given in [19, Lemma 1 and Theorem 2] in the finite-dimensional setting.

It is worthwhile to compare (5.20) with other variational expressions for the relative entropy. For simplicity assume that $\psi, \varphi \in M_*^+$ are such that ψ is a state and φ is faithful. For each $H \in M_{\mathrm{sa}}$ let φ^H be the *perturbed functional* (a faithful functional in M_*^+) introduced in [7, 33]. Then the variational expression of $D(\psi\|\varphi)$ due to Petz [107] is

$$D(\psi\|\varphi) = \sup_{H \in M_{\mathrm{sa}}} \{\psi(H) - \log \varphi^H(1)\}. \tag{5.21}$$

From Araki's *Golden–Thompson inequality* [8] we have

$$\varphi^H(1) \le \varphi(e^H). \tag{5.22}$$

(Indeed, in the finite-dimensional case, when φ has the density matrix e^{H_0}, the density of the perturbed φ^H is e^{H_0+H}, so (5.22) reduces to the usual Golden–Thompson inequality $\mathrm{Tr}\, e^{H_0+H} \le \mathrm{Tr}(e^{H_0}e^H)$.) Since

$$\log \varphi^H(1) \le \log \varphi(e^H) \le \varphi(e^H) - 1,$$

we conclude by (5.20)–(5.22) that

$$\begin{aligned} D(\psi\|\varphi) &\ge \sup_{H \in M_{\mathrm{sa}}} \{\psi(H) - \log \varphi(e^H)\} \\ &\ge \sup_{H \in M_{\mathrm{sa}}} \{\psi(H) + 1 - \varphi(e^H)\} = D^{\mathrm{pr}}(\psi\|\varphi). \end{aligned}$$

Let E_H be the spectral resolution of $H \in M_{\mathrm{sa}}$, and $dp(\lambda) := d\psi(E_H(\lambda))$, $dq(\lambda) := d\varphi(E_H(\lambda))$ on $(\mathbb{R}, \mathcal{B}(\mathbb{R}))$ supported on $[-\|H\|, \|H\|]$, where p is a probability measure. By Proposition 5.3 we have

$$\begin{aligned} D^{\mathrm{pr}}(\psi\|\varphi) &\geq D(p\|q) \\ &\geq \int_{-\|H\|}^{\|H\|} \lambda\, dp(\lambda) - \log \int_{-\|H\|}^{\|H\|} e^{\lambda}\, dq(\lambda) \\ &= \psi(H) - \log\varphi(e^H), \end{aligned}$$

where the above latter inequality is well-known for the classical relative entropy. Therefore, in addition to (5.20) we observe that

$$D^{\mathrm{pr}}(\psi\|\varphi) = \sup_{H \in M_{\mathrm{sa}}} \{\psi(H) - \log\varphi(e^H)\},$$

which was shown in the finite-dimensional case in [60, (1.7)] and [101, Proposition 7.13] but does not seem explicitly given so far in the von Neumann algebra setting.

Example 5.11 Here is one more example. Consider

$$g_\lambda(t) := \frac{t}{t+\lambda} \in \mathcal{F}_{\mathrm{cc}}^{\nearrow}(0,\infty), \quad f_\lambda := -g_\lambda, \quad \text{where } 0 < \lambda < \infty.$$

We have $g_\lambda(0^+) = 0$, $g_\lambda(\infty) = 1$, $g'_\lambda(0^+) = 1/\lambda$, and

$$g_\lambda^\flat(t) = -\{1 - (\lambda t)^{1/2}\}^2, \qquad 0 < t \leq 1/\lambda,$$

so that

$$(g_\lambda^\flat)^{-1}(t) = \frac{1}{\lambda}\{1 - (-t)^{1/2}\}^2, \qquad -1 < t < 0,$$

which is operator convex on $(-1, 0)$. Hence by Theorem 5.7 (2) and case (4) of Theorem 5.8 we have

$$\begin{aligned} S_{f_\lambda}^{\mathrm{pr}}(\psi\|\varphi) = S_{f_\lambda}^{\mathrm{meas}}(\psi\|\varphi) &= \sup_{A \in M_{(0,1/\lambda)}} \Big\{-\psi(A) - \varphi(\{1 - (\lambda A)^{1/2}\}^2)\Big\} \\ &= \sup_{A \in M_{(0,1)}} \Big\{-\frac{1}{\lambda}\psi(\{1 - A^{1/2}\}^2) - \varphi(A)\Big\} \\ &= \sup\big\{-\psi(X^2) - \varphi(Y^2) : X, Y \in M_+,\ \lambda^{1/2}X + Y = 1\big\}. \end{aligned}$$

5.3 Optimal Measurements

In the finite-dimensional case, it is known [58, Proposition 4.17] that the supremum in (5.1) is attained, and that if f is operator convex, then the supremum in (5.2) is attained as well. This problem of attainability of (projective) measurements for (5.3) and (5.6) does not seem easy in the infinite-dimensional case. In this section we consider the attainability problem or the existence of optimal measurements for the (projectively) measured f-divergences S_f^{meas} and S_f^{pr}, based on the variational expressions obtained in the previous section.

The next lemma extending [55, Lemma 5.4] will be useful.

Lemma 5.12 *Let $-\infty < a < b < \infty$ and T, T_n ($n \in \mathbb{N}$) be self-adjoint bounded operators on a Hilbert space $\mathcal{H}$ such that $a1 \le T_n \le b1$ for all $n \in \mathbb{N}$ and $T_n \to T$ in WOT (the weak operator topology).*

(1) *Assume that f is an operator convex function on $[a, b]$. Then for any $\xi \in \mathcal{H}$,*

$$\langle \xi, f(T)\xi \rangle \le \liminf_{n\to\infty} \langle \xi, f(T_n)\xi \rangle.$$

In fact, the function $T \mapsto \langle \xi, f(T)\xi \rangle$ is lower semicontinuous in WOT on $B(\mathcal{H})_{[a,b]}$.

(2) *Assume that f is an operator concave function on $[a, b]$. Then for any $\xi \in \mathcal{H}$,*

$$\langle \xi, f(T)\xi \rangle \ge \limsup_{n\to\infty} \langle \xi, f(T_n)\xi \rangle.$$

In fact, the function $T \mapsto \langle \xi, f(T)\xi \rangle$ is upper semicontinuous in WOT on $B(\mathcal{H})_{[a,b]}$.

Proof The proofs of the two assertions are similar, so we prove (2) only. Let E_λ be a net of finite-dimensional orthogonal projections on $\mathcal{H}$ such that $E_\lambda \nearrow 1$. Set $T_{n,\lambda} := E_\lambda T_n E_\lambda + a(I - E_\lambda)$ and $T_\lambda := E_\lambda T E_\lambda + a(I - E_\lambda)$ for every $n \in \mathbb{N}$ and λ. For each λ, since $\|T_{n,\lambda} - T_\lambda\| \to 0$ as $n \to \infty$, it follows that $\|f(T_{n,\lambda}) - f(T_\lambda)\| \to 0$ as $n \to \infty$. By Choi [25, Theorem 2.1] (also [51, Theorem 2.5.7]) we have

$$f(T_{n,\lambda}) \ge E_\lambda f(T_n) E_\lambda + f(a)(I - E_\lambda),$$

so that

$$\begin{aligned} \langle \xi, f(T_\lambda)\xi \rangle &= \lim_{n\to\infty} \langle \xi, f(T_{n,\lambda})\xi \rangle \\ &\ge \limsup_{n\to\infty} \langle \xi, E_\lambda f(T_n) E_\lambda \xi \rangle + f(a)\langle \xi, (I - E_\lambda)\xi \rangle. \end{aligned} \tag{5.23}$$

Note that

$$\begin{aligned}&|\langle \xi, E_\lambda f(T_n)E_\lambda \xi\rangle - \langle \xi, f(T_n)\xi\rangle| \\ &\quad \le |\langle E_\lambda \xi - \xi, f(T_n)E_\lambda \xi\rangle| + |\langle \xi, f(T_n)(E_\lambda \xi - \xi)\rangle| \le 2K\|\xi\|\,\|E_\lambda \xi - \xi\|,\end{aligned}$$

where $K := \sup_n \|f(T_n)\| < +\infty$. Therefore, $\langle \xi, E_\lambda f(T_n)E_\lambda \xi\rangle \to \langle \xi, f(T_n)\xi\rangle$ as $\lambda \to$"∞" uniformly for n, so it follows that

$$\begin{aligned}\limsup_{n\to\infty}\langle \xi, f(T_n)\xi\rangle &= \lim_\lambda \limsup_{n\to\infty}\langle \xi, E_\lambda f(T_n)E_\lambda \xi\rangle \\ &\le \lim_\lambda \{\langle \xi, f(T_\lambda)\xi\rangle - f(a)\langle \xi, (I - E_\lambda)\xi\rangle\} \quad \text{(by (5.23))} \\ &= \lim_\lambda \langle \xi, f(T_\lambda)\xi\rangle = \langle \xi, f(T)\xi\rangle,\end{aligned}$$

where the last equality follows since $T_\lambda \to T$ in SOT (the strong operator topology) implies that $f(T_\lambda) \to f(T)$ in SOT, see [71] (or [122, Theorem A.2]).

For the latter statement, we may note that the above argument is valid also when $\{T_\nu\}$ is a net in $B(\mathcal{H})_{[a,b]}$ with $T_\nu \to T$ in WOT.

Theorem 5.13 *Assume that $f \in \mathcal{F}_{\mathrm{cv}}^{\nearrow}(0,\infty)$ and either $f^\sharp$ is operator convex on $(f'(0^+),\infty)$ or $(f^\sharp)^{-1}$ is operator concave on $(-f(0^+),\infty)$. Then for every $\psi, \varphi \in M_*^+$ such that $\psi \le \lambda\varphi$ for some $\lambda \in (0,\infty)$, there exists an $H \in M_{\mathrm{sa}}$ such that*

$$S_f^{\mathrm{meas}}(\psi\|\varphi) = S_f^{\mathrm{pr}}(\psi\|\varphi) = S_f(\mathcal{E}_H(\psi)\|\mathcal{E}_H(\varphi)).$$

That is, the supremum of expression (5.3) *is attained by a projective measurement induced by H.*

Proof Let $f \in \mathcal{F}_{\mathrm{cv}}^{\nearrow}(0,\infty)$. Assume that $f^\sharp$ is operator convex on (a,∞), where $a := f'(0^+)$. From $(f^\sharp)'(\infty) = \infty$ (since $f^\sharp \in \mathcal{F}_{\mathrm{cv}}^{\nearrow}(0,\infty)$ by Lemma 5.6(1)), there is a $t_0 \in (a,\infty)$ such that $(f^\sharp)'(t_0) > \lambda$. For any $A \in M_{(a,\infty)}$ with the spectral decomposition $A = \int_a^\infty t\,dE_A(t)$, define an $A_0 \in M_{[a,t_0]}$ by

$$A_0 := \int_a^{t_0} t\,dE_A(t) + t_0 E_A((t_0,\infty)).$$

We then have

$$\begin{aligned}&\{\psi(A_0) - \varphi(f^\sharp(A_0))\} - \{\psi(A) - \varphi(f^\sharp(A))\} \\ &= -\int_{(t_0,\infty)} (t - t_0)\,d\psi(E_A(t)) + \int_{(t_0,\infty)} (f^\sharp(t) - f^\sharp(t_0))\,d\varphi(E_A(t)) \\ &\ge \int_{(t_0,\infty)} (t - t_0)\left(\frac{f^\sharp(t) - f^\sharp(t_0)}{t - t_0} - \lambda\right) d\varphi(E_A(t))\end{aligned}$$

thanks to $\psi \le \lambda\varphi$. Since

$$\frac{f^\sharp(t) - f^\sharp(t_0)}{t - t_0} \ge (f^\sharp)'(t_0) > \lambda, \qquad t > t_0,$$

it follows that

$$\psi(A_0) - \varphi(f^\sharp(A_0)) \ge \psi(A) - \varphi(f^\sharp(A)).$$

By Theorem 5.7 (1) and case (1) of Theorem 5.8, this implies that

$$S_f^{\mathrm{meas}}(\psi\|\varphi) = S_t^{\mathrm{pr}}(\psi\|\varphi) = \sup_{A \in M_{[a,t_0]}} \{\psi(A) - \varphi(f^\sharp(A))\},$$

where $f^\sharp(0) := f^\sharp(0^+)$ when $a = 0$. Noting that $\varphi(f^\sharp(A)) = \langle\Phi, f^\sharp(A)\Phi\rangle$, where Φ is the vector representative of φ (in the standard form), we find by Lemma 5.12 (1) that the function $A \in M_{[a,t_0]} \mapsto \psi(A) - \varphi(f^\sharp(A))$ is upper semicontinuous in WOT. Since $M_{[a,t_0]}$ is compact in WOT, there exists an $H \in M_{[a,t_0]}$ such that

$$S^{\mathrm{pr}}(\psi\|\varphi) = \psi(H) - \varphi(f^\sharp(H)).$$

Letting $p := d\psi(E_H(t))$ and $q := d\varphi(E_H(t))$ we have

$$\begin{aligned} S^{\mathrm{pr}}(\psi\|\varphi) &= \int_a^{t_0} t\,dp(t) - \int_a^{t_0} f^\sharp(t)\,dq(t) \\ &= \int_a^{t_0} \left(t\,\frac{dp}{dq}(t) - f^\sharp(t)\right) dq(t) \\ &\le \int_a^{t_0} f\left(\frac{dp}{dq}\right) dq = S_f(p\|q) \le S_f^{\mathrm{pr}}(\psi\|\varphi), \end{aligned}$$

where the last inequality follows from Proposition 5.3. Thus, we obtain

$$S_f^{\mathrm{pr}}(\psi\|\varphi) = S_f(p\|q) = S_f(\mathcal{E}_H(\psi)\|\mathcal{E}_H(\varphi)),$$

as desired.

Next, assume that $(f^\sharp)^{-1}$ is operator concave on (b, ∞), where $b := -f(0^+)$. Note that $(f^\sharp)^{-1}(t) \nearrow \infty$ as $b < t \nearrow \infty$ by Theorem 5.7 (1) and

$$\left[(f^\sharp)^{-1}\right]'(t) = \frac{1}{(f^\sharp)'((f^\sharp)^{-1}(t))} \longrightarrow \frac{1}{(f^\sharp)'(\infty)} = 0 \quad \text{as } t \to \infty.$$

So there is a $t_0 \in (b, \infty)$ such that $\left[(f^\sharp)^{-1}\right]'(t_0) < 1/\lambda$. Now, the proof proceeds as in the previous case. For any $A \in M_{(b,\infty)}$ with the spectral decomposition $A = \int_b^\infty t\,dE_A(t)$, define an $A_0 \in M_{[b,t_0]}$ by

$$A_0 := \int_b^{t_0} t\,dE_A(t) + t_0 E_A((t_0, \infty)).$$

Then

$$\begin{aligned}
&\left\{\psi((f^\sharp)^{-1}(A_0)) - \varphi(A_0)\right\} - \left\{\psi((f^\sharp)^{-1}(A)) - \varphi(A)\right\} \\
&\quad = -\int_{(t_0,\infty)} \left((f^\sharp)^{-1}(t) - (f^\sharp)^{-1}(t_0)\right) d\psi(E_A(t)) + \int_{(t_0,\infty)} (t - t_0)\, d\varphi(E_A(t)) \\
&\quad \ge \int_{(t_0,\infty)} \lambda(t - t_0)\left(\frac{1}{\lambda} - \frac{(f^\sharp)^{-1}(t) - (f^\sharp)^{-1}(t_0)}{t - t_0}\right) d\varphi(E_A(t)) \ge 0,
\end{aligned}$$

since

$$\frac{(f^\sharp)^{-1}(t) - (f^\sharp)^{-1}(t_0)}{t - t_0} \le \left[(f^\sharp)^{-1}\right]'(t_0) < \frac{1}{\lambda}, \qquad t > t_0.$$

Therefore,

$$\psi((f^\sharp)^{-1}(A_0)) - \varphi(A_0) \ge \psi((f^\sharp)^{-1}(A)) - \varphi(A).$$

By Theorem 5.7 (1) and case (1) of Theorem 5.8, this implies that

$$S_f^{\mathrm{meas}}(\psi\|\varphi) = S_f^{\mathrm{pr}}(\psi\|\varphi) = \sup_{A \in M_{[b,t_0]}} \left\{\psi((f^\sharp)^{-1}(A)) - \varphi(A)\right\},$$

where $(f^\sharp)^{-1}$ is operator concave on $[b, \infty)$ with $(f^\sharp)^{-1}(b) := a$. By Lemma 5.12 (2) there exists an $H_0 \in M_{[b,t_0]}$ such that $S_f^{\mathrm{pr}}(\psi\|\varphi) = \psi((f^\sharp)^{-1}(H_0)) - \varphi(H_0)$. Let $H := (f^\sharp)^{-1}(H_0)$; then $H \in M_{[a,t_1]}$ with $t_1 := (f^\sharp)^{-1}(t_0)$ and $H_0 = f^\sharp(H)$, so we have $S_f^{\mathrm{pr}}(\psi\|\varphi) = \psi(H) - \varphi(f^\sharp(H))$. Hence, similarly to the previous case, we have $S_f^{\mathrm{pr}}(\psi\|\varphi) = S_f(\mathcal{E}_H(\psi)\|\mathcal{E}_H(\varphi))$.

Theorem 5.14 *Assume that $g \in \mathcal{F}_{\mathrm{cc}}^{\nearrow}(0, \infty)$ and either $g^\flat$ is operator concave on $(0, g'(0^+))$ or $(g^\flat)^{-1}$ is operator convex on $(-g(\infty), -g(0^+))$. Let $f := -g$ and $\psi, \varphi \in M_*^+$. Assume that one of the following conditions (1)–(4) is satisfied:*

(1) $g(\infty) < \infty$, $g'(0^+) < \infty$ *(with no condition on ψ, φ),*
(2) $g(\infty) = \infty$, $g'(0^+) < \infty$, *and $\psi \le \lambda\varphi$ for some $\lambda > 0$,*
(3) $g(\infty) < \infty$, $g'(0^+) = \infty$, *and $\varphi \le \lambda\psi$ for some $\lambda > 0$,*
(4) $g(\infty) = \infty$, $g'(0^+) = \infty$, *and $\lambda^{-1}\varphi \le \psi \le \lambda\varphi$ for some $\lambda > 1$.*

Then there exists an $H \in M_{\mathrm{sa}}$ *such that*

$$S_f^{\mathrm{meas}}(\psi\|\varphi) = S_f^{\mathrm{pr}}(\psi\|\varphi) = S_f(\mathcal{E}_H(\psi)\|\mathcal{E}_H(\varphi)).$$

Proof Let $g \in \mathcal{F}_{\mathrm{cc}}^{\nearrow}(0,\infty)$ and $f := -g$. We may assume that g is non-constant; the conclusion is obviously true when g and hence f are constant.

Case (1). Assume that $g^\flat$ is operator concave on $(0,a)$, where $a := g'(0^+)$. Since $g^\flat(0^+) = -g(\infty) > -\infty$ (by Lemma 5.6 (2)) and $a = g'(0^+) < \infty$, $g^\flat$ is an operator concave function on $[0,a]$, where $g^\flat(0) = -g(\infty)$ and $g^\flat(a) = -g(0^+) < \infty$ (since $g'(0^+) < \infty$). By Theorem 5.7 (2) and case (3) of Theorem 5.8, we have

$$S_f^{\mathrm{meas}}(\psi\|\varphi) = S_f^{\mathrm{pr}}(\psi\|\varphi) = \sup_{A\in M_{[0,a]}} \{-\psi(A) + \varphi(g^\flat(A))\}.$$

Hence by Lemma 5.12 (2), there exists an $H \in M_{[0,a]}$ such that $S_f^{\mathrm{pr}}(\psi\|\varphi) = -\psi(H) + \varphi(g^\flat(H))$, and the remaining proof is similar to that of Theorem 5.13. Next, assume that $(g^\flat)^{-1}$ is operator convex on (c,b), where $b := -g(0^+)$ and $c := -g(\infty)$, so $(g^\flat)^{-1}$ extends to an operator convex function on $[c,b]$ with $(g^\flat)^{-1}(c) = 0$ and $(g^\flat)^{-1}(b) = g'(0^+)$. By Theorem 5.7 (2) and case (3) of Theorem 5.8 we have

$$S_f^{\mathrm{meas}}(\psi\|\varphi) = S_f^{\mathrm{pr}}(\psi\|\varphi) = \sup_{A\in M_{[c,b]}} \{-\psi((g^\flat)^{-1}(A)) + \varphi(A)\}.$$

Hence the proof proceeds as above by using Lemma 5.12 (1).

Case (2). Assume that $g^\flat$ is operator concave on $(0,a)$, where $a := g'(0^+) < \infty$. Since $g^\flat(0^+) = -g(\infty) = -\infty$ and hence $(g^\flat)'(0^+) = \infty$, there is a $t_0 \in (0,a)$ such that $(g^\flat)'(t_0) > \lambda$. For any $A \in M_{(0,a)}$ with the spectral decomposition $A = \int_0^a t\,dE_A(t)$, define an $A_0 \in M_{[t_0,a]}$ by

$$A_0 := t_0 E_A((0,t_0)) + \int_{[t_0,a]} t\,dE_A(t).$$

Then

$$\begin{aligned}
&\{-\psi(A_0) + \varphi(g^\flat(A_0))\} - \{-\psi(A) + \varphi(g^\flat(A))\} \\
&= -\int_{(0,t_0)} (t_0 - t)\,d\psi(E_A(t)) + \int_{(0,t_0)} (g^\flat(t_0) - g^\flat(t))\,d\varphi(E_A(t)) \\
&\ge \int_{(0,t_0)} (t_0 - t)\biggl(\frac{g^\flat(t_0) - g^\flat(t)}{t_0 - t} - \lambda\biggr)\,d\varphi(E_A(t)) \ge 0,
\end{aligned}$$

since

$$\frac{g^\flat(t_0)-g^\flat(t)}{t_0-t} \ge (g^\flat)'(t_0) > \lambda, \qquad 0<t<t_0.$$

Therefore, by Theorem 5.7 (2) and case (3) of Theorem 5.8, we have

$$S_f^{\mathrm{meas}}(\psi\|\varphi) = S_f^{\mathrm{pr}}(\psi\|\varphi) = \sup_{A\in M_{[t_0,a]}} \{-\psi(A)+\varphi(g^\flat(A))\}.$$

Since $g^\flat$ is operator concave on $[t_0,a]$, it follows from Lemma 5.12 (2) that the above supremum is attained by some $H \in M_{[t_0,a]}$, for which we have

$$S_f^{\mathrm{pr}}(\psi\|\varphi) = S_f(\mathcal{E}_H(\psi)\|\mathcal{E}_H(\varphi))$$

similarly to the proof of Theorem 5.13.
Next, assume that $(g^\flat)^{-1}$ is operator convex on $(-\infty,b)$, where $b := -g(0^+) < \infty$. Since $g^\flat(0^+) = -g(\infty) = -\infty$ and hence $(g^\flat)'(0^+) = \infty$, note that

$$\big[(g^\flat)^{-1}\big]'(t) = \frac{1}{(g^\flat)'((g^\flat)^{-1}(t))} \longrightarrow \frac{1}{(g^\flat)'(0^+)} = 0 \quad \text{as } b>t\to-\infty.$$

So there is a $t_0 \in (-\infty,b)$ such that $\big[(g^\flat)^{-1}\big]'(t_0) < 1/\lambda$. For any $A \in M_{(-\infty,b)}$ with $A = \int_{-\infty}^b t\,dE_A(t)$, define an $A_0 \in M_{[t_0,b]}$ by

$$A_0 := t_0E_A((-\infty,t_0)) + \int_{t_0}^b t\,dE_A(t).$$

Then

$$\begin{aligned}
&\{-\psi((g^\flat)^{-1}(A_0)+\varphi(A_0)\} - \{-\psi((g^\flat)^{-1}(A))+\varphi(A)\}\\
&= -\int_{(-\infty,t_0)} ((g^\flat)^{-1}(t_0)-(g^\flat)^{-1}(t))\,d\psi(E_A(t)) + \int_{(-\infty,t_0)} (t_0-t)\,d\varphi(E_A(t))\\
&\ge \int_{(-\infty,t_0)} (t_0-t)\bigg(\frac{1}{\lambda} - \frac{(g^\flat)^{-1}(t_0)-(g^\flat)^{-1}(t)}{t_0-t}\bigg)\,d\psi(E_A(t)) \ge 0,
\end{aligned}$$

since

$$\frac{(g^\flat)^{-1}(t_0)-(g^\flat)^{-1}(t)}{t_0-t} \le \big[(g^\flat)^{-1}\big]'(t_0) < \frac{1}{\lambda}, \qquad -\infty<t<t_0.$$

Hence we have

$$S_f^{\rm meas}(\psi\|\varphi) = S_f^{\rm pr}(\psi\|\varphi) = \sup_{A\in M_{[t_0,b]}} \{-\psi((g^\flat)^{-1}(A)) + \varphi(A)\}$$
$$= -\psi((g^\flat)^{-1}(H_0)) + \varphi(H_0)$$

for some $H_0 \in M_{[t_0,b]}$. Letting $H := (g^\flat)^{-1}(H_0) \in M_{[t_1,a]}$, where $t_1 := g^\flat(t_0)$, we have

$$S_f^{\rm pr}(\psi\|\varphi) = -\psi(H) + \varphi(g^\flat(H)) = S_f(\mathcal{E}_H(\psi)\|\mathcal{E}_H(\varphi))$$

as before.

Case (3). Assume that $g^\flat$ is operator concave on $(0,\infty)$, where $g'(0^+) = \infty$. Since $g^\flat(0^+) = -g(\infty) > -\infty$, $g^\flat$ extends to $[0,\infty)$ with $g^\flat(0) = -g(\infty)$. Since $(g^\flat)'(\infty) = 0$ (due to $g^\flat \in \mathcal{F}_{\rm cc}^{\nearrow}(0,\infty)$), there is a $t_0 \in (0,\infty)$ such that $(g^\flat)'(t_0) < 1/\lambda$. For any $A \in M_{++}$ with $A = \int_0^\infty t\,dE_A(t)$, define an $A_0 \in M_{[0,t_0]}$ by

$$A_0 := \int_0^{t_0} t\,dE_A(t) + t_0 E_A((t_0,\infty)).$$

Then, from $\varphi \le \lambda\psi$ we have

$$\{-\psi(A_0) + \varphi(g^\flat(A_0))\} - \{-\psi(A) + \varphi(g^\flat(A))\}$$
$$= \int_{(t_0,\infty)} (t-t_0)\,d\psi(E_A(t)) - \int_{(t_0,\infty)} (g^\flat(t) - g^\flat(t_0))\,d\varphi(E_A(t))$$
$$\ge \int_{(t_0,+\infty)} (t-t_0)\left(\frac{1}{\lambda} - \frac{g^\flat(t)-g^\flat(t_0)}{t-t_0}\right) d\varphi(E_A(t)) \ge 0,$$

since

$$\frac{g^\flat(t)-g^\flat(t_0)}{t-t_0} \le (g^\flat)'(t_0) < \frac{1}{\lambda}, \qquad t > t_0.$$

Hence by Theorem 5.7 (2) and case (3) of Theorem 5.8, we have

$$S_f^{\rm meas}(\psi\|\varphi) = S_f^{\rm pr}(\psi\|\varphi) = \sup_{A\in M_{[0,t_0]}} \{-\psi(A) + \varphi(g^\flat(A))\},$$

and the remaining proof is the same as before.

Next, assume that $(g^\flat)^{-1}$ is operator convex on (c, b), where $b := -g(0^+) \in (-\infty, \infty]$ and $c := -g(\infty) > -\infty$. Since $g^\flat(\infty) = b$ and $(g^\flat)'(\infty) = 0$, note that

$$\left[(g^\flat)^{-1}\right]'(t) = \frac{1}{(g^\flat)'((g^\flat)^{-1}(t))} \longrightarrow \frac{1}{(g^\flat)'(\infty)} = \infty \quad \text{as } c < t \nearrow b.$$

So there is a $t_0 \in (c, b)$ such that $\left[(g^\flat)^{-1}\right]'(t_0) > \lambda$. For any $A \in M_{(c,b)}$ with $A = \int_c^b t\, dE_A(t)$, define an $A_0 \in M_{[c,t_0]}$ by

$$A_0 := \int_c^{t_0} t\, dE_A(t) + t_0 E_A((t_0, b)).$$

Then

$$\begin{aligned}
&\{-\psi((g^\flat)^{-1}(A_0)) + \varphi(A_0)\} - \{-\psi((g^\flat)^{-1}(A)) + \varphi(A)\} \\
&= \int_{(t_0,b)} ((g^\flat)^{-1}(t) - (g^\flat)^{-1}(t_0))\, d\psi(E_A(t)) + \int_{(t_0,b)} (t - t_0)\, d\varphi(E_A(t)) \\
&\ge \int_{(t_0,b)} (t_0 - t)\left(\frac{(g^\flat)^{-1}(t) - (g^\flat)^{-1}(t_0)}{t - t_0} - \lambda\right)(t - t_0)\, d\psi(E_A(t)) \ge 0,
\end{aligned}$$

since

$$\frac{(g^\flat)^{-1}(t) - (g^\flat)^{-1}(t_0)}{t - t_0} \ge [(g^\flat)^{-1}]'(t_0) > \lambda, \qquad t_0 < t < b.$$

Since $(g^\flat)^{-1}$ is operator convex on $[c, t_0]$ with $(g^\flat)^{-1}(c) = 0$, we have

$$\begin{aligned}
S_f^{\mathrm{meas}}(\psi\|\varphi) = S_f^{\mathrm{pr}}(\psi\|\varphi) &= \sup_{A \in M_{[c,t_0]}} \{-\psi((g^\flat)^{-1}(A)) + \varphi(A)\} \\
&= -\psi((g^\flat)^{-1}(H_0)) + \varphi(H_0)
\end{aligned}$$

for some $H_0 \in M_{[c,t_0]}$, and the remaining proof is the same as before.

Case (4). When $g^\flat$ is operator concave on $(0, \infty)$, we may combine the proofs of cases (2) and (3). In this case, we can choose t_0, t_1 with $0 < t_0 < t_1 < \infty$ such that $(g^\flat)'(t_0) > \lambda$ and $(g^\flat)'(t_1) < 1/\lambda$. Then for any $A \in M_{++}$, define an $A_0 \in M_{[t_0,t_1]}$ by

$$A_0 := t_0 E_A((0, t_0)) + \int_{[t_0,t_1]} t\, dE_A(t)) + t_1 E_A((t_1, \infty)).$$

Then the proof proceeds as before, while we omit the details. When $(g^{\flat})^{-1}$ is operator convex on $(-\infty, b)$ where $b := -g(0^+) \in (-\infty, \infty]$, the proof is again the combination of those of cases (2) and (3); we omit the details.

Remark 5.15 As in Theorem 5.8, when the transpose $\widetilde{g}$ is in $\mathcal{F}_{cc}^{\nearrow}(0, \infty)$ and the assumption in Theorem 5.14 is satisfied for $\widetilde{g}$, the same attainability result for S_f^{pr} as in Theorem 5.14 holds under exchanging the roles of ψ, φ.

Examples 5.16

(1) When $1 < \alpha < \infty$, consider $f_\alpha(t) := t^\alpha \in \mathcal{F}_{cv}^{\nearrow}(0, \infty)$. From Example 5.9 (1), $(f_\alpha^{\sharp})^{-1}$ is operator concave on $(0, \infty)$. Hence by Theorem 5.13, if $\psi \le \lambda\varphi$ for some $\lambda > 0$, then there exists an $H \in M_{\mathrm{sa}}$ such that

$$S_{f_\alpha}^{\mathrm{meas}}(\psi\|\varphi) = S_{f_\alpha}^{\mathrm{pr}}(\psi\|\varphi) = S_{f_\alpha}(\mathcal{E}_H(\psi)\|\mathcal{E}_H(\varphi)),$$

so that

$$D_\alpha^{\mathrm{meas}}(\psi\|\varphi) = D_\alpha^{\mathrm{pr}}(\psi\|\varphi) = D_\alpha(\mathcal{E}_H(\psi)\|\mathcal{E}_H(\varphi)).$$

(2) When $0 < \alpha < 1$, consider $g_\alpha(t) := t^\alpha \in \mathcal{F}_{cc}^{\nearrow}(0, \infty)$ and $f_\alpha := -g_\alpha$. Then $g_\alpha(0^+) = 0$, $g_\alpha(\infty) = \infty$, and $g_\alpha'(0^+) = \infty$. From Example 5.9 (2), when $1/2 \le \alpha < 1$, $\widetilde{g}_\alpha^{\flat}$ is operator concave on $(0, \widetilde{g}'_\alpha(0^+)) = (0, \infty)$. When $0 < \alpha \le 1/2$, $(\widetilde{g}_\alpha^{\flat})^{-1}$ is operator convex on $(-\widetilde{g}_\alpha(\infty), -\widetilde{g}_\alpha(0^+)) = (-\infty, 0)$. Hence by case (4) of Theorem 5.14 and Remark 5.15, if $\lambda^{-1}\varphi \le \psi \le \lambda\varphi$ for some $\lambda > 1$, there exists an $H \in M_{\mathrm{sa}}$ such that

$$S_{f_\alpha}^{\mathrm{meas}}(\psi\|\varphi) = S_{f_\alpha}^{\mathrm{pr}}(\psi\|\varphi) = S_{f_\alpha}(\mathcal{E}_H(\psi)\|\mathcal{E}_H(\varphi)),$$

so that

$$D_\alpha^{\mathrm{meas}}(\psi\|\varphi) = D_\alpha^{\mathrm{pr}}(\psi\|\varphi) = D_\alpha(\mathcal{E}_H(\psi)\|\mathcal{E}_H(\varphi)).$$

(3) Consider $g_0(t) := \log t \in \mathcal{F}_{cc}^{\nearrow}(0, \infty)$ and $f_0 := -g_0$. Then $g_0(0^+) = -\infty$, $g_0(\infty) = \infty$, and $g_0'(0^+) = \infty$. From Example 5.10, $g_0^{\flat}$ is operator concave on $(0, g_0'(0^+)) = (0, \infty)$. Hence by case (4) of Theorem 5.14, for every $\psi, \varphi \in M_*^+$ with $\lambda^{-1}\varphi \le \psi \le \lambda\varphi$, there exists an $H \in M_{\mathrm{sa}}$ such that

$$D^{\mathrm{meas}}(\psi\|\varphi) = D^{\mathrm{pr}}(\psi\|\varphi) = D(\mathcal{E}_H(\psi)\|\mathcal{E}_H(\varphi)).$$

(4) For each $\lambda \in (0, \infty)$, consider $g_\lambda(t) := \frac{t}{t+\lambda} \in \mathcal{F}_{cc}^{\nearrow}(0, \infty)$ and $f_\lambda := -g_\lambda$. Then $g_\lambda(0^+) = 0$, $g_\lambda(\infty) = 1$, and $g_\lambda'(0^+) = 1/\lambda$. From Example 5.11, $(g_\lambda^{\flat})^{-1}$ is operator convex on $(-g_\lambda(\infty), -g_\lambda(0^+)) = (-1, 0)$. Hence by case (1) of

Theorem 5.14, for every $\psi, \varphi \in M_*^+$ (with no condition), there exists an $H \in M_{\mathrm{sa}}$ such that

$$S_{f_\lambda}^{\mathrm{meas}}(\psi\|\varphi) = S_{f_\lambda}^{\mathrm{pr}}(\psi\|\varphi) = S_{f_\lambda}(\mathcal{E}_H(\psi)\|\mathcal{E}_H(\varphi)).$$

In the two examples below we will give the descriptions of optimal measurements for the particular Rényi divergences D_α when $\alpha = 2$ and $1/2$. Similar discussions on optimal measurements for those Rényi divergences in the finite-dimensional case are found in [91]. In the following, let $\psi, \varphi \in M_*^+$ with $\psi(1) = 1$ for simplicity.

Example 5.17 (Optimal Measurement for D_2) Let $f_2(t) := t^2$ and so $f_2^\sharp(t) = t^2/4$ as given in Example 5.9 (1). By (5.17) we have

$$Q_2^{\mathrm{meas}}(\psi\|\varphi) = S_{f_2}^{\mathrm{meas}}(\psi\|\varphi) = \sup_{x \in M_+} \{2\psi(x) - \varphi(x^2)\}.$$

Assume that $\psi \le \lambda\varphi$ for some $\lambda > 0$. In this case, we may assume that φ is faithful. Then by the linear Radon–Nikodym theorem due to Sakai [116, 1.24.4], there exists an $h_0 \in M_+$ such that

$$\psi(x) = \frac{1}{2}\varphi(h_0 x + x h_0), \qquad x \in M. \tag{5.24}$$

For every $h \in M_{sa}$ we find that

$$\begin{aligned}
&2\psi(h_0 + h) - \varphi((h_0 + h)^2) \\
&\quad = 2\psi(h_0) - \varphi(h_0^2) + \{2\psi(h) - \varphi(h_0 h + h h_0)\} - \varphi(h^2) \\
&\quad \le 2\psi(h_0) - \varphi(h_0^2).
\end{aligned}$$

Therefore,

$$Q_2^{\mathrm{meas}}(\psi\|\varphi) = 2\psi(h_0) - \varphi(h_0^2) = 2\psi(h_0) - \psi(h_0) = \psi(h_0).$$

Consequently, the optimal measurement for D_2 is induced from the spectral resolution of h_0.

According to Kosaki [76, Theorem 1.6], a linear Radon–Nikodym derivative h_0 satisfying (5.24) exists if and only if $\widetilde{\psi} \le \lambda\varphi$ for some $\lambda > 0$, where

$$\widetilde{\psi} := \int_{-\infty}^{\infty} \frac{1}{\cosh \pi t}\, \psi \circ \sigma_t^\varphi \, dt \tag{5.25}$$

(of course, $\psi \le \lambda\varphi \implies \widetilde{\psi} \le \lambda\varphi$), and in this case, h_0 is determined as

$$h_0 = [D\widetilde{\psi} : D\varphi]_{-i/2}^* [D\widetilde{\psi} : D\varphi]_{-i/2},$$

where $[D\widetilde{\psi} : D\varphi]_t$, $t \in \mathbb{R}$, is Connes' cocycle derivative and $[D\widetilde{\psi} : D\varphi]_{-i/2}$ is the analytic continuation at $t = -i/2$, see Lemma A.58 of Sect. A.7. Therefore,

$$D_2^{\mathrm{meas}}(\psi\|\varphi) = \log \psi\big([D\widetilde{\psi} : D\varphi]_{-i/2}^*[D\widetilde{\psi} : D\varphi]_{-i/2}\big).$$

Assume that $M = B(\mathcal{H})$ on a finite-dimensional Hilbert space $\mathcal{H}$ and $\psi, \varphi \in B(\mathcal{H})_*^+$ have the density operators D_ψ, D_φ. Then we note that $\widetilde{\psi}$ in (5.25) has the density

$$D_{\widetilde{\psi}} = \int_{-\infty}^{\infty} \frac{1}{\cosh \pi t} D_\varphi^{-it} D_\psi D_\varphi^{it}\, dt.$$

Therefore,

$$\begin{aligned} h_0 &= (D_{\widetilde{\psi}}^{1/2} D_\varphi^{-1/2})^*(D_{\widetilde{\psi}}^{1/2} D_\varphi^{-1/2}) = D_\varphi^{-1/2} D_{\widetilde{\psi}} D_\varphi^{-1/2} \\ &= \int_{-\infty}^{\infty} \frac{1}{\cosh \pi t} D_\varphi^{-it} D_\varphi^{-1/2} D_\psi D_\varphi^{-1/2} D_\varphi^{-it}\, dt \end{aligned}$$

and

$$Q_2^{\mathrm{meas}}(\psi\|\varphi) = \mathrm{Tr}\, D_\psi h_0 = \int_{-\infty}^{\infty} \frac{1}{\cosh \pi t} \mathrm{Tr}\, D_\psi D_\varphi^{-it} D_\varphi^{-1/2} D_\psi D_\varphi^{-1/2} D_\varphi^{-it}\, dt.$$

Since (5.24) is nothing but the familiar Lyapunov equation $D_\psi = \frac{1}{2}(h_0 D_\varphi + D_\varphi h_0)$, the alternative well-known formula of h_0 is

$$h_0 = 2\int_0^{\infty} e^{-tD_\varphi} D_\psi e^{-tD_\varphi}\, dt.$$

Example 5.18 (Optimal Measurement for $D_{1/2}$) We first show that for every $\psi, \varphi \in M_*^+$ (with $\psi(1) = 1$),

$$D_{1/2}^{\mathrm{meas}}(\psi\|\varphi) = \widetilde{D}_{1/2}(\psi\|\varphi) = -2\log F(\psi, \varphi). \tag{5.26}$$

Indeed, the latter equality is (3.29). By Example 5.9 (2) we note that

$$Q_{1/2}^{\mathrm{meas}}(\psi\|\varphi) = \frac{1}{2} \inf_{x\in M_{++}} \{\psi(x^{-1}) + \varphi(x)\}.$$

On the other hand, by (3.34) of Lemma 3.19 we have

$$\widetilde{Q}_{1/2}(\psi\|\varphi) = \frac{1}{2} \inf_{x\in M_{++}} \{\psi(x^{-1}) + \varphi(x)\}$$

so that (5.26) holds, which also says that the measured version of the fidelity F is the same as F itself and $F(\psi,\varphi)=\frac{1}{2}\sup_{x\in M_{++}}\{\psi(x^{-1})+\varphi(x)\}$. This and other similar expressions for F were given in [2, 3].

Now, assume that $h_\psi^\delta\le\lambda h_\varphi^\delta$ for some $\delta,\lambda>0$, or equivalently, $s(\psi)\le s(\varphi)$ and Connes' cocycle derivative $[D\psi:D\varphi]_t$ has the analytic continuation to the strip $-\delta/2\le\operatorname{Im}z\le 0$, see Lemma A.58 of Sect. A.7. Then by the $p=1$ case of Lemma A.59 (Kosaki's extension [81] of Sakai's quadratic Radon–Nikodym theorem), there exists a unique $k_1\in(s(\varphi)Ms(\varphi))_+$ such that $h_\psi=k_1h_\varphi k_1$, i.e., $\psi(x)=\varphi(k_1xk_1)$ for all $x\in M$. Moreover, in this case, k_1 satisfies $(h_\varphi^{1/2}h_\psi h_\varphi^{1/2})^{1/2}=h_\varphi^{1/2}k_1h_\varphi^{1/2}$ so that

$$F(\psi,\varphi)=\operatorname{tr}(h_\varphi^{1/2}h_\psi h_\varphi^{1/2})^{1/2}=\operatorname{tr}h_\varphi k_1=\varphi(k_1). \tag{5.27}$$

With the spectral decomposition $k_1=\int_0^r t\,dE_{k_1}(t)$ with $r:=\|k_1\|$, let $p:=d\psi(E_{k_1}(t))$ and $q:=d\varphi(E_{k_1}(t))$. Then $p\ll q$ (absolutely continuous) since $s(\psi)\le s(\varphi)$, and we have

$$\begin{aligned}Q_{1/2}^{\mathrm{meas}}(\psi\|\varphi)&\le\int_0^r\left(\frac{dp}{dq}\right)^{1/2}dq=\lim_{\varepsilon\searrow0}\int_\varepsilon^r\left(\frac{dp}{dq}\right)^{1/2}dq\\&\le\lim_{\varepsilon\searrow0}\frac{1}{2}\int_\varepsilon^r\left(t^{-1}\frac{dp}{dq}(t)+t\right)dq(t)\\&=\lim_{\varepsilon\searrow0}\frac{1}{2}\left(\int_\varepsilon^r t^{-1}\,dp(t)+\int_0^r t\,dq(t)\right).\end{aligned}$$

Furthermore, since $\int_0^r t\,dq(t)=\varphi(k_1)$ and

$$\begin{aligned}\int_\varepsilon^r t^{-1}\,dp(t)&=\psi\left(\int_\varepsilon^r t^{-1}\,dE_{k_1}(t)\right)=\varphi\left(k_1\left(\int_\varepsilon^r t^{-1}\,dE_{k_1}(t)\right)k_1\right)\\&=\varphi\left(\int_\varepsilon^r t\,dE_{k_1}(t)\right)\longrightarrow\varphi(k_1)\quad\text{as }\varepsilon\searrow0,\end{aligned}$$

it follows that

$$D_{1/2}^{\mathrm{meas}}(\psi\|\varphi)\ge D_{1/2}(p\|q)=-2\log\int_0^r\left(\frac{dp}{dq}\right)^{1/2}dq\ge-2\log\varphi(k_1). \tag{5.28}$$

By (5.26)–(5.28) we find that

$$D_{1/2}^{\mathrm{meas}}(\psi\|\varphi)=-2\log F(\psi,\varphi)=-2\log\varphi(k_1)$$

and the optimal measurement for $D_{1/2}$ is induced from the spectral resolution of k_1.

In the finite-dimensional case with the density operators D_ψ, D_φ with φ faithful, the optimal measurement is given by the spectral decomposition of

$$k_1 := D_\varphi^{-1/2}(D_\varphi^{1/2} D_\psi D_\varphi^{1/2})^{1/2} D_\varphi^{-1/2} = D_\psi \# D_\varphi^{-1},$$

the geometric mean of D_ψ and D_φ^{-1}, as explained in [100, Sec. 9.2.2].

Problems 5.19

(1) An example of an operator convex function on $(0, \infty)$ for which $S_f^{\mathrm{pr}}(\psi\|\varphi) < S_f^{\mathrm{meas}}(\psi\|\varphi)$ for some $\psi, \varphi \in M_*^+$ is not known. Such an example might exist in the finite-dimensional setting if any.

(2) In the attainability results in Theorems 5.13 and 5.14 (as well as Examples 5.17 and 5.18) we have proved the existence of von Neumann measurements induced by bounded self-adjoint operators in M, under such an assumption as $\psi \le \lambda\varphi$ or more strongly $\lambda^{-1}\varphi \le \psi \le \lambda\varphi$ for some $\lambda > 0$. These dominance assumptions are rather too strong in the infinite-dimensional setting, so it is desirable to remove them by considering measurements induced by unbounded self-adjoint operators. However, the problem seems difficult because the compactness argument has been used in the proof of the theorems.

(3) It is also interesting to find a more explicit form of optimal measurements, as in Examples 5.17 and 5.18, in the attainability results of Theorems 5.13 and 5.14. Since an optimal measurement is given as the maximizer of a certain strictly convex function on a certain set $M_{[a,b]}$, the maximizer might explicitly be specified.

Chapter 6
Reversibility and Quantum Divergences

6.1 Petz' Recovery Map

Let M and N be von Neumann algebras, whose standard forms are $(M, \mathcal{H}, J, \mathcal{P})$ and $(N, \mathcal{H}_0, J_0, \mathcal{P}_0)$, respectively. For convenience, we first summarize basic properties of positive linear maps between von Neumann algebras, although those have already been used in previous chapters. For a positive linear map $\gamma : N \to M$, γ is *unital* if $\gamma(1) = 1$. The γ is *normal* if $\gamma(y_j) \nearrow \gamma(y)$ for any net $\{y_j\}$ in N_+ with $y_j \nearrow y \in N_+$, or equivalently, γ is continuous with respect to the σ-weak topologies (i.e., the $\sigma(N, N_*)$, $\sigma(M, M_*)$-topologies) on N and M. We call γ a *Schwarz map* if $\gamma(y)^*\gamma(y) \le \gamma(y^*y)$ for all $y \in N$. The γ is said to be *n-positive* if $\gamma \otimes \mathrm{id}_n : N \otimes \mathbb{M}_n \to M \otimes \mathbb{M}_n$ is positive, where $\mathbb{M}_n$ is the $n \times n$ matrix algebra and $(\gamma \otimes \mathrm{id}_n)([y_{ij}]) := [\gamma(y_{ij})]$ for $[y_{ij}]_{i,j=1}^n \in \mathbb{M}_n(N) = N \otimes \mathbb{M}_n$. Furthermore, γ is said to be *completely positive (CP)* if it is n-positive for all $n \ge 1$. As is well-known,

$$\text{CP} \implies \text{2-positive} \implies \text{Schwarz map} \implies \text{simply positive.}$$

In quantum information in the finite-dimensional setting, quantum channels (implementing quantum operations) are typically *TPCP* (trace-preserving CP) maps, whose dual maps are unital CP maps. But we may consider, depending on problems, weaker notions of positivity such as Schwarz positivity or 2-positivity or even plain positivity for quantum operations. In the von Neumann algebra setting, normality is essential for a channel $\gamma : N \to M$, which means that γ is the dual of the predual map $\gamma_* : M_* \to N_*$.

The next theorem was proved in [1, 106] under the assumption that both of φ and $\varphi \circ \gamma$ are faithful, but the faithfulness of φ is not essential.

Theorem 6.1 ([1, 106]) *Let $\gamma : N \to M$ be a unital normal positive map, and $\varphi \in M_*^+$. Assume that $\varphi \circ \gamma$ is faithful. Let $\Phi \in \mathcal{P}$ and $\Phi_0 \in \mathcal{P}_0$ be the vector representatives of φ and $\varphi \circ \gamma$, respectively. Then there exists a unique unital normal*

F. Hiai, *Quantum f-Divergences in von Neumann Algebras*,
Mathematical Physics Studies, https://doi.org/10.1007/978-981-33-4199-9_6

positive map $\beta : M \to N$ such that

$$\langle J_0\beta(x)\Phi_0, y\Phi_0\rangle = \langle Jx\Phi, \gamma(y)\Phi\rangle \tag{6.1}$$

for all $x \in M$ and all $y \in N$.

Proof First, assuming that for each $x \in M$, $\beta(x) \in N$ exists so that (6.1) holds for all $y \in N$, we prove that $\beta : M \to N$ is a unital normal positive map. Since Φ_0 is a cyclic vector for N (i.e., $N\Phi_0 := \{y\Phi_0 : y \in N\}$ is dense in $\mathcal{H}_0$), it is obvious that $\beta(x) \in N$ is uniquely determined by (6.1) and $\beta : M \to N$ is a linear map. Since $\langle J\Phi, \gamma(y)\Phi\rangle = (\varphi \circ \gamma)(y) = \langle J_0\Phi_0, y\Phi_0\rangle$ for all $y \in N$, we have $\beta(1) = 1$. For every $x \in M_+$ and $y \in N$, it follows from (6.1) that

$$\begin{aligned}\langle (J_0\beta(x)J_0)y\Phi_0, y\Phi_0\rangle &= \langle J_0\beta(x)\Phi_0, y^*y\Phi_0\rangle = \langle Jx\Phi, \gamma(y^*y)\Phi\rangle \\ &= \langle (JxJ)\gamma(y^*y)^{1/2}\Phi, \gamma(y^*y)^{1/2}\Phi\rangle \ge 0,\end{aligned}$$

showing that $J_0\beta(x)J_0 \ge 0$ and so $\beta(x) \in N_+$. Hence β is a unital positive map. To prove the normality of β, let $\{x_j\} \subset M_+$ be a net with $x_j \nearrow x \in M_+$; then $\beta(x_j) \nearrow y_0$ for some $y_0 \in N_+$. For every $y \in N$ we have

$$\langle J_0y_0\Phi_0, y\Phi_0\rangle = \lim_j \langle J_0\beta(x_j)\Phi_0, y\Phi_0\rangle = \lim_j \langle Jx_j\Phi, \gamma(y)\Phi\rangle = \langle Jx\Phi, \gamma(y)\Phi\rangle,$$

which implies that $y_0 = \beta(x)$ so that $\beta(x_j) \nearrow \beta(x)$. Hence β is normal.

It remains to prove, for each $x \in M$, the existence of $\beta(x)$ satisfying (6.1) for all $y \in N$. For this, by linearity we may assume that $x \in M_+$. Define $\psi_x \in N_*$ by $\psi_x(y) := \langle Jx\Phi, \gamma(y)\Phi\rangle$, $y \in N$. For any $y \in N_+$ one has

$$\psi_x(y) = \langle \gamma(y)^{1/2}(JxJ)\gamma(y)^{1/2}\Phi, \Phi\rangle \le \|x\|\langle \gamma(y)\Phi, \Phi\rangle = \|x\|(\varphi \circ \gamma)(y)$$

so that $0 \le \psi_x \le \|x\|(\varphi \circ \gamma)$. Hence by Lemma A.24 of Sect. A.3 there exists a $B \in N$ such that $\Psi_x = B\Phi_0$, where $\Psi_x \in \mathcal{P}_0$ is the vector representative of ψ_x. Since $\Psi_x = J_0\Psi_x = J_0B\Phi_0 = J_0BJ_0\Phi_0$, one has for every $y \in N$,

$$\begin{aligned}\psi_x(y) &= \langle B\Phi_0, yJ_0BJ_0\Phi_0\rangle = \langle J_0B^*J_0B\Phi_0, y\Phi_0\rangle \\ &= \langle J_0B^*\Psi_x, y\Phi_0\rangle = \langle J_0B^*B\Phi_0, y\Phi_0\rangle.\end{aligned}$$

Letting $\beta(x) := B^*B \in N_+$ one obtains $\psi_x(y) = \langle J_0\beta(x)\Phi_0, y\Phi_0\rangle$, $y \in N$, which gives (6.1).

Definition 6.2 For $\gamma : N \to M$ and $\varphi \in M_*^+$ as in Theorem 6.1, we refer to $\beta : M \to N$ satisfying (6.1) as *Petz' recovery map* of γ with respect to φ, and denote it by γ_φ^*. The γ_φ^* is defined by (6.1) independently of the choice of the standard forms of M, N due to Theorem A.19 of Sect. A.3.

Proposition 6.3 *Let $\gamma : N \to M$ and $\varphi \in M_*^+$ be as in Theorem 6.1.*

(1) $\varphi \circ \gamma \circ \gamma_\varphi^* = \varphi$.
(2) *Assume that φ is faithful as well as $\varphi \circ \gamma$. Then $(\gamma_\varphi^*)_{\varphi\circ\gamma}^* = \gamma$, i.e., γ is Petz' recovery map of γ_φ^* with respect to $\varphi \circ \gamma$.*
(3) *If γ is 2-positive (resp., CP), then γ_φ^* is 2-positive (resp., CP).*

Proof We write β for γ_φ^*.

(1) Letting $y = 1$ in (6.1) gives

$$\varphi(x) = \langle Jx\Phi, \Phi\rangle = \langle J_0\beta(x)\Phi_0, \Phi_0\rangle = \varphi \circ \gamma(\beta(x)), \qquad x \in M,$$

so that $\varphi = \varphi \circ \gamma \circ \beta$.
(2) immediately follows from (1) and (6.1) since the roles of γ and β are symmetric in (6.1).
(3) Assume that γ is 2-positive. Consider the von Neumann algebra $M^{(2)} := M \otimes \mathbb{M}_2$, the tensor product of M with the 2×2 matrix algebra $\mathbb{M}_2(\mathbb{C})$, whose standard form is described in Example A.20 of Sect. A.3. In fact, the standard representation of $M^{(2)}$ is given in (A.5) on $\mathcal{H}^{(2)} := \mathcal{H} \oplus \mathcal{H} \oplus \mathcal{H} \oplus \mathcal{H}$ with the standard involution $J^{(2)}$ in (A.7). Similarly, we take the standard $\mathcal{H}_0^{(2)}$ and $J_0^{(2)}$ for $N^{(2)} := N \otimes \mathbb{M}_2$. Set $\varphi^{(2)} \in (M^{(2)})_*^+$ by $\varphi^{(2)}\left(\begin{bmatrix} x_{11} & x_{12} \\ x_{21} & x_{22}\end{bmatrix}\right) := \varphi(x_{11}) + \varphi(x_{22})$. Since γ is 2-positive, we have $\varphi^{(2)} \circ (\gamma \otimes \mathrm{id}_2) \in (N^{(2)})_*^+$, which is given as $\varphi^{(2)} \circ (\gamma \otimes \mathrm{id}_2)\left(\begin{bmatrix} y_{11} & y_{12} \\ y_{21} & y_{22}\end{bmatrix}\right) = \varphi \circ \gamma(y_{11}) + \varphi \circ \gamma(y_{22})$. Moreover, note that the vector representatives of $\varphi^{(2)}$ and $\varphi^{(2)} \circ (\gamma \otimes \mathrm{id}_2)$ are $\Phi^{(2)} = \begin{bmatrix} \Phi & 0 \\ 0 & \Phi\end{bmatrix}$ $(= \Phi \oplus 0 \oplus 0 \oplus \Phi)$ and $\Phi_0^{(2)} = \begin{bmatrix} \Phi_0 & 0 \\ 0 & \Phi_0\end{bmatrix}$ $(= \Phi_0 \oplus 0 \oplus 0 \oplus \Phi_0)$, respectively, see Example A.20. Then for $x = \begin{bmatrix} x_{11} & x_{12} \\ x_{21} & x_{22}\end{bmatrix} \in M^{(2)}$ and $y = \begin{bmatrix} y_{11} & y_{12} \\ y_{21} & y_{22}\end{bmatrix} \in N^{(2)}$, with $\beta = \gamma_\varphi^*$, we have

$$\begin{aligned}
\langle J_0^{(2)}(\beta \otimes \mathrm{id}_2)(x)\Phi_0^{(2)}, y\Phi_0^{(2)}\rangle &= \sum_{i,j=1}^{2} \langle J_0\beta(x_{ij})\Phi_0, y_{ij}\Phi_0\rangle \\
&= \sum_{i,j=1}^{2} \langle Jx_{ij}\Phi, \gamma(y_{ij})\Phi\rangle \\
&= \langle J^{(2)}x\Phi^{(2)}, (\gamma \otimes \mathrm{id}_2)(y)\Phi^{(2)}\rangle,
\end{aligned}$$

so that $\beta \otimes \mathrm{id}_2$ is Petz' recovery map of $\gamma \otimes \mathrm{id}_2$ with respect to $\varphi^{(2)}$. Hence $\beta \otimes \mathrm{id}_2$ is positive (as seen from the proof of Theorem 6.1), i.e., β is 2-positive.

For any $n \geq 1$, if γ is n-positive, then one can prove that so is β, in a similar way by taking the standard forms of $M \otimes \mathbb{M}_n$ and $N \otimes \mathbb{M}_n$ (while the details are omitted). Thus, if γ is CP, then so is β.

Remark 6.4 Petz' recovery map γ_φ^* of a Schwarz map γ is not necessarily a Schwarz map, as shown by [65, Proposition 2] even in the finite-dimensional setting. This is the reason why we often need to assume in the reversibility problem that a quantum operation is 2-positive rather than a Schwarz map.

Let M_0 be a von Neumann subalgebra of M and $(M_0, \mathcal{H}_0, J_0, \mathcal{P}_0)$ be the standard form of M_0. For a faithful $\varphi \in M_*^+$ let $\Phi_0 \in \mathcal{P}_0$ be the vector representative of $\varphi|_{M_0}$ as well as $\Phi \in \mathcal{P}$ of φ. Then the *generalized conditional expectation* $E_\varphi = E_\varphi^{M_0} : M \to M_0$ (due to Accardi and Cecchini [1]) is given as

$$E_\varphi(x) := J_0 V_\varphi^* J x J V_\varphi J_0, \qquad x \in M,$$

which is also defined by

$$E_\varphi(x)\Phi_0 = J_0 V_\varphi^* J x \Phi, \qquad x \in M,$$

where $V_\varphi : \mathcal{H}_0 \to \mathcal{H}$ is the isometry defined by $V_\varphi y \Phi_0 := y\Phi$, $y \in M_0$. In this case, since

$$\langle J_0 E_\varphi(x)\Phi_0, y\Phi_0 \rangle = \langle Jx\Phi, V_\varphi y \Phi_0 \rangle = \langle Jx\Phi, y\Phi \rangle, \qquad x \in M,\ y \in M_0,$$

we have the following:

Proposition 6.5 *Let M_0 be a von Neumann subalgebra of M. If $\varphi \in M_*^+$ is faithful, then Petz' recovery map of the inclusion map $M_0 \hookrightarrow M$ with respect to φ is the generalized conditional expectation $E_\varphi : M \to M_0$.*

We can extend Petz' recovery map γ_φ^* to arbitrary $\varphi \in M_*^+$ as follows:

Proposition 6.6 *Let γ be as in Theorem 6.1. For every $\varphi \in M_*^+$ let $\Phi \in \mathcal{P}$ and $\Phi_0 \in \mathcal{P}_0$ be the vector representatives of φ and $\varphi \circ \gamma$, respectively. Let $e := s(\varphi) \in M$ and $e_0 := s(\varphi \circ \gamma) \in N$. Then there exists a unique unital (i.e., $\beta(1) = e_0$) normal positive map $\beta : M \to e_0 N e_0$ such that* (6.1) *holds for all $x \in M$ and all $y \in e_0 N e_0$. Furthermore, $\beta(1-e) = 0$ and $\beta(x) = \beta(exe)$ for all $x \in M$, and we have:*

(1) $\varphi \circ \gamma \circ \beta = \varphi$.
(2) *If γ is 2-positive (resp., CP), then β is 2-positive (resp., CP).*

Proof Let $e_0' := J_0 e_0 J_0 = s_{N'}(\varphi \circ \gamma)$ and $q_0 := e_0 e_0'$. By Proposition A.16 note that $e_0 N e_0 \cong q_0 N q_0$ (by $y \in e_0 N e_0 \mapsto y q_0 \in q_0 N q_0$) and the standard form of $q_0 N q_0$ is $(q_0 N q_0, q_0 \mathcal{H}_0, q_0 J_0 q_0, q_0 \mathcal{P}_0)$. Moreover, $\Phi_0 = q_0 \Phi_0 \in q_0 \mathcal{P}_0$. Although $\gamma|_{e_0 N e_0} : e_0 N e_0 \to M$ is not necessarily unital, we can apply the proof of Theorem 6.1 to $\gamma|_{e_0 N e_0}$, so we have a unique normal map $\beta : M \to e_0 N e_0$ such that

$$\langle (q_0 J_0 q_0)(\beta(x)q_0)\Phi_0, (yq_0)\Phi_0\rangle = \langle Jx\Phi, \gamma(y)\Phi\rangle, \qquad x \in M,\ y \in e_0 N e_0.$$

Since $q_0 J_0 q_0 = e_0 e_0' J_0$, the above LHS is equal to

$$\langle J_0\beta(x)\Phi_0, e_0 e_0' y\Phi_0\rangle = \langle J_0\beta(x)\Phi_0, y\Phi_0\rangle.$$

Hence β is determined by (6.1) for all $x \in M$ and $y \in e_0 N e_0$. Then $\beta(1) = e_0$ follows since $\langle J\Phi, \gamma(y)\Phi\rangle = (\varphi \circ \gamma)(y) = \langle J_0 e_0 \Phi_0, y\Phi_0\rangle$ for $y \in e_0 N e_0$. Moreover, for every $x \in M$, since $Jx(1-e)\Phi = 0$, one has $\beta(x(1-e)) = 0$ and $\beta((1-e)x) = \beta(x^*(1-e))^* = 0$. Therefore, $\beta(1-e) = 0$ and $\beta(x) = \beta(exe)$ for all $x \in M$.

Furthermore, since $\varphi \circ \gamma(e_0) = \varphi \circ \gamma(1) = \varphi(1)$, one has $\varphi(1 - \gamma(e_0)) = 0$ so that $(1 - \gamma(e_0))\Phi = 0$. Letting $y = e_0$ in (6.1) gives the equality in (1) as in the proof of Proposition 6.3 (1). The assertion in (2) is seen similarly to the proof of Proposition 6.3 (3).

Remark 6.7 Here we give a slightly different description of β than that given in Proposition 6.6. In the situation of the proposition define

$$\hat{\gamma} := e\gamma(\cdot)e|_{e_0 N e_0} : e_0 N e_0 \longrightarrow eMe. \tag{6.2}$$

Since $\varphi(1 - \gamma(e_0)) = 0$ as seen in the proof of Proposition 6.6, $e(1 - \gamma(e_0))e = 0$ so that $\hat{\gamma}(e_0) = e$. Hence $\hat{\gamma}$ is a unital normal positive map, so we can define Petz' recovery map $\hat{\beta}$ of $\hat{\gamma}$ with respect to $\varphi|_{eMe}$ (Definition 6.2). Now we show that β is essentially the same as $\hat{\beta}$ in such a way that

$$\beta(x) = \beta(exe) = \hat{\beta}(exe), \qquad x \in M. \tag{6.3}$$

Indeed, consider the standard form $(qMq, q\mathcal{H}, qJq, q\mathcal{P})$ of $eMe \cong qMq$ where $q := eJeJ$, as well as $(q_0 N q_0, q_0\mathcal{H}_0, q_0 J_0 q_0, q_0\mathcal{P}_0)$ of $e_0 N e_0 \cong q_0 N q_0$ in the proof of Proposition 6.6. Then $\hat{\beta}$ is determined by

$$\langle (q_0 J_0 q_0)(\hat{\beta}(x)q_0)\Phi_0, (yq_0)\Phi_0\rangle = \langle (qJq)(xq)\Phi, \hat{\gamma}(y)\Phi\rangle, \quad x \in eMe,\ y \in e_0 N e_0,$$

which can easily be reduced to

$$\langle J_0\hat{\beta}(x)\Phi_0, y\Phi_0\rangle = \langle Jx\Phi, \gamma(y)\Phi\rangle, \qquad x \in eMe,\ y \in e_0 N e_0.$$

Hence for every $x \in M$ and $y \in e_0 N e_0$ we have

$$\langle J_0\hat{\beta}(exe)\Phi_0, y\Phi_0\rangle = \langle Jexe\Phi, \gamma(y)\Phi\rangle = \langle Jx\Phi, JeJ\gamma(y)\Phi\rangle = \langle Jx\Phi, \gamma(y)\Phi\rangle,$$

showing (6.3) by Proposition 6.6.

We denote $\beta : M \to e_0 N e_0$ in Proposition 6.6 by the same γ_φ^* as in Definition 6.2, which may be called again Petz' recovery map of γ with respect to φ. This extended version will be used later in Sects. 6.4, 7.1 and 8.1.

Example 6.8 Here we consider the finite-dimensional case $M = B(\mathcal{H})$ and $N = B(\mathcal{K})$ with $\dim \mathcal{H}$, $\dim \mathcal{K} < \infty$. Let $T : B(\mathcal{H}) \to B(\mathcal{K})$ be a trace-preserving positive map, so the dual map $\gamma = T^* : B(\mathcal{K}) \to B(\mathcal{H})$ is a unital positive map. For every $\varphi \in B(\mathcal{H})_*^+$ with density D_φ, the density of $\varphi \circ \gamma \in B(\mathcal{K})_+^*$ is $T(D_\varphi)$ and $E_0 = s(\varphi \circ \gamma)$ is the support projection of $T(D_\varphi)$. Let $\beta : B(\mathcal{H}) \to E_0 B(\mathcal{K}) E_0$ be as given in Proposition 6.6. Recall that the standard representation of $B(\mathcal{H})$ is the left multiplication on $B(\mathcal{H})$ with the Hilbert–Schmidt inner product and $J = {}^*$. Hence for every $X \in B(\mathcal{H})$, $\beta(X)$ is given as

$$\operatorname{Tr} T(D_\varphi)^{1/2} \beta(X) T(D_\varphi)^{1/2} Y = \operatorname{Tr} T(D_\varphi^{1/2} X D_\varphi^{1/2}) Y, \quad Y \in E_0 B(\mathcal{K}) E_0.$$

Since $E_0 T(D_\varphi^{1/2}) E_0 = T(D_\varphi^{1/2})$ and $E_0 T(D_\varphi^{1/2} X D_\varphi^{1/2}) E_0 = T(D_\varphi^{1/2} X D_\varphi^{1/2})$, we have

$$T(D_\varphi)^{1/2} \beta(X) T(D_\varphi)^{1/2} = T(D_\varphi^{1/2} X D_\varphi^{1/2}). \tag{6.4}$$

Therefore,

$$\beta(X) = T(D_\varphi)^{-1/2} T(D_\varphi^{1/2} X D_\varphi^{1/2}) T(D_\varphi)^{-1/2},$$

where $T(D_\varphi)^{-1/2}$ is defined under restriction to the support $E_0 \mathcal{K}$. This expression of β is the same as in [58, (3.19)].

6.2 A Technical Lemma

The next notion of multiplicative domains plays an important role not only in operator algebras but also in quantum information.

Definition 6.9 Let $\gamma : N \to M$ be a unital normal Schwarz map between von Neumann algebras. The *multiplicative domain* of γ is defined as

$$\mathcal{M}_\gamma := \{y \in N : \gamma(y^* y) = \gamma(y)^* \gamma(y),\ \gamma(y y^*) = \gamma(y) \gamma(y^*)\}.$$

The next result was first shown by Choi [25] when γ is a unital 2-positive map. See [63, Lemma 3.9] for the case of a Schwarz map.

Proposition 6.10 *Let $\gamma : N \to M$ be as in Definition 6.9. For any $y \in N$, $\gamma(y^*y) = \gamma(y)^*\gamma(y)$ if and only if $\gamma(zy) = \gamma(z)\gamma(y)$ for all $z \in N$. Consequently,*

$$\mathcal{M}_\gamma = \{y \in N : \gamma(zy) = \gamma(z)\gamma(y),\ \gamma(yz) = \gamma(y)\gamma(z) \text{ for all } z \in N\},$$

and hence $\mathcal{M}_\gamma$ is a von Neumann subalgebra of N.

When γ is a unital normal Schwarz map from M into itself, we also consider the fixed-point set

$$\mathcal{F}_\gamma := \{x \in M : \gamma(x) = x\}.$$

The set $\mathcal{F}_\gamma$ is not generally a subalgebra of M, and there are no general inclusion relations between $\mathcal{F}_\gamma$ and $\mathcal{M}_\gamma$, see [58, Appendix B]. But we have the next result, which seems to have been first observed in [86] (see also [14, Theorem 2.3], [24, Lemma 3.4] and [65, Theorem 1 (i)]). We give a proof for convenience.

Lemma 6.11 *Let γ be a unital normal Schwarz map from M into itself. Assume that there exists a faithful $\varphi \in M_*^+$ such that $\varphi \circ \gamma = \varphi$. Then*

$$\mathcal{F}_\gamma = \{x \in M : \gamma(zx) = \gamma(z)x,\ \gamma(xz) = x\gamma(z) \text{ for all } z \in M\} \subset \mathcal{M}_\gamma,$$

and hence $\mathcal{F}_\gamma$ is a von Neumann subalgebra of M.

Proof Since the latter inclusion assertion is obvious by Proposition 6.10, it suffices to prove the first equality. The inclusion $\supset$ is obvious. Conversely, let $x \in \mathcal{F}_\gamma$. Then $x^*x = \gamma(x)^*\gamma(x) \le \gamma(x^*x)$ and $\varphi(\gamma(x^*x) - x^*x) = \varphi(x^*x) - \varphi(x^*x) = 0$, implying $\gamma(x^*x) = x^*x = \gamma(x)^*\gamma(x)$. Similarly, $\gamma(xx^*) = xx^* = \gamma(x)\gamma(x)^*$. Therefore, $x \in \mathcal{M}_\gamma$ and by Proposition 6.10 we have $\gamma(zx) = \gamma(z)x$ and $\gamma(xz) = x\gamma(z)$ for all $z \in M$.

The next lemma due to Petz [106] will play a crucial role in the next section.

Lemma 6.12 ([106]) *Let $\gamma : N \to M$ be a unital normal 2-positive map, and $\varphi \in M_*^+$. Assume that both of φ and $\varphi \circ \gamma$ are faithful. Let σ_t^φ be the modular automorphism group of M associated with φ and $\sigma_t^{\varphi\circ\gamma}$ be that of N associated with $\varphi \circ \gamma$ (see Sect. A.2). Then for any $v \in N$ the following conditions are equivalent:*

(i) $\gamma(v^*v) = \gamma(v)^*\gamma(v)$ *and* $\gamma(\sigma_t^{\varphi\circ\gamma}(v)) = \sigma_t^\varphi(\gamma(v))$ *for all* $t \in \mathbb{R}$;
(ii) $\gamma_\varphi^* \circ \gamma(v) = v$.

Moreover, $N_1 := \mathcal{F}_{\gamma_\varphi^*\circ\gamma}$ and $M_1 := \mathcal{F}_{\gamma\circ\gamma_\varphi^*}$ are von Neumann subalgebras of N and M respectively, and $\gamma|_{N_1}$ is an isomorphism from N_1 onto M_1, whose inverse is $\gamma_\varphi^*|_{M_1}$.

Proof (i) $\Longrightarrow$ (ii). Let $\Phi \in \mathcal{P}$ and $\Phi_0 \in \mathcal{P}_0$ be as in Theorem 6.1. Recall that the operator $x\Phi \mapsto x^*\Phi$ ($x \in M$) is closable and its closure has the polar decomposition $J\Delta^{1/2}$ with Δ the modular operator with respect to φ. Similarly, the closure of $y\Phi_0 \mapsto y^*\Phi_0$ ($y \in N$) is $J_0\Delta_0^{1/2}$ with the modular operator Δ_0 with respect to $\varphi \circ \gamma$. Then $\sigma_t^\varphi(x) = \Delta^{it}x\Delta^{-it}$ ($x \in M$) and $\sigma_t^{\varphi\circ\gamma}(y) = \Delta_0^{it}y\Delta_0^{-it}$ ($y \in N$), see Sect. A.2. Since γ is a Schwarz map, one has

$$\|\gamma(y)\Phi\|^2 = \varphi(\gamma(y)^*\gamma(y)) \le \varphi \circ \gamma(y^*y) = \|y\Phi_0\|^2, \qquad y \in N,$$

so that a contraction $V_\varphi : \mathcal{H}_0 \to \mathcal{H}$ can be defined by extending $y\Phi_0 \mapsto \gamma(y)\Phi$ ($y \in N$). By the second assumption of (i) one has $\gamma(\sigma_t^{\varphi\circ\gamma}(v^*)) = \sigma_t^\varphi(\gamma(v^*))$ so that $V_\varphi\sigma_t^{\varphi\circ\gamma}(v^*)\Phi_0 = \sigma_t^\varphi(\gamma(v^*))\Phi$ for all $t \in \mathbb{R}$. This means that $V_\varphi\Delta_0^{it}v^*\Phi_0 = \Delta^{it}\gamma(v^*)\Phi$ for all $t \in \mathbb{R}$. Hence, in view of Theorem B.1 of Appendix B, the analytic continuation of those at $z = -i/2$ gives

$$V_\varphi\Delta_0^{1/2}v^*\Phi_0 = \Delta^{1/2}\gamma(v^*)\Phi,$$

that is,

$$V_\varphi J_0 v\Phi_0 = J\gamma(v)\Phi, \tag{6.5}$$

which implies that

$$\begin{aligned}\|V_\varphi J_0 v\Phi_0\|^2 &= \|\gamma(v)\Phi\|^2 = \varphi(\gamma(v)^*\gamma(v)) \\ &= \varphi(\gamma(v^*v)) = \|v\Phi_0\|^2 = \|J_0 v\Phi_0\|^2,\end{aligned}$$

where we have used the first assumption of (i). Letting $\zeta_0 := J_0 v\Phi_0 \in \mathcal{H}_0$, since V_φ is a contraction, one has

$$\|V_\varphi^*V_\varphi\zeta_0 - \zeta_0\|^2 = \|V_\varphi^*V_\varphi\zeta_0\|^2 - 2\|V_\varphi\zeta_0\|^2 + \|\zeta_0\|^2 = \|V_\varphi^*V_\varphi\zeta_0\|^2 - \|\zeta_0\|^2 \le 0,$$

so that $V_\varphi^*V_\varphi\zeta_0 = \zeta_0$. This and (6.5) give

$$J_0 v\Phi_0 = V_\varphi^*V_\varphi J_0 v\Phi_0 = V_\varphi^* J\gamma(v)\Phi. \tag{6.6}$$

Therefore,

$$\langle J_0 v\Phi_0, y\Phi_0\rangle = \langle V_\varphi^* J\gamma(v)\Phi, y\Phi_0\rangle = \langle J\gamma(v)\Phi, \gamma(y)\Phi\rangle, \qquad y \in N, \tag{6.7}$$

which, compared with (6.1), is equivalent to the equality in (ii).

Before proving (ii) $\Longrightarrow$ (i), we prove the latter assertion. By Proposition 6.3 (1) and (2),

$$\varphi \circ (\gamma \circ \gamma_\varphi^*) = \varphi, \qquad \varphi \circ \gamma \circ (\gamma_\varphi^* \circ \gamma) = \varphi \circ \gamma. \tag{6.8}$$

Since γ is 2-positive, so is γ_φ^* by Proposition 6.3 (3). Hence $\gamma \circ \gamma_\varphi^*$ and $\gamma_\varphi^* \circ \gamma$ are 2-positive. Since φ and $\varphi \circ \gamma$ are faithful by assumption, it follows from Lemma 6.11 that the fixed-point sets M_1 and N_1 are von Neumann subalgebras of M and N, respectively. Since $\gamma(y) = \gamma \circ \gamma_\varphi^* \circ \gamma(y)$ for all $y \in N_1$ and $\gamma_\varphi^*(x) = \gamma_\varphi^* \circ \gamma \circ \gamma_\varphi^*(x)$ for all $x \in M_1$, it is immediate to see that $\gamma|_{N_1}$ maps N_1 to M_1 bijectively and $\gamma_\varphi^*|_{M_1}$ is the inverse of $\gamma|_{N_1}$. Next, assume that $v \in N_1$. Then

$$\begin{aligned}\langle J_0 v\Phi_0, y\Phi_0\rangle &= \langle J_0 \gamma_\varphi^*(\gamma(v))\Phi_0, y\Phi_0\rangle \\ &= \langle J\gamma(v)\Phi, \gamma(y)\Phi\rangle = \langle V_\varphi^* J\gamma(v)\Phi, y\Phi_0\rangle, \qquad y \in N,\end{aligned}$$

so that (6.7) holds, which in turn gives the equality $J_0 v\Phi_0 = V_\varphi^* J\gamma(v)\Phi$ in (6.6). Hence

$$\varphi \circ \gamma(v^*v) = \|v\Phi_0\|^2 = \|V_\varphi^* J\gamma(v)\Phi\|^2 \le \|\gamma(v)\Phi\|^2 = \varphi(\gamma(v)^*\gamma(v)).$$

Since γ is a Schwarz map and φ is faithful, $\gamma(v^*v) = \gamma(v)^*\gamma(v)$ holds. Since $v^* \in N_1$, $\gamma(vv^*) = \gamma(v)\gamma(v)^*$ holds as well, so that $v \in \mathcal{M}_\gamma$. Thus, by Proposition 6.10 we have $\gamma(yv) = \gamma(y)\gamma(v)$ and $\gamma(vy) = \gamma(v)\gamma(y)$ for all $y \in N$, showing that $\gamma|_{N_1}$ is an isomorphism.

(ii) $\Longrightarrow$ (i). Assume (ii), i.e., $v \in N_1$. The first equality in (i) was already shown above. As for the second, since $(\varphi \circ \gamma)|_{N_1} = (\varphi|_{M_1}) \circ (\gamma|_{N_1})$, we find by using the KMS condition (see Sect. A.2) that[1]

$$\gamma(\sigma_t^{(\varphi\circ\gamma)|_{N_1}}(y)) = \sigma_t^{\varphi|_{M_1}}(\gamma(y)), \qquad y \in N_1,\ t \in \mathbb{R}. \tag{6.9}$$

The mean ergodic theorem says that the norm-limit

$$E(x) := \lim_{n\to\infty} \frac{1}{n}\sum_{k=0}^{n-1} (\gamma \circ \gamma_\varphi^*)^k(x)$$

exists for every $x \in M$ and E is a norm one projection from M onto M_1. By Lemma 6.11 with (6.8) it is immediate that E is a conditional expectation onto M_1 such that $\varphi \circ E = \varphi$. (This is also a consequence of Tomiyama's theorem [131].) The normality of E is also immediate from $\varphi \circ E = \varphi$ since φ is faithful. Therefore, we have $\sigma_t^{\varphi|_{M_1}} = \sigma_t^\varphi|_{M_1}$ $(t \in \mathbb{R})$ by Takesaki's theorem (see Theorem A.10 of Sect. A.2). Similarly, replacing φ and $\gamma \circ \gamma_\varphi^*$ in the above argument with $\varphi \circ \gamma$ and $\gamma_\varphi^* \circ \gamma$, we have $\sigma_t^{(\varphi\circ\gamma)|_{N_1}} = \sigma_t^{\varphi\circ\gamma}|_{N_1}$ $(t \in \mathbb{R})$. Inserting these into (6.9) gives the second equality in (i).

[1] More explicitly speaking, by the KMS condition based on Theorem A.7, one can easily see that if $\gamma : N \to M$ is a $*$-isomorphism between von Neumann algebras and $\omega \in M_*^+$ is faithful, then $\sigma_t^{\omega\circ\gamma} = \gamma^{-1} \circ \sigma_t^\omega \circ \gamma$ for all $t \in \mathbb{R}$.

Remark 6.13 In the above proof of Lemma 6.12 we need the 2-positivity of γ to guarantee γ_φ^* being a Schwarz map (see Remark 6.4).

6.3 Preservation of Connes' Cocycle Derivatives

Throughout this section, let f be an operator convex function on $(0, \infty)$, having the integral expression in (2.5). We write μ_f for the representing measure in (2.5) to specify its dependence on f, and denote its support by $\operatorname{supp}\mu_f$, i.e., the topological support of μ_f consisting of $t \in [0, \infty)$ such that $\mu_f((t-\varepsilon, t+\varepsilon)) > 0$ for any $\varepsilon > 0$. Let M, N be von Neumann algebras and $\gamma : N \to M$ be a unital normal *Schwarz* map. Furthermore, let $\psi, \varphi \in M_*^+$ and assume that both of φ and $\varphi \circ \gamma$ are faithful (but ψ is general). There are two main ingredients of the discussions below. One is the standard f-divergence $S_f(\psi\|\varphi)$ and another is Connes' cocycle derivative $[D\psi : D\varphi]_t$ $(t \in \mathbb{R})$, whose concise survey is in Sect. A.7.

In the section we prove two lemmas stating that the preservation of a standard f-divergence implies that of Connes' cocycle derivatives. Those lemmas are the most substantial parts of the proof of the reversibility/sufficiency theorems in Sect. 6.4.

Lemma 6.14 *Assume that* $\operatorname{supp}\mu_f$ *has a limit point in* $(0, \infty)$*. If*

$$S_f(\psi \circ \gamma \| \varphi \circ \gamma) = S_f(\psi\|\varphi) < +\infty, \tag{6.10}$$

then we have

$$\gamma([D(\psi \circ \gamma) : D(\varphi \circ \gamma)]_t) = [D\psi : D\varphi]_t, \qquad t \in \mathbb{R}. \tag{6.11}$$

Proof We are working in the standard forms $(M, \mathcal{H}, J, \mathcal{P})$ and $(N, \mathcal{H}_0, J_0, \mathcal{P}_0)$ as before. Write $\psi_0 := \psi \circ \gamma$ and $\varphi_0 := \varphi \circ \gamma$. Let $\Psi, \Phi \in \mathcal{P}$ and $\Psi_0, \Phi_0 \in \mathcal{P}_0$ be the vector representatives of ψ, φ and ψ_0, φ_0, respectively. Below we divide the proof into several steps.

Step 1. A contraction $V_\varphi : \mathcal{H}_0 \to \mathcal{H}$ is defined by extending $y\Phi_0 \mapsto \gamma(y)\Phi$ $(y \in N)$ as in the first part of the proof of Lemma 6.12. For every $y \in N$ one has

$$\begin{aligned}\|\Delta_{\psi,\varphi}^{1/2} V_\varphi(y\Phi_0)\|^2 &= \|\Delta_{\psi,\varphi}^{1/2}\gamma(y)\Phi\|^2 = \|\gamma(y)^*\Psi\|^2 \\ &= \psi(\gamma(y)\gamma(y)^*) \le \psi(\gamma(yy^*)) = \psi_0(yy^*) \\ &= \|y^*\Psi_0\|^2 = \|\Delta_{\psi_0,\varphi_0}^{1/2}(y\Phi_0)\|^2. \end{aligned} \tag{6.12}$$

Note that the operator $\Delta_{\psi,\varphi}^{1/2} V_\varphi|_{N\Phi_0}$ is closable. Indeed, let $y_n \in N$ ($n \in \mathbb{N}$) be such that $y_n\Phi_0 \to 0$ and $\Delta_{\psi,\varphi}^{1/2} V_\varphi(y_n\Phi_0) = \Delta_{\psi,\varphi}^{1/2}\gamma(y_n)\Phi \to \zeta \in \mathcal{H}$. Since

$$\|\gamma(y_n)\Phi\|^2 = \varphi(\gamma(y_n)^*\gamma(y_n)) \le \varphi(\gamma(y_n^* y_n)) = \|y_n\Phi_0\|^2,$$

we have $\gamma(y_n)\Phi \to 0$. Since $\Delta_{\psi,\varphi}^{1/2}$ is a closed operator, $\zeta = 0$ follows. Thus, $\Delta_{\psi,\varphi}^{1/2} V_\varphi|_{N\Phi_0}$ has the closure T and (6.12) means that $\|T(y\Phi_0)\| \le \|\Delta_{\psi_0,\varphi_0}^{1/2}(y\Phi_0)\|$ for all $y \in N$. Since $N\Phi_0$ is a core of $\Delta_{\psi_0,\varphi_0}^{1/2}$, we see by Proposition B.5 of Appendix B that

$$T^*T \le \Delta_{\psi_0,\varphi_0} \tag{6.13}$$

in the sense of Proposition B.4. Let $\Delta_{\psi,\varphi} = \int_0^\infty t\,dE_{\psi,\varphi}(t)$ be the spectral decomposition of $\Delta_{\psi,\varphi}$, and for each $n \in \mathbb{N}$ let $H_n := \int_0^n t\,dE_{\psi,\varphi}(t)$. Since $N\Phi_0$ is a core of T and

$$\|H_n^{1/2} V_\varphi(y\Phi_0)\| = \|H_n^{1/2}(\gamma(y)\Phi)\| \le \|\Delta_{\psi,\varphi}^{1/2}(\gamma(y)\Phi)\| = \|T(y\Phi_0)\|, \quad y \in N,$$

we have $V_\varphi^* H_n V_\varphi \le T^*T$ by Proposition B.5 again. This and (6.13) show that $V_\varphi^* H_n V_\varphi \le \Delta_{\psi_0,\varphi_0}$. Therefore, for every $s > 0$,

$$(s1 + \Delta_{\psi_0,\varphi_0})^{-1} \le (s1 + V_\varphi^* H_n V_\varphi)^{-1} \le V_\varphi^*(s1 + H_n)^{-1} V_\varphi,$$

where the latter inequality above is a consequence of Hansen's inequality [49] since $t \mapsto (s+t)^{-1}$ is an operator monotone decreasing function on $[0,\infty)$. Letting $n \to \infty$ gives

$$(s1 + \Delta_{\psi_0,\varphi_0})^{-1} \le V_\varphi^*(s1 + \Delta_{\psi,\varphi})^{-1} V_\varphi, \qquad s > 0. \tag{6.14}$$

Step 2. Write the expression in (2.5) as

$$f(t) = a + b(t-1) + cf_2(t) + \int_{[0,\infty)} g_s(t)\,d\mu_f(s), \qquad t \in (0,\infty), \tag{6.15}$$

where $f_2(t) := (t-1)^2$ and $g_s(t) := (t-1)^2(t+s)^{-1}$ for $s \in [0,\infty)$. Note that

$$S_f(\psi\|\varphi) = (a-b)\varphi(1) + b\psi(1) + cS_{f_2}(\psi\|\varphi) + \int_{[0,\infty)} S_{g_s}(\psi\|\varphi)\,d\mu_f(s) < +\infty, \tag{6.16}$$

$$S_f(\psi_0\|\varphi_0) = (a-b)\varphi_0(1) + b\psi_0(1) + cS_{f_2}(\psi_0\|\varphi_0) + \int_{[0,\infty)} S_{g_s}(\psi_0\|\varphi_0)\,d\mu_f(s), \tag{6.17}$$

and

$$\varphi_0(1) = \varphi(1), \qquad \psi_0(1) = \psi(1),$$
$$S_{f_2}(\psi_0\|\varphi_0) \le S_{f_2}(\psi\|\varphi), \qquad S_{g_s}(\psi_0\|\varphi_0) \le S_{g_s}(\psi\|\varphi) \tag{6.18}$$

by the monotonicity property in Theorem 2.7 (iv). Therefore, assumption (6.10) implies that

$$S_{g_s}(\psi_0\|\varphi_0) = S_{g_s}(\psi\|\varphi) \quad \text{for } \mu_f\text{-a.e. } s \in [0,\infty).$$

Write

$$g_s(t) = \frac{((t+s)-(1+s))^2}{t+s} = t-(2+s)+(1+s)^2 h_s(t), \qquad t \in (0,\infty),$$

where $h_s(t) := (t+s)^{-1}$ for $s \in [0,\infty)$. Since

$$S_{g_s}(\psi\|\varphi) = \psi(1) - (2+s)\varphi(1) + (1+s)^2 S_{h_s}(\psi\|\varphi), \tag{6.19}$$

$$S_{g_s}(\psi_0\|\varphi_0) = \psi_0(1) - (2+s)\varphi_0(1) + (1+s)^2 S_{h_s}(\psi_0\|\varphi_0), \tag{6.20}$$

we have

$$S_{h_s}(\psi_0\|\varphi_0) = S_{h_s}(\psi\|\varphi) \quad \text{for } \mu_f\text{-a.e. } s \in [0,\infty),$$

that is,

$$\langle \Phi_0, (s1+\Delta_{\psi_0,\varphi_0})^{-1}\Phi_0\rangle = \langle \Phi, (s1+\Delta_{\psi,\varphi})^{-1}\Phi\rangle \quad \text{for } \mu_f\text{-a.e. } s \in [0,\infty). \tag{6.21}$$

We easily see that

$$\langle \Phi, (z1+\Delta_{\psi,\varphi})^{-1}\Phi\rangle = \int_0^\infty \frac{1}{z+t}\,d\|E_{\psi,\varphi}(t)\Phi\|^2$$

is analytic in $\{z \in \mathbb{C} : \mathrm{Re}\, z > 0\}$, and similarly for $\langle \Phi_0, (z1+\Delta_{\psi_0,\varphi_0})^{-1}\Phi_0\rangle$. From the assumption on $\mathrm{supp}\,\mu_f$ and (6.21) it follows that the set of $s > 0$ for which equality (6.21) holds has a limit point in $(0,\infty)$. Thus, the coincidence theorem for analytic functions yields that for all $s > 0$,

$$\langle \Phi_0, (s1+\Delta_{\psi_0,\varphi_0})^{-1}\Phi_0\rangle = \langle \Phi, (s1+\Delta_{\psi,\varphi})^{-1}\Phi\rangle$$

and hence

$$\langle \Phi_0, \left[V_\varphi^*(s1 + \Delta_{\psi,\varphi})^{-1}V_\varphi - (s1 + \Delta_{\psi_0,\varphi_0})^{-1}\right]\Phi_0\rangle = 0$$

thanks to $V_\varphi \Phi_0 = \Phi$. By (6.14) this implies that

$$V_\varphi^*(s1 + \Delta_{\psi,\varphi})^{-1}V_\varphi \Phi_0 = (s1 + \Delta_{\psi_0,\varphi_0})^{-1}\Phi_0,$$

and therefore

$$V_\varphi^*(s1 + \Delta_{\psi,\varphi})^{-1}\Phi = (s1 + \Delta_{\psi_0,\varphi_0})^{-1}\Phi_0, \qquad s > 0. \tag{6.22}$$

Step 3. Let $C_0[0, \infty)$ denote the Banach space of continuous complex functions ϕ on the locally compact space $[0, \infty)$ vanishing at infinity (i.e., $\lim_{t\to\infty} \phi(t) = 0$) with the sup-norm. Noting that $h_s \in C_0[0, \infty)$ for $s \in (0, \infty)$, we define $\mathfrak{A}$ to be the norm-closed complex linear span of h_s, $s \in (0, \infty)$, in $C_0[0, \infty)$. Since

$$h_{s_1}h_{s_2} = \frac{1}{s_2 - s_1}(h_{s_1} - h_{s_2}) \quad \text{for } s_1 \neq s_2,$$

$$h_s^2 = \lim_{\varepsilon \searrow 0} \frac{1}{\varepsilon}(h_s - h_{s+\varepsilon}) \text{ (in the norm)} \quad \text{for } s > 0,$$

we see that $\mathfrak{A}$ is a closed subalgebra of $C_0[0, \infty)$. Obviously, $h_s(t) > 0$ for all $t \in [0, \infty)$, and $h_s(t_1) \neq h_s(t_2)$ for every $t_1, t_2 \in [0, \infty)$ with $t_1 \neq t_2$. Hence the Stone–Weierstrass theorem implies that $\mathfrak{A} = C_0[0, \infty)$. Since

$$\phi \in C_0[0, \infty) \longmapsto V_\varphi^*\phi(\Delta_{\psi,\varphi})\Phi, \ \phi(\Delta_{\psi_0,\varphi_0})\Phi_0 \in \mathcal{H}_0$$

are bounded linear maps with respect to the sup-norm on $C_0[0, \infty)$ and the norm on $\mathcal{H}_0$, it follows from (6.22) that

$$V_\varphi^*\phi(\Delta_{\psi,\varphi})\Phi = \phi(\Delta_{\psi_0,\varphi_0})\Phi_0, \qquad \phi \in C_0[0, \infty). \tag{6.23}$$

Step 4. For every $s > 0$ one has by (6.22)

$$\begin{aligned}\|V_\varphi^*(s1 + \Delta_{\psi,\varphi})^{-1}\Phi\|^2 &= \langle \Phi_0, (s1 + \Delta_{\psi_0,\varphi_0})^{-2}\Phi_0\rangle \\ &= \langle \Phi_0, V_\varphi^*(s1 + \Delta_{\psi,\varphi})^{-2}\Phi\rangle \quad \text{(by (6.23))} \\ &= \langle \Phi, (s1 + \Delta_{\psi,\varphi})^{-2}\Phi\rangle \quad (\text{since } V_\varphi\Phi_0 = \Phi) \\ &= \|(s1 + \Delta_{\psi,\varphi})^{-1}\Phi\|^2.\end{aligned}$$

Letting $\zeta := (s1 + \Delta_{\psi,\varphi})^{-1}\Phi \in \mathcal{H}$, since the above means $\|V_\varphi^*\zeta\|^2 = \|\zeta\|^2$, one has

$$\|V_\varphi V_\varphi^*\zeta - \zeta\|^2 = \|V_\varphi V_\varphi^*\zeta\|^2 - 2\|V_\varphi^*\zeta\|^2 + \|\zeta\|^2 = \|V_\varphi V_\varphi^*\zeta\|^2 - \|\zeta\|^2 \le 0,$$

so that $V_\varphi V_\varphi^*\zeta = \zeta$, i.e.,

$$V_\varphi V_\varphi^*(s1 + \Delta_{\psi,\varphi})^{-1}\Phi = (s1 + \Delta_{\psi,\varphi})^{-1}\Phi.$$

By (6.22) this implies that

$$V_\varphi(s1 + \Delta_{\psi_0,\varphi_0})^{-1}\Phi_0 = (s1 + \Delta_{\psi,\varphi})^{-1}\Phi, \qquad s > 0.$$

Hence, similarly to Step 3 we have

$$V_\varphi\phi(\Delta_{\psi_0,\varphi_0})\Phi_0 = \phi(\Delta_{\psi,\varphi})\Phi, \qquad \phi \in C_0[0,\infty). \tag{6.24}$$

Step 5. For each $t \in \mathbb{R}$ set

$$\phi(x) := x^{it} \quad (x > 0), \qquad \phi(0) := 0. \tag{6.25}$$

For $n \in \mathbb{N}$ let

$$\kappa_n(x) := \begin{cases} nx & (0 \le x \le 1/n), \\ 1 & (1/n \le x \le n), \\ 1 - n(x-n) & (n \le x \le n + \frac{1}{n}), \\ 0 & (x \ge n + \frac{1}{n}), \end{cases} \tag{6.26}$$

and $\phi_n(x) := \kappa_n(x)\phi(x)$ for $x \in [0,\infty)$. Let $s(\psi), s(\psi_0)$ be the support projections of ψ, ψ_0, respectively. Note that the support projection of $\Delta_{\psi,\varphi}$ is $s(\psi)$ and that of $\Delta_{\psi_0,\varphi_0}$ is $s(\psi_0)$, because φ, φ_0 are faithful. Then it is immediate to verify that $\phi_n \in C_0[0,\infty)$ and

$$\phi_n(\Delta_{\psi,\varphi}) \longrightarrow \phi(\Delta_{\psi,\varphi}) = s(\psi)\Delta_{\psi,\varphi}^{it} \quad \text{strongly},$$
$$\phi_n(\Delta_{\psi_0,\varphi_0}) \longrightarrow \phi(\Delta_{\psi_0,\varphi_0}) = s(\psi_0)\Delta_{\psi_0,\varphi_0}^{it} \quad \text{strongly}.$$

By (6.24), $V_\varphi\phi_n(\Delta_{\psi_0,\varphi_0})\Phi_0 = \phi_n(\Delta_{\psi,\varphi})\Phi$. Letting $n \to \infty$ gives

$$V_\varphi(s(\psi_0)\Delta_{\psi_0,\varphi_0}^{it}\Phi_0) = s(\psi)\Delta_{\psi,\varphi}^{it}\Phi.$$

Since $\Delta_{\varphi}^{-it}\Phi = \Phi$ and $\Delta_{\varphi_0}^{-it}\Phi_0 = \Phi_0$, where $\Delta_\varphi = \Delta_{\varphi,\varphi}$ is the modular operator associated with φ, we have

$$V_\varphi(s(\psi_0)\Delta_{\psi_0,\varphi_0}^{it}\Delta_{\varphi_0}^{-it}\Phi_0) = s(\psi)\Delta_{\psi,\varphi}^{it}\Delta_{\varphi}^{-it}\Phi. \tag{6.27}$$

From Proposition A.47 and Remark A.48 we write

$$[D\psi : D\varphi]_t = s(\psi)\Delta_{\psi,\varphi}^{it}\Delta_{\varphi}^{-it}, \quad [D\psi_0 : \varphi_0]_t = s(\psi_0)\Delta_{\psi_0,\varphi_0}^{it}\Delta_{\varphi_0}^{-it}, \qquad t \in \mathbb{R}.$$

(Normally, $s(\psi)$ and $s(\psi_0)$ are removed in the above formulas, because the support projection of $\Delta_{\psi,\varphi}$ is $s(\psi)$, see Proposition A.22 (1), and $\Delta_{\psi,\varphi}^{it}$ is defined with restriction on $s(\psi)\mathcal{H}$.) Hence (6.27) means that

$$V_\varphi([D\psi_0 : D\varphi_0]_t\Phi_0) = [D\psi : D\varphi]_t\Phi,$$

that is,

$$\gamma([D\psi_0 : D\varphi_0]_t)\Phi = [D\psi : D\varphi]_t\Phi.$$

Since Φ is separating for M (i.e., $x \in M$, $x\Phi = 0 \implies x = 0$), we arrive at (6.11).

Lemma 6.15 *Assume that γ is 2-positive. If* (6.11) *holds, then we have*

$$\gamma_\varphi^*([D\psi : D\varphi]_t) = [D(\psi \circ \gamma) : D(\varphi \circ \gamma)]_t, \qquad t \in \mathbb{R}, \tag{6.28}$$

$$\gamma \circ \gamma_\varphi^*([D\psi : D\varphi]_t) = [D\psi : D\varphi]_t, \qquad t \in \mathbb{R}, \tag{6.29}$$

$$\psi \circ \gamma \circ \gamma_\varphi^* = \psi. \tag{6.30}$$

Proof We write $\varphi_0 := \varphi \circ \gamma$, $\psi_0 := \psi \circ \gamma$, $u_t := [D\psi : D\varphi]_t$ and $v_t := [D\psi_0 : D\varphi_0]_t$ ($t \in \mathbb{R}$). The assumption says that $\gamma(v_t) = u_t$ for all $t \in \mathbb{R}$. From properties of Connes' cocycle derivatives in Theorem A.52 (i) we have

$$\varphi(\gamma(v_t)^*\gamma(v_t)) = \varphi(u_t^*u_t) = \varphi(\sigma_t^\varphi(u_0)) = \varphi(u_0) = \varphi(\gamma(v_0)) = \varphi_0(v_0),$$
$$\varphi(\gamma(v_t^*v_t)) = \varphi_0(v_t^*v_t) = \varphi_0(\sigma_t^{\varphi_0}(v_0)) = \varphi_0(v_0), \qquad t \in \mathbb{R}.$$

Since γ is 2-positive (hence a Schwarz map) and φ is faithful, the above equalities imply that

$$\gamma(v_t^*v_t) = \gamma(v_t)^*\gamma(v_t), \qquad t \in \mathbb{R}. \tag{6.31}$$

From the cocycle identity of v_t, u_t (Theorem A.52 (ii)) and Proposition 6.10 we find that

$$\begin{aligned}\gamma(\sigma_s^{\varphi_0}(v_t)) &= \gamma(v_s^* v_{s+t}) = \gamma(v_s)^*\gamma(v_{s+t}) \\ &= u_s^* u_{s+t} = \sigma_s^{\varphi}(u_t) = \sigma_s^{\varphi}(\gamma(v_t)), \qquad s,t \in \mathbb{R}. \end{aligned} \tag{6.32}$$

By (6.31) and (6.32) it follows from Lemma 6.12 that $v_t \in N_1$ so that $u_t = \gamma(v_t) \in M_1$ and $\gamma_\varphi^*(u_t) = v_t$ for all $t \in \mathbb{R}$, where N_1, M_1 are as given in Lemma 6.12. Therefore, $\gamma \circ \gamma_\varphi^*(u_t) = \gamma(v_t) = u_t$ for all $t \in \mathbb{R}$, so (6.28) and (6.29) have been shown. Moreover, since $\gamma|_{N_1}$ and $\gamma_\varphi^*|_{M_1}$ are $*$-isomorphisms by Lemma 6.12, note that $v_t \in \mathcal{M}_\gamma$ and $u_t \in \mathcal{M}_{\gamma_\varphi^*}$ (see Definition 6.9).

Now, let Φ, Φ_0 be the vector representatives of φ, φ_0, respectively, as in Theorem 6.1. For every $x \in M$, $y \in N$ and $s, t \in \mathbb{R}$, since $\Delta_{\psi_0,\varphi_0}^{it} = v_t \Delta_{\varphi_0}^{it}$ and $\Delta_{\psi,\varphi}^{it} = u_t \Delta_\varphi^{it}$ (Proposition A.47 and Remark A.48), we obtain

$$\begin{aligned}&\langle J_0 y \Delta_{\psi_0,\varphi_0}^{is} \Phi_0, \gamma_\varphi^*(x) \Delta_{\psi_0,\varphi_0}^{it} \Phi_0\rangle \\ &= \langle J_0 y v_s \Phi_0, \gamma_\varphi^*(x) v_t \Phi_0\rangle = \langle J_0 y v_s \Phi_0, \gamma_\varphi^*(x)\gamma_\varphi^*(u_t)\Phi_0\rangle \\ &= \langle J_0 y v_s \Phi_0, \gamma_\varphi^*(x u_t)\Phi_0\rangle \qquad \text{(by Proposition 6.10, since } u_t \in \mathcal{M}_{\gamma_\varphi^*}) \\ &= \langle J_0 \gamma_\varphi^*(x u_t)\Phi_0, y v_s \Phi_0\rangle = \langle J x u_t \Phi, \gamma(y v_s)\Phi\rangle \qquad \text{(by (6.1))} \\ &= \langle J\gamma(y v_s)\Phi, x u_t \Phi\rangle = \langle J\gamma(y) u_s \Phi, x u_t \Phi\rangle \qquad \text{(since } v_s \in \mathcal{M}_\gamma) \\ &= \langle J\gamma(y)\Delta_{\psi,\varphi}^{is}\Phi, x\Delta_{\psi,\varphi}^{it}\Phi\rangle.\end{aligned}$$

By analytic continuation of the above both sides twice at $s = -i/2$ and then at $t = -i/2$, it follows that

$$\langle J_0 y \Delta_{\psi_0,\varphi_0}^{1/2}\Phi_0, \gamma_\varphi^*(x)\Delta_{\psi_0,\varphi_0}^{1/2}\Phi_0\rangle = \langle J\gamma(y)\Delta_{\psi,\varphi}^{1/2}\Phi, x\Delta_{\psi,\varphi}^{1/2}\Phi\rangle,$$

that is,

$$\langle J_0 y \Psi_0, \gamma_\varphi^*(x)\Psi_0\rangle = \langle J\gamma(y)\Psi, x\Psi\rangle, \qquad x \in M,\ y \in N, \tag{6.33}$$

where Ψ, Ψ_0 are the vector representatives of ψ, ψ_0, respectively. Taking $y = 1$ in (6.33) yields $\psi_0 \circ \gamma_\varphi^*(x) = \psi(x)$ for all $x \in M$, showing (6.30).

Remark 6.16 Assume in Lemma 6.15 that ψ is faithful too. Then from Theorem A.52 (i) one has $u_0 = s(\psi) = 1$ and $s(\psi_0) = v_0 = \gamma_\varphi^*(u_0) = 1$. Hence $\psi_0 = \psi \circ \gamma$ is also faithful automatically. Therefore, equality (6.33) means that γ_φ^* is Petz' recovery map of γ with respect to ψ, so we have $\gamma_\psi^* = \gamma_\varphi^*$.

6.4 Reversibility via Standard f-Divergences

Throughout the section, assume that $\gamma : N \to M$ is a unital normal 2-*positive* map between von Neumann algebras. For given $\psi, \varphi \in M_*^+$ we say that γ is *reversible* for $\{\psi, \varphi\}$ if there exists a unital normal 2-positive map $\beta : M \to N$ such that

$$\psi \circ \gamma \circ \beta = \psi \quad \text{and} \quad \varphi \circ \gamma \circ \beta = \varphi.$$

When M_0 is a von Neumann subalgebra of M, we say that M_0 is *sufficient* for $\{\psi, \varphi\}$ if the inclusion map $M_0 \hookrightarrow M$ is reversible for $\{\psi, \varphi\}$, i.e., there exists a unital normal 2-positive map $\beta : M \to M_0$ such that $\psi \circ \beta = \psi$ and $\varphi \circ \beta = \varphi$.

Let $\Psi, \Phi \in \mathcal{P}$ be the vector representatives of ψ, φ, respectively, in the standard form $(M, \mathcal{H}, J, \mathcal{P})$. The *transition probability* of $\psi, \varphi \in M_*^+$ is

$$P(\psi, \varphi) := \langle \Psi, \Phi \rangle = -S_{-t^{1/2}}(\psi \| \varphi), \tag{6.34}$$

which already appeared in (3.30). Note that $P(\psi, \varphi)$ has the (reverse) monotonicity property under unital normal Schwarz maps (by Theorem 2.7 (iv)).

The next theorem was originally proved by Petz [106, Theorem 5] under the assumption that φ, $\varphi \circ \gamma$, ψ and $\psi \circ \gamma$ are all faithful. The faithfulness assumption of ψ and $\psi \circ \gamma$ was later removed in Jenčová and Petz [68, Theorem 3].

Theorem 6.17 ([68, 106]) *Let $\psi, \varphi \in M_*^+$ and assume that φ and $\varphi \circ \gamma$ are faithful. Then the following conditions (i)–(v) are equivalent:*

(i) $\psi \circ \gamma \circ \gamma_\varphi^* = \psi$ *($\varphi \circ \gamma \circ \gamma_\varphi^* = \varphi$ is automatic by Proposition 6.3 (1));*
(ii) *γ is reversible for $\{\psi, \varphi\}$;*
(iii) $P(\psi, \varphi) = P(\psi \circ \gamma, \varphi \circ \gamma)$*;*
(iv) $[D\psi : D\varphi]_t = \gamma([D(\psi \circ \gamma) : D(\varphi \circ \gamma)]_t)$ *for all $t \in \mathbb{R}$;*
(v) $[D\psi : D\varphi]_t = \gamma \circ \gamma_\varphi^*([D\psi : D\varphi]_t)$ *for all $t \in \mathbb{R}$.*

Moreover, if ψ and $\psi \circ \gamma$ are faithful too, then the above (i)–(v) are equivalent to the following:

(vi) $\gamma_\psi^* = \gamma_\varphi^*$.

Proof (i) $\Longrightarrow$ (ii) is obvious. In view of (6.34), (ii) $\Longrightarrow$ (iii) is immediate from the monotonicity property of the standard f-divergence.

(iii) $\Longrightarrow$ (iv). Condition (iii) is rephrased as

$$S_{-t^{1/2}}(\psi \| \varphi) = S_{-t^{1/2}}(\psi \circ \gamma \| \varphi \circ \gamma),$$

whose value is finite. Recall the well-known integral expression of $-t^{1/2}$

$$-t^{1/2} = -t + \frac{1}{\pi} \int_{(0,\infty)} \left(\frac{t}{1+s} - \frac{t}{t+s} \right) s^{-1/2}\, ds,$$

which is rewritten[2] in the form of (2.5) as

$$-t^{1/2} = -1 - \frac{1}{2}(t-1) + \frac{1}{\pi}\int_{(0,\infty)} \frac{(t-1)^2}{t+s}\frac{s^{1/2}}{(1+s)^2}\,ds,$$

so that the support of the representing measure of $-t^{1/2}$ is $(0,\infty)$. Hence (iii) $\Longrightarrow$ (iv) is a special case of Lemma 6.14.

(iv) $\Longrightarrow$ (i) and (iv) $\Longrightarrow$ (v) are contained in Lemma 6.15.

(v) $\Longrightarrow$ (iii). Let $\mathfrak{M}$ be the von Neumann subalgebra of M generated by $\{[D\psi : D\varphi]_t : t \in \mathbb{R}\}$. From the cocycle identity of $[D\psi : D\varphi]_t$ (see Theorem A.52 (ii)), $\mathfrak{M}$ is globally invariant under the modular automorphism σ^φ. Hence by Theorem A.10 (Sect. A.2) says that there exists the conditional expectation E from M onto $\mathfrak{M}$ with respect to φ. Furthermore, since $\sigma_t^\varphi|_{\mathfrak{M}} = \sigma_t^{\varphi|_{\mathfrak{M}}}$ for all $t \in \mathbb{R}$ by Theorem A.10, it follows that $[D\psi : D\varphi]_t$ is a $\sigma^{\varphi|_{\mathfrak{M}}}$-cocycle in $\mathfrak{M}$. Therefore, Connes' inverse theorem (Theorem A.54) implies that there exists a (unique) normal semifinite weight $\tilde{\psi}$ on $\mathfrak{M}$ such that

$$[D\psi : D\varphi]_t = [D\tilde{\psi} : D(\varphi|_{\mathfrak{M}})]_t, \qquad t \in \mathbb{R}. \tag{6.35}$$

On the other hand, one has

$$\begin{aligned}[D(\tilde{\psi}\circ E) : D\varphi]_t &= [D(\tilde{\psi}\circ E) : D((\varphi|_{\mathfrak{M}})\circ E)]_t \\ &= [D\tilde{\psi} : D(\varphi|_{\mathfrak{M}})]_t, \qquad t \in \mathbb{R},\end{aligned} \tag{6.36}$$

due to Proposition A.55. Combining (6.35) and (6.36) implies by Proposition A.53 (3)[3] that $\psi = \tilde{\psi}\circ E$ and hence $\psi\circ E = \psi$.

Now, let $N_1 := \mathcal{F}_{\gamma_\varphi^*\circ\gamma}$ and $M_1 := \mathcal{F}_{\gamma\circ\gamma_\varphi^*}$ (see Lemma 6.12) and assume (v), that is, $\mathfrak{M} \subset M_1$. Then one has

$$\begin{aligned}P(\psi|_{M_1}, \varphi|_{M_1}) &\le P(\psi|_{\mathfrak{M}}, \varphi|_{\mathfrak{M}}) \\ &\le P(\psi, \varphi) \quad (\text{since } \psi = (\psi|_{\mathfrak{M}})\circ E,\ \varphi = (\varphi|_{\mathfrak{M}})\circ E) \\ &\le P(\psi\circ\gamma, \varphi\circ\gamma) \le P((\psi\circ\gamma)|_{N_1}, (\varphi\circ\gamma)|_{N_1})\end{aligned} \tag{6.37}$$

---

[2]This can be checked by a direct computation

$$\begin{aligned}&\int_0^\infty \left[\left(\frac{t}{1+s} - \frac{t}{t+s}\right)s^{-1/2} - \frac{(t-1)^2}{t+s}\frac{s^{1/2}}{(1+s)^2}\right]ds \\ &\quad = (t-1)\int_0^\infty \frac{ds}{(1+s)^2 s^{1/2}} = (t-1)\int_0^\infty \frac{2ds}{(1+s^2)^2} = \frac{\pi}{2}(t-1).\end{aligned}$$

[3]This holds also for normal semifinite weights.

$$= P((\psi|_{M_1}) \circ (\gamma|_{N_1}), (\varphi|_{M_1}) \circ (\gamma|_{N_1}))$$
$$= P(\psi|_{M_1}, \varphi|_{M_1}),$$

where the last equality holds since $\gamma|_{N_1} : N_1 \to M_1$ is an isomorphism (Lemma 6.12). Therefore, (iii) follows.

Finally, assume that ψ and $\psi \circ \gamma$ are also faithful. Then (iv) $\Longrightarrow$ (vi) was shown in the proof of Lemma 6.15 as noted in Remark 6.16. (vi) $\Longrightarrow$ (i) is obvious by Proposition 6.3 (1).

The following is the specialization of Theorem 6.17 to the case of a von Neumann subalgebra. The corollary was first given in [106] under the assumption that both of φ, ψ are faithful. The faithfulness assumption of ψ was removed in [68, Theorem 1].

Corollary 6.18 ([68, 106]) *Let M_0 be a von Neumann subalgebra of M, and $\psi, \varphi \in M_*^+$ with φ faithful. Then the following conditions (i)–(v) are equivalent:*

(i) $\psi \circ E_\varphi = \psi$, *where $E_\varphi : M \to M_0$ is the generalized conditional expectation with respect to φ (see the paragraph just before Proposition 6.5);*
(ii) M_0 *sufficient for* $\{\psi, \varphi\}$;
(iii) $P(\psi, \varphi) = P(\psi|_{M_0}, \varphi|_{M_0})$;
(iv) $[D\psi : D\varphi]_t = [D(\psi|_{M_0}) : D(\varphi|_{M_0})]_t$ *for all* $t \in \mathbb{R}$;
(v) $[D\psi : D\varphi]_t \in M_0$ *for all* $t \in \mathbb{R}$.

Moreover, if ψ is faithful too, then the above (i)–(v) are also equivalent to the following:

(vi) $E_\psi = E_\varphi$, *where $E_\psi : M \to M_0$ is the generalized conditional expectation with respect to ψ.*

Proof In view of Proposition 6.5, all the conditions of the corollary, except (v), exactly correspond to those in Theorem 6.17 in this specialized case. Condition (v) of Theorem 6.17 means that $[D\psi, D\varphi]_t = E_\varphi([D\psi, D\varphi]_t)$ for all $t \in \mathbb{R}$, which obviously implies (v) of the corollary. Conversely, assume that (v) of the corollary holds, and let $\mathfrak{M}$ be the von Neumann subalgebra of M generated by $\{[D\psi : d\varphi]_t : t \in \mathbb{R}\}$; then $\mathfrak{M} \subset M_0$. In a similar way to the proof of (v) $\Longrightarrow$ (iii) of Theorem 6.17, there exists the conditional expectation E from M to $\mathfrak{M}$ with respect to φ, and we have $\psi \circ E = \psi$. We hence have

$$P(\psi|_{\mathfrak{M}}, \varphi|_{\mathfrak{M}}) \le P(\psi, \varphi) \le P(\psi|_{M_0}, \varphi|_{M_0}) \le P(\psi|_{\mathfrak{M}}, \varphi|_{\mathfrak{M}}),$$

where the first inequality above is as in (6.37). Therefore, (iii) of the corollary follows.

We now present the main reversibility theorem via standard f-divergences.

Theorem 6.19 *Let $\psi, \varphi \in M_*^+$ with $\psi \neq 0$ and assume that $s(\psi) \le s(\varphi)$. Then the following conditions are equivalent:*

(i) $\psi \circ \gamma \circ \gamma_\varphi^* = \psi$ *($\varphi \circ \gamma \circ \gamma_\varphi^* = \varphi$ is automatic), where γ_φ^* is Petz' recovery map in the extended sense of Proposition 6.6;*
(ii) *γ is reversible for $\{\psi, \varphi\}$;*
(iii) *$S_f(\psi\|\varphi) = S_f(\psi \circ \gamma\|\varphi \circ \gamma)$ for every operator convex function f on $(0, \infty)$;*
(iv) *$S_f(\psi\|\varphi) = S_f(\psi \circ \gamma\|\varphi \circ \gamma) < +\infty$ for some operator convex function f on $(0, \infty)$ such that $\operatorname{supp} \mu_f$ has a limit point in $(0, \infty)$;*
(v) *$D_\alpha(\psi\|\varphi) = D_\alpha(\psi \circ \gamma\|\varphi \circ \gamma)$ for some $\alpha \in (0, 1)$, where $D_\alpha(\psi\|\varphi)$ is the α-Rényi divergence (see Sect. 3.1);*
(vi) *$P(\psi, \varphi) = P(\psi \circ \gamma, \varphi \circ \gamma)$.*

Proof Write $\psi_0 := \psi \circ \gamma$, $\varphi_0 := \varphi \circ \gamma$, $e := s(\varphi)$ $(\in M)$ and $e_0 := s(\varphi_0)$ $(\in N)$.

(i) $\Longrightarrow$ (ii). Let $\beta := \gamma_\varphi^* : M \to e_0Ne_0$ be given in Proposition 6.6. Assume (i); then $\psi \circ \gamma \circ \beta = \psi$ as well as $\varphi \circ \gamma \circ \beta = \varphi$ by Proposition 6.6 (1). We extend β to a unital map $\widetilde{\beta} : M \to N$ when $e_0 \neq 1$. Choose a normal state ρ on M and define

$$\widetilde{\beta}(x) := \beta(x) + \rho(x)(1 - e_0), \qquad x \in M.$$

Then it is clear that $\widetilde{\beta}$ is a unital normal 2-positive map. Since $s(\psi) \le s(\varphi)$ implies that $s(\psi_0) \le s(\varphi_0) = e_0$, we have for every $x \in M$,

$$\psi \circ \gamma \circ \widetilde{\beta}(x) = \psi \circ \gamma \circ \beta(x) = \psi(x)$$

as well as $\varphi \circ \gamma \circ \widetilde{\beta}(x) = \varphi(x)$. Hence (ii) holds.

(ii) $\Longrightarrow$ (iii) is immediate from the monotonicity property of S_f. (iii) $\Longrightarrow$ (iv) is obvious. When $\alpha \in (0, 1)$, note that $-t^\alpha$ is an operator convex function on $(0, \infty)$, the support of whose representing measure is $(0, \infty)$ similarly to that of $-t^{1/2}$ mentioned in the proof of (iv) $\Longrightarrow$ (v) of Theorem 6.17. By the definition of D_α note that condition (v) is equivalent to $S_{-f_\alpha}(\psi\|\varphi) = S_{-f_\alpha}(\psi \circ \gamma\|\varphi \circ \gamma)$, whose value is always finite. Hence it is clear that (iii) $\Longrightarrow$ (v) and (v) $\Longrightarrow$ (iv). Moreover, (vi) is equivalent to (v) with $\alpha = 1/2$. After all, it only remains to prove that (iv) implies (i).

(iv) $\Longrightarrow$ (i). Let $\hat{\gamma} : e_0Ne_0 \to eMe$ be defined by (6.2), which is a unital normal 2-positive map. Since $s(\psi) \le s(\varphi) = e$ and $s(\psi_0) \le s(\varphi_0) = e_0$, note that

$$\psi_0|_{e_0Ne_0} = (\psi|_{eMe}) \circ \hat{\gamma}, \qquad \varphi_0|_{e_0Ne_0} = (\varphi|_{eMe}) \circ \hat{\gamma}. \tag{6.38}$$

Hence, for any operator convex function f on $(0, \infty)$, we have

$$S_f(\psi_0\|\varphi_0) = S_f(\psi_0|_{e_0Ne_0}\|\varphi_0|_{e_0Ne_0}) = S_f((\psi|_{eMe}) \circ \hat{\gamma}\|(\varphi|_{eMe}) \circ \hat{\gamma}) \tag{6.39}$$

as well as

$$S_f(\psi\|\varphi) = S_f(\psi|_{eMe}\|\varphi|_{eMe}). \tag{6.40}$$

Now, assume (iv). Then by (6.39) and (6.40), for f in condition (iv) we obtain

$$S_f(\psi|_{eMe}\|\varphi|_{eMe}) = S_f((\psi|_{eMe})\circ\hat{\gamma}\|(\varphi|_{eMe})\circ\hat{\gamma}) < +\infty.$$

Since $\varphi|_{eMe}$ and $(\varphi|_{eMe})\circ\hat{\gamma}$ are faithful, it follows from Lemmas 6.14 and 6.15 that

$$\psi|_{eMe} = (\psi|_{eMe})\circ\hat{\gamma}\circ\hat{\beta}, \tag{6.41}$$

where $\hat{\beta} : eMe \to e_0Ne_0$ is Petz' recovery map of $\hat{\gamma}$ with respect to $\varphi|_{eMe}$. For every $x \in M$, since $\gamma_\varphi^*(x) = \hat{\beta}(exe)$ as given in (6.3) of Remark 6.7, we find by (6.38) and (6.41) that

$$\begin{aligned}\psi_0\circ\gamma_\varphi^*(x) &= (\psi|_{eMe})\circ\hat{\gamma}\circ\gamma_\varphi^*(x) = (\psi|_{eMe})\circ\hat{\gamma}\circ\hat{\beta}(exe)\\ &= \psi(exe) = \psi(x),\end{aligned}$$

showing (i).

For every $\psi, \varphi \in M_*^+$, it is easy to see that γ is reversible for $\{\psi, \varphi\}$ if and only if γ is reversible for $\{\psi, \psi+\varphi\}$. Hence the next theorem follows from Theorem 6.19. This way of presentation of the reversibility theorem was given in [68, Theorem 3] for a family $\{\psi_\theta\}$ in M_*^+.

Theorem 6.20 *For every $\psi, \varphi \in M_*^+$ with $\psi \neq 0$ the following conditions are equivalent:*

(i) $\psi\circ\gamma\circ\gamma_{\psi+\varphi}^* = \psi$;
(ii) γ *is reversible for* $\{\psi, \varphi\}$;
(iii) $S_f(\psi\|\psi+\varphi) = S_f(\psi\circ\gamma\|(\psi+\varphi)\circ\gamma)$ *for every operator convex function* f *on* $(0, \infty)$;
(iv) $S_f(\psi\|\psi+\varphi) = S_f(\psi\circ\gamma\|(\psi+\varphi)\circ\gamma) < +\infty$ *for some operator convex function* f *on* $(0, \infty)$ *such that* $\operatorname{supp}\mu_f$ *has a limit point in* $(0, \infty)$;
(v) $D_\alpha(\psi\|\psi+\varphi) = D_\alpha(\psi\circ\gamma\|(\psi+\varphi)\circ\gamma)$ *for some* $\alpha \in (0, 1)$;
(vi) $P(\psi, \psi+\varphi) = P(\psi\circ\gamma, (\psi+\varphi)\circ\gamma)$.

Remark 6.21 When $M = B(\mathcal{H})$ with $\dim\mathcal{H} < \infty$, the assumption on $\operatorname{supp}\mu_f$ in condition (iv) of Theorems 6.19 and 6.20 can be relaxed to $|\operatorname{supp}\mu_f| \geq (\dim\mathcal{H})^2$, see [63, Theorem 5.1]. But the support condition on μ_f cannot completely be removed, so f cannot be a general nonlinear operator convex function in (iv) of the theorems. In fact, some counter-examples in the finite-dimensional case are known when $f(t) = t^2$ (see [70], [58, Example 4.8]) and when $f(t) = (1+t)^{-1}$ (see [65, Example 1]), in which S_f is preserved under a CP map γ and yet reversibility fails to hold.

Problem 6.22 It is interesting to find whether or not Theorem 6.19 holds true without the assumption $s(\psi) \le s(\varphi)$ on supports. In particular, assume that $s(\psi)$ and $s(\varphi)$ are orthogonal. In this case, condition (v) (also (vi)) implies that $s(\psi \circ \gamma)$ and $s(\varphi \circ \gamma)$ are also orthogonal. Then one can easily see that γ is reversible for $\{\psi, \varphi\}$. So it seems that the problem is not so hopeless.

6.5 Reversibility via Sandwiched Rényi Divergences

The reversibility via the sandwiched Rényi divergence $\widetilde{D}_\alpha$ (discussed in Sect. 3.3) in the von Neumann algebra setting has recently been obtained by Jenčová [66, 67], which is reported below without proofs. This reversibility theorem was proved in [66] for the case $\alpha > 1$ and in [67] for the case $1/2 < \alpha < 1$.

Theorem 6.23 ([66, 67]) *Let $\gamma : N \to M$ be a unital normal 2-positive map between von Neumann algebras. Let $\psi, \varphi \in M_*^+$ with $\psi \ne 0$ and $s(\psi) \le s(\varphi)$. Let $\alpha \in (1/2, \infty) \setminus \{1\}$. If*

$$\widetilde{D}_\alpha(\psi \circ \gamma \| \varphi \circ \gamma) = \widetilde{D}_\alpha(\psi \| \varphi) < +\infty$$

then γ is reversible for $\{\psi, \varphi\}$.

Remark 6.24 Recall the two limit cases $\widetilde{D}_{1/2}(\psi \| \varphi) = -2 \log(F(\psi, \varphi)/\psi(1))$ in (3.29) and $\lim_{\alpha \to \infty} \widetilde{D}_\alpha(\psi \| \varphi) = D_{\max}(\psi \| \varphi)$ in Theorem 3.16 (2). In these limit cases, the reversibility as in the above theorem for a CP map γ fails to hold even in the finite-dimensional case, see [96, Corollary A.9] (also [58, Remark 5.15]).

Chapter 7
Reversibility and Measurements

7.1 Approximation of Connes' Cocycle Derivatives and Approximate Reversibility

This chapter is concerned with the approximate reversibility (sufficiency) for a sequence of quantum operations $\alpha_k : M_k \to M$ (or quantum channels with input M and outputs M_k). Our main problem is to characterize the approximate reversibility of $(\alpha_k : M_k \to M)_{k=1}^{\infty}$ for $\psi, \varphi \in M_*^+$ in terms of the convergence $S_f(\psi \circ \alpha_k \| \varphi \circ \alpha_k) \to S_f(\psi \| \varphi)$. In particular, we are concerned with the case where α_k's are measurement operations with commutative M_k's (or quantum-classical channels). The problem was formerly investigated by Petz [108] (also [101, Chap. 9]), and we will revisit Petz' results with some refinements.

Throughout the chapter, let f be an operator convex function on $(0, \infty)$, and $\psi, \varphi \in M_*^+$ be as before. The aim of this section is to prove the approximate versions of the two lemmas of Sect. 6.3 and of the main result of Sect. 6.4. The first lemma is the approximate version of Lemma 6.14, where the equalities in (6.10) and (6.11) are replaced with the convergences in (7.1) and (7.2). The lemma improves [108, Lemma 3.1] (also [101, Lemma 9.7]) where the case $S_f = S$ (the relative entropy) was treated under an additional assumption that $\lambda^{-1}\varphi \le \psi \le \lambda\varphi$ for some $\lambda > 0$.

Lemma 7.1 *Let $\alpha_k : M_k \to M$ ($k \in \mathbb{N}$) be a sequence of unital normal Schwarz maps with von Neumann algebras M_k. Assume that* $\operatorname{supp}\mu_f$ *has a limit point in $(0, \infty)$ and that $\psi, \varphi \in M_*^+$ and α_k ($k \in \mathbb{N}$) are all faithful. If*

$$\lim_k S_f(\psi \circ \alpha_k \| \varphi \circ \alpha_k) = S_f(\psi \| \varphi) < +\infty, \tag{7.1}$$

F. Hiai, *Quantum f-Divergences in von Neumann Algebras*,
Mathematical Physics Studies, https://doi.org/10.1007/978-981-33-4199-9_7

then we have

$$\lim_k \alpha_k([D(\psi \circ \alpha_k) : D(\varphi \circ \alpha_k)]_t) = [D\psi : D\varphi]_t \ \textit{strongly*}, \quad t \in \mathbb{R}. \tag{7.2}$$

Proof We are working in the standard forms $(M, \mathcal{H}, J, \mathcal{P})$ and $(M_k, \mathcal{H}_k, J_k, \mathcal{P}_k)$. Write $\varphi_k := \varphi \circ \alpha_k$ and $\psi_k := \psi \circ \alpha_k$, and let Φ, Φ_k be the vector representatives of φ, φ_k, respectively. We show first the strong convergence in (7.2), whose proof is divided into several steps similarly to that of Lemma 6.14. The convergence will be strengthened into the strong* one in the final step.

Step 1. For each k, a contraction $V_k : \mathcal{H}_k \to \mathcal{H}$ is defined by extending $y\Phi_k \mapsto \alpha_k(y)\Phi$ $(y \in M_k)$. As in Step 1 of the proof of Lemma 6.14, replacing $\gamma, N, \psi_0, \varphi_0$ with $\alpha_k, M_k, \psi_k, \varphi_k$, we have

$$(s1 + \Delta_{\psi_k,\varphi_k})^{-1} \le V_k^*(s1 + \Delta_{\psi,\varphi})^{-1}V_k, \qquad s > 0. \tag{7.3}$$

Step 2. Writing f as in (6.15) we have equalities (6.16), (6.17) and inequalities (6.18), where ψ_0, φ_0 are replaced with ψ_k, φ_k for any k. Therefore, assumption (7.1) implies that

$$\int_{[0,\infty)} \left[S_{g_s}(\psi\|\varphi) - S_{g_s}(\psi_k\|\varphi_k)\right] d\mu_f(s) \longrightarrow 0$$

(as well as $S_{f_2}(\psi_k\|\varphi_k) \to S_{f_2}(\psi\|\varphi)$ if $c > 0$), and so we can choose a subsequence $\{k(l)\}$ of $\{k\}$ such that

$$S_{g_s}(\psi_{k(l)}\|\varphi_{k(l)}) \longrightarrow S_{g_s}(\psi\|\varphi) \quad \text{for } \mu_f\text{-a.e. } s \in [0, \infty).$$

Thus, we may assume that

$$S_{g_s}(\psi_k\|\varphi_k) \longrightarrow S_{g_s}(\psi\|\varphi) \quad \text{for } \mu_f\text{-a.e. } s \in [0, \infty). \tag{7.4}$$

(Indeed, suppose that (7.2) does not hold; then there are a $t_0 \in \mathbb{R}$ and a neighborhood $\mathcal{V}$ of $[D\psi : D\varphi]_{t_0}$ in the strong topology, and a subsequence $\{m_k\}$ of $\{k\}$ such that $\alpha_k([D(\psi \circ \alpha_{m_k}) : D(\varphi \circ \alpha_{m_k})]_t) \notin \mathcal{V}$ for all k. From the above argument we can choose a subsequence $\{m_{k(l)}\}$ of $\{m_k\}$ such that

$$S_{g_s}(\psi_{m_{k(l)}}\|\varphi_{m_{k(l)}}) \longrightarrow S_{g_s}(\psi\|\varphi) \quad \text{for } \mu_f\text{-a.e. } s \in [0, \infty). \tag{7.5}$$

Then a contradiction occurs since the proof below under (7.5) shows the strong convergence in (7.2) for the subsequence $\{m_{k(l)}\}$.)

With $h_s(t) := (t + s)^{-1}$, $s \in [0, \infty)$, we have (6.19) and (6.20), where ψ_0, φ_0 are replaced with ψ_k, φ_k for any k, so that we have by (7.4)

$$S_{h_s}(\psi_k\|\varphi_k) \longrightarrow S_{h_s}(\psi\|\varphi) \quad \text{for } \mu_f\text{-a.e. } s \in [0, \infty),$$

that is,

$$\langle \Phi_k, (s1 + \Delta_{\psi_k,\varphi_k})^{-1}\Phi_k\rangle \longrightarrow \langle \Phi, (s1 + \Delta_{\psi,\varphi})^{-1}\Phi\rangle \quad \text{for } \mu_f\text{-a.e. } s \in [0,\infty). \tag{7.6}$$

Note that $\langle \Phi, (z1 + \Delta_{\psi,\varphi})^{-1}\Phi\rangle$ and $\langle \Phi_k, (z1 + \Delta_{\psi_k,\varphi_k})^{-1}\Phi_k\rangle$ are analytic in $\{z \in \mathbb{C} : \mathrm{Re}\, z > 0\}$, as seen just after (6.21). From the assumption on $\mathrm{supp}\, \mu_f$ and (7.6) it follows that the set of $s > 0$ for which the convergence in (7.6) holds has a limit point in $(0, \infty)$. Moreover, it is clear that the set of analytic functions $\langle \Phi_k, (z1 + \Delta_{\psi_k,\varphi_k})^{-1}\Phi_k\rangle$ on $\{z \in \mathbb{C} : \mathrm{Re}\, z > 0\}$ is locally bounded, i.e., uniformly bounded on each compact subset of $\{z \in \mathbb{C} : \mathrm{Re}\, z > 0\}$. Thus, Vitali's theorem for analytic functions (see, e.g., [115, Sec. 7.3, p. 156]) yields that for all $s > 0$,

$$\langle \Phi_k, (s1 + \Delta_{\psi_k,\varphi_k})^{-1}\Phi_k\rangle \longrightarrow \langle \Phi, (s1 + \Delta_{\psi,\varphi})^{-1}\Phi\rangle$$

and hence

$$\langle \Phi_k, \left[V_k^*(s1 + \Delta_{\psi,\varphi})^{-1}V_k - (s1 + \Delta_{\psi_k,\varphi_k})^{-1}\right]\Phi_k\rangle \longrightarrow 0,$$

thanks to $V_k\Phi_k = \Phi$. By (7.3) this implies that

$$\|V_k^*(s1 + \Delta_{\psi,\varphi})^{-1}V_k\Phi_k - (s1 + \Delta_{\psi_k,\varphi_k})^{-1}\Phi_k\| \longrightarrow 0,$$

and therefore

$$\|V_k^*(s1 + \Delta_{\psi,\varphi})^{-1}\Phi - (s1 + \Delta_{\psi_k,\varphi_k})^{-1}\Phi_k\| \longrightarrow 0, \qquad s > 0. \tag{7.7}$$

Step 3. As shown in Step 3 of the proof of Lemma 6.14, note that the linear span of h_s, $s \in [0, \infty)$, is dense in $C_0[0, \infty)$ with the sup-norm. Since

$$\phi \in C_0[0,\infty) \longmapsto V_k^*\phi(\Delta_{\psi,\varphi})\Phi,\ \phi(\Delta_{\psi_k,\varphi_k})\Phi_k \in \mathcal{H}_k \quad (k \in \mathbb{N})$$

are uniformly bounded linear maps with respect to the sup-norm on $C_0[0, \infty)$ and the norm on $\mathcal{H}_k$, it follows from (7.7) that

$$\|V_k^*\phi(\Delta_{\psi,\varphi})\Phi - \phi(\Delta_{\psi_k,\varphi_k})\Phi_k\| \longrightarrow 0, \qquad \phi \in C_0[0,\infty). \tag{7.8}$$

Step 4. For every $s > 0$ one has

$$\begin{aligned}
\lim_k \|V_k^*(s1 + \Delta_{\psi,\varphi})^{-1}\Phi\|^2 &= \lim_k \langle \Phi_k, (s1 + \Delta_{\psi_k,\varphi_k})^{-2}\Phi_k\rangle \quad \text{(by (7.7))} \\
&= \langle \Phi, (s1 + \Delta_{\psi,\varphi})^{-2}\Phi\rangle \quad \text{(by (7.8) and } V_k\Phi_k = \Phi) \\
&= \|(s1 + \Delta_{\psi,\varphi})^{-1}\Phi\|^2.
\end{aligned}$$

Letting $\zeta := (s1 + \Delta_{\psi,\varphi})^{-1}\Phi \in \mathcal{H}$, since the above means that $\lim_k \|V_k^*\zeta\|^2 = \|\zeta\|^2$, one has

$$\|V_k V_k^*\zeta - \zeta\|^2 = \|V_k V_k^*\zeta\|^2 - 2\|V_k^*\zeta\|^2 + \|\zeta\|^2 \le 2\|\zeta\|^2 - 2\|V_k^*\zeta\|^2 \longrightarrow 0$$

so that $\|V_k V_k^*\zeta - \zeta\| \to 0$, i.e.,

$$\|V_k V_k^*(s1 + \Delta_{\psi,\varphi})^{-1}\Phi - (s1 + \Delta_{\psi,\varphi})^{-1}\Phi\| \longrightarrow 0.$$

By (7.7) this implies that

$$\|V_k(s1 + \Delta_{\psi_k,\varphi_k})^{-1}\Phi_k - (s1 + \Delta_{\psi,\varphi})^{-1}\Phi\| \longrightarrow 0, \qquad s > 0.$$

Hence, similarly to Step 3 we have

$$\|V_k\phi(\Delta_{\psi_k,\varphi_k})\Phi_k - \phi(\Delta_{\psi,\varphi})\Phi\| \longrightarrow 0, \qquad \phi \in C_0[0,\infty). \tag{7.9}$$

Step 5. Here we may assume that φ is a state, so $\|\Phi\| = 1$ and $\|\Phi_k\| = 1$ for all k. For each $t \in \mathbb{R}$ let $\phi(x)$, $x \ge 0$, be the same as (6.25). For $n \in \mathbb{N}$ with $n \ge 2$ let κ_n be the same as (6.26) and define χ_n by

$$\chi_n(x) := \begin{cases} 0 & (0 \le x \le 1/n), \\ nx - 1 & (1/n \le x \le 2/n), \\ 1 & (2/n \le x \le n - \frac{1}{n}), \\ n(n-x) & (n - \frac{1}{n} \le x \le n), \\ 0 & (x \ge n). \end{cases}.$$

Moreover, let $\phi_n(x) := \kappa_n(x)\phi(x)$ for $x \in [0,\infty)$. Then it is obvious that $\kappa_n, \chi_n, \phi_n \in C_0[0,\infty)$ and $\chi_n \le 1_{[1/n,n]} \le \kappa_n$. Since $s(\psi) = s(\varphi) = 1$ and $s(\psi_k) = s(\varphi_k) = 1_{M_k}$, note that $\Delta_{\psi,\varphi}$ and $\Delta_{\psi_k,\varphi_k}$ are non-singular and so $\|\chi_n(\Delta_{\psi,\varphi})\Phi - \Phi\| \to 0$ as $n \to \infty$. For any $\varepsilon \in (0,1)$ one can choose an n_0 such that

$$\|\chi_{n_0}(\Delta_{\psi,\varphi})\Phi\| \ge 1 - \varepsilon. \tag{7.10}$$

By (7.9) there exists a k_0 such that

$$\|V_k\chi_{n_0}(\Delta_{\psi_k,\varphi_k})\Phi_k - \chi_{n_0}(\Delta_{\psi,\varphi})\Phi\| \le \varepsilon, \qquad k \ge k_0. \tag{7.11}$$

It then follows from (7.10) and (7.11) that for every $n \geq n_0$ and $k \geq k_0$,

$$\begin{aligned}\|\chi_n(\Delta_{\psi_k,\varphi_k})\Phi_k\| &\geq \|\chi_{n_0}(\Delta_{\psi_k,\varphi_k})\Phi_k\| \geq \|V_k\chi_{n_0}(\Delta_{\psi_k,\varphi_k})\Phi_k\| \\ &\geq \|\chi_{n_0}(\Delta_{\psi,\varphi})\Phi\| - \|V_k\chi_{n_0}(\Delta_{\psi_k,\varphi_k})\Phi_k - \chi_{n_0}(\Delta_{\psi,\varphi})\Phi\| \\ &\geq 1 - 2\varepsilon.\end{aligned}$$

Therefore, for every $n \geq n_0$ and $k \geq k_0$,

$$\begin{aligned}\|(1-\kappa_n)(\Delta_{\psi_k,\varphi_k})\Phi_k\|^2 &\leq \|(1 - E_{\psi_k\varphi_k}([1/n, n])\Phi_k\|^2 \\ &= 1 - \|E_{\psi_k,\varphi_k}([1/n, n])\Phi_k\|^2 \\ &\leq 1 - \|\chi_n(\Delta_{\psi_k,\varphi_k})\Phi_k\|^2 \\ &\leq 1 - (1-2\varepsilon)^2 \leq 4\varepsilon, \end{aligned} \tag{7.12}$$

where $\Delta_{\psi_k,\varphi_k} = \int_0^\infty t\,dE_{\psi_k,\varphi_k}(t)$ is the spectral decomposition of $\Delta_{\psi_k,\varphi_k}$. Furthermore, one can choose an $n_1 \geq n_0$ such that $\|(1-\kappa_{n_1})(\Delta_{\psi,\varphi})\Phi\| \leq \varepsilon$. From this and (7.12) it follows that for every $k \geq k_0$,

$$\begin{aligned}\|V_k\Delta_{\psi_k,\varphi_k}^{it}\Phi_k - \Delta_{\psi,\varphi}^{it}\Phi\| &\leq \|V_k\phi(\Delta_{\psi_k,\varphi_k})\Phi_k - V_k\phi_{n_1}(\Delta_{\psi_k,\varphi_k})\Phi_k\| \\ &\quad + \|V_k\phi_{n_1}(\Delta_{\psi_k,\varphi_k})\Phi_k - \phi_{n_1}(\Delta_{\psi,\varphi})\Phi\| \\ &\quad + \|\phi_{n_1}(\Delta_{\psi,\varphi})\Phi - \phi(\Delta_{\psi,\varphi})\Phi\| \\ &\leq \|\phi(\Delta_{\psi_k,\varphi_k})(1-\kappa_{n_1})(\Delta_{\psi_k,\varphi_k})\Phi_k\| \\ &\quad + \|V_k\phi_{n_1}(\Delta_{\psi_k,\varphi_k})\Phi_k - \phi_{n_1}(\Delta_{\psi,\varphi})\Phi\| \\ &\quad + \|\phi(\Delta_{\psi,\varphi})(1-\kappa_{n_1})(\Delta_{\psi,\varphi})\Phi\| \\ &\leq 2\varepsilon^{1/2} + \|V_k\phi_{n_1}(\Delta_{\psi_k,\varphi_k})\Phi_k - \phi_{n_1}(\Delta_{\psi,\varphi})\Phi\| + \varepsilon.\end{aligned}$$

Therefore, by (7.9) there exists a $k_1 \geq k_0$ such that

$$\|V_k\Delta_{\psi_k,\varphi_k}^{it}\Phi_k - \Delta_{\psi,\varphi}^{it}\Phi\| \leq 2(\varepsilon^{1/2} + \varepsilon), \qquad k \geq k_1,$$

which implies that

$$\|V_k\Delta_{\psi_k,\varphi_k}^{it}\Phi_k - \Delta_{\psi,\varphi}^{it}\Phi\| \longrightarrow 0 \quad \text{as } k \to \infty.$$

Since $\Delta_{\varphi_k}^{-it}\Phi_k = \Phi_k$ and $\Delta_{\varphi}^{-it}\Phi = \Phi$, the above implies by Proposition A.47 (and Remark A.48) that

$$\|V_k[D\psi_k : D\varphi_k]_t\Phi_k - [D\psi : D\varphi]_t\Phi\| \longrightarrow 0 \quad \text{as } k \to \infty.$$

Since $[D\psi_k : D\varphi_k]_t \in M_k$ so that $V_k[D\psi_k : D\varphi_k]_t\Phi_k = \alpha_k([D\psi_k : D\varphi_k]_t)\Phi$, it follows that

$$\lim_{k\to\infty} \|\alpha_k([D\psi_k : D\varphi_k]_t)\Phi - [D\psi : D\varphi]_t\Phi\| = 0$$

for each $t \in \mathbb{R}$. This shows that the strong convergence in (7.2) holds since $\overline{M'\Phi} = \mathcal{H}$.

Step 6. We replace f with the transpose $\widetilde{f}$. It is immediate to see that $d\mu_{\widetilde{f}}(s) = sd\mu_f(s^{-1})$ for $s \in (0,\infty)$; hence $\widetilde{f}$ satisfies the same assumption as f. Moreover, by Proposition 2.3 (5), condition (7.1) is rewritten as

$$\lim_k S_{\widetilde{f}}(\varphi\circ\alpha_k\|\psi\circ\alpha_k) = S_{\widetilde{f}}(\varphi\|\psi) < +\infty.$$

Hence one can apply the assertion proved above to $\widetilde{f}$ with exchanging the roles of ψ, φ, so that one has

$$\lim_k \alpha_k([D(\varphi\circ\alpha_k) : D(\psi\circ\alpha_k)]_t) = [D\varphi : D\psi]_t \ \text{ strongly}, \quad t\in\mathbb{R}.$$

Therefore, the strong* convergence in (7.2) has been shown in view of Proposition A.53 (1).

The next lemma is the approximate version of Lemma 6.15, which was given in [108] (also [101, Chap. 9]). We will make the proof more readable than that in [101, 108] while essentially the same.

Lemma 7.2 *Let $\alpha_k : M_k \to M$ and $\psi, \varphi \in M_*^+$ be as in Lemma 7.1. Assume that α_k's are all 2-positive and (7.2) holds. Let $\alpha_{k,\varphi}^* : M \to M_k$ be Petz' recovery map of α_k with respect to φ. If $\lambda^{-1}\varphi \le \psi \le \lambda\varphi$ for some $\lambda > 0$, then*

$$\lim_k \alpha_k\circ\alpha_{k,\varphi}^*([D\psi : D\varphi]_t) = [D\psi : D\varphi]_t \quad \textit{strongly*}, \qquad t\in\mathbb{R}, \tag{7.13}$$

$$\lim_k \psi\circ\alpha_k\circ\alpha_{k,\varphi}^* = \psi \quad \textit{in } \sigma(M_*, M). \tag{7.14}$$

To prove the lemma, we first prepare three technical lemmas.

Lemma 7.3 *Let $V_k : \mathcal{H}_k \to \mathcal{K}_k$ ($k \in \mathbb{N}$) be a sequence of contractions between Hilbert spaces. Let $\xi_k \in \mathcal{H}_k$ and $\eta_k \in \mathcal{K}_k$. If $\|V_k\xi_k - \eta_k\| \to 0$ and $\lim_k \|\xi_k\| = \lim_k \|\eta_k\| < +\infty$, then $\|V_k^*\eta_k - \xi_k\| \to 0$.*

Proof We have

$$\begin{aligned}\|V_k^*\eta_k - \xi_k\|^2 &= \|V_k^*\eta_k\|^2 - 2\mathrm{Re}\,\langle V_k^*\eta_k, \xi_k\rangle + \|\xi_k\|^2 \\ &\le \|\eta_k\|^2 - 2\mathrm{Re}\,\langle \eta_k, V_k\xi_k\rangle + \|\xi_k\|^2.\end{aligned}$$

Since

$$\left|\langle \eta_k, V_k\xi_k\rangle - \|\eta_k\|^2\right| = |\langle \eta_k, V_k\xi_k - \eta_k\rangle| \le \|\eta_k\| \, \|V_k\xi_k - \eta_k\| \longrightarrow 0,$$

we have $\langle \eta_k, V_k\xi_k\rangle \to \lim_k \|\eta_k\|^2$. Therefore,

$$\limsup_k \|V_k^*\eta_k - \xi_k\|^2 \le \lim_k \|\eta_k\|^2 - 2\lim_k \|\eta_k\|^2 + \lim_k \|\xi_k\|^2 = 0,$$

as asserted.

Lemma 7.4 *Let $\alpha_k : A_k \to B(\mathcal{H})$ be a sequence of unital Schwarz maps, where A_k's are unital C^*-algebras. Let $u_k, a_k \in A_k$ be given such that u_k's are unitaries and $\sup_k \|a_k\| < +\infty$. Assume that $\alpha_k(u_k) \to u$ in the strong* operator topology and $\alpha_k(a_k) \to a$ in the weak operator topology for some $u, a \in B(\mathcal{H})$, where u is a unitary. Then $\alpha_k(a_k u_k) \to au$ and $\alpha_k(u_k a_k) \to ua$ in the weak operator topology.*

Remark 7.5 The lemma is in [108, Lemma 3.2] and [101, Lemma 9.8] without proof, where the convergence $\alpha_k(a_k u_k) \to au$ in the strong* topology is claimed if $\alpha_k(u_k) \to u$ and $\alpha_k(a_k) \to a$ in the strong* topology. In fact, we are not able to see this assertion in the strong* convergence, and the above modification will be enough for discussions below.

Proof (Lemma 7.4) For every $\lambda \in \mathbb{R}$ we have

$$\begin{aligned}
&\lambda\{\alpha_k(a_k)\alpha_k(u_k) + \alpha_k(u_k)^*\alpha_k(a_k)^*\} \\
&\quad = \alpha_k(a_k + \lambda u_k^*)\alpha_k(a_k^* + \lambda u_k) - \alpha_k(a_k)\alpha_k(a_k)^* - \lambda^2\alpha_k(u_k)^*\alpha_k(u_k) \\
&\quad \le \alpha_k((a_k + \lambda u_k^*)(a_k^* + \lambda u_k)) - \alpha_k(a_k)\alpha_k(a_k)^* - \lambda^2\alpha_k(u_k)^*\alpha_k(u_k) \\
&\quad = \alpha_k(a_k a_k^*) - \alpha_k(a_k)\alpha_k(a_k)^* + \lambda\{\alpha_k(a_k u_k) + \alpha_k(u_k^* a_k^*)\} \\
&\qquad + \lambda^2\{\alpha_k(u_k^* u_k) - \alpha_k(u_k)^*\alpha_k(u_k)\}.
\end{aligned} \tag{7.15}$$

For any vector $\xi \in \mathcal{H}$ the above implies that for every $\lambda > 0$,

$$\begin{aligned}
&\langle \xi, \{\alpha_k(a_k)\alpha_k(u_k) + \alpha_k(u_k)^*\alpha_k(a_k)^*\}\xi\rangle \\
&\quad \le \frac{1}{\lambda}\langle \xi, \{\alpha_k(a_k a_k^*) - \alpha_k(a_k)\alpha_k(a_k)^*\}\xi\rangle + \langle \xi, \{\alpha_k(a_k u_k) + \alpha_k(u_k^* a_k^*)\}\xi\rangle \\
&\qquad + \lambda\langle \xi, \{1 - \alpha_k(u_k)^*\alpha_k(u_k)\}\xi\rangle.
\end{aligned} \tag{7.16}$$

Since $\alpha_k(u_k)^*\alpha_k(u_k) \to u^*u = 1$ strongly, one can choose $\lambda_k > 0$, $k \in \mathbb{N}$, such that

$$\lambda_k \longrightarrow +\infty, \qquad \lambda_k\langle \xi, \{1 - \alpha_k(u_k)^*\alpha_k(u_k)\}\xi\rangle \longrightarrow 0.$$

Then using (7.16) to $\lambda = \lambda_k$ and letting $k \to \infty$ one has

$$\langle \xi, (au + u^* a^*)\xi \rangle \le \liminf_k \langle \xi, \{\alpha_k(a_k u_k) + \alpha_k(u_k^* a_k^*)\}\xi \rangle. \tag{7.17}$$

Similarly, for every $\lambda < 0$,

$$\begin{aligned}
&\langle \xi, \{\alpha_k(a_k)\alpha_k(u_k) + \alpha_k(u_k^*)\alpha_k(a_k^*)\}\xi \rangle \\
&\quad \ge \frac{1}{\lambda} \langle \xi, \{\alpha_k(a_k a_k^*) - \alpha_k(a_k)\alpha_k(a_k)^*\}\xi \rangle + \langle \xi, \{\alpha_k(a_k u_k) + \alpha_k(u_k^* a_k^*)\}\xi \rangle \\
&\qquad + \lambda \langle \xi, \{1 - \alpha_k(u_k)^* \alpha_k(u_k)\}\xi \rangle.
\end{aligned}$$

Taking $\lambda_k < 0, k \in \mathbb{N}$, with $\lambda_k \to -\infty$ and $\lambda_k \langle \xi, \{1 - \alpha_k(u_k)^* \alpha_k(u_k)\}\xi \rangle \to 0$, one has

$$\langle \xi, (au + u^* a^*)\xi \rangle \ge \limsup_k \langle \xi, \{\alpha_k(a_k u_k) + \alpha_k(u_k^* a_k^*)\}\xi \rangle. \tag{7.18}$$

From (7.17) and (7.18) it follows that

$$\alpha_k(a_k u_k) + \alpha_k(u_k^* a_k^*) \longrightarrow au + u^* a^* \text{ weakly.} \tag{7.19}$$

Next, replace u_k with iu_k in (7.15) to obtain

$$\begin{aligned}
&\lambda i \{\alpha_k(a_k)\alpha_k(u_k) - \alpha_k(u_k)^* \alpha_k(a_k)^*\} \\
&\quad \le \alpha_k(a_k a_k^*) - \alpha_k(a_k)\alpha_k(a_k)^* + \lambda i \{\alpha_k(a_k u_k) - \alpha_k(u_k^* a_k^*)\} \\
&\qquad + \lambda^2 \{\alpha_k(u_k^* u_k) - \alpha_k(u_k)^* \alpha_k(u_k)\}.
\end{aligned}$$

From this, the same argument as above gives

$$\alpha_k(a_k u_k) - \alpha_k(u_k^* a_k^*) \longrightarrow au - u^* a^* \text{ weakly.} \tag{7.20}$$

Combining (7.19) and (7.20) yields that $\alpha_k(a_k u_k) \to au$ weakly. By replacing u_k, a_k with u_k^*, a_k^* in the above argument we have $\alpha_k(a_k^* u_k^*) \to a^* u^*$ weakly so that $\alpha_k(u_k a_k) \to ua$ weakly.

Lemma 7.6 *Let $A \subset B(\mathcal{H})$ be a unital C^*-algebra. Let $\beta_k : A \to B(\mathcal{H})$ be a sequence of unital Schwarz maps and $\Phi \in \mathcal{H}$ be such that $\langle \Phi, \beta_k(a)\Phi \rangle = \langle \Phi, a\Phi \rangle$ for all k and $a \in A$. If $u, v \in A$ are unitaries and $\beta_k(u) \to u$ and $\beta_k(v) \to v$ in the strong* topology, then*

$$\lim_k \langle v\Phi, \beta_k(a)u\Phi \rangle = \langle v\Phi, au\Phi \rangle, \qquad a \in A.$$

Proof It suffices to show that $\langle v\Phi, au\Phi\rangle$ is a unique limit point of $\{\langle v\Phi, \beta_k(a)u\Phi\rangle\}_{k=1}^\infty$. For this we may and do assume that $\mathcal{H}$ is a separable Hilbert space. Indeed, for each $a \in A$, let A_0 be the C^*-subalgebra of A generated by $1, u, v, a, \beta_k(a)$ $(k \in \mathbb{N})$, and B_0 be the C^*-subalgebra of $B(\mathcal{H})$ generated by A_0, $\beta_k(A_0)$ $(k \in \mathbb{N})$. Then $\mathcal{H}_0 := \overline{B_0\Phi}$ is a separable subspace of $\mathcal{H}$ and we may consider $\beta_k(x)|_{\mathcal{H}_0}$ $(x \in A_0)$. Since the weak topology is metrizable on bounded subsets of $B(\mathcal{H})$, by taking a subsequence we can assume that $\beta_k(a) \to b \in B(\mathcal{H})$ in the weak topology. The repeated use of Lemma 7.4 implies that $\beta_k(v^*au) \to v^*bu$ in the weak topology. Therefore,

$$\lim_k \langle v\Phi, \beta_k(a)u\Phi\rangle = \langle \Phi, v^*bu\Phi\rangle = \lim_k \langle \Phi, \beta_k(v^*au)\Phi\rangle = \langle v\Phi, au\Phi\rangle,$$

as asserted.

Proof (Lemma 7.2) For each k let $V_k : \mathcal{H}_k \to \mathcal{H}$ be a contraction as given in Step 1 of the proof of Lemma 7.1. Let φ_k, ψ_k and Φ, Φ_k be as in the proof of Lemma 7.1. Set $u_t := [D\psi : D\varphi]_t$ and $u_{k,t} := [D\psi_k : D\varphi_k]_t$ $(t \in \mathbb{R})$. We divide the proof into several steps to make it readable.

Step 1. For each k we show that $\|\Delta_\varphi^{1/4}\alpha_k(y)\Phi\| \le \|\Delta_{\varphi_k}^{1/4} y\Phi_k\|$ for all $y \in M_k$. For every $y \in M_k$ we have

$$\begin{aligned}\|\Delta_\varphi^{1/2}\alpha_k(y)\Phi\|^2 &= \|\alpha_k(y)^*\Phi\|^2 = \varphi(\alpha_k(y)\alpha_k(y)^*)\\ &\le \varphi(\alpha_k(yy^*)) = \|y^*\Phi_k\|^2 = \|\Delta_{\varphi_k}^{1/2} y\Phi_k\|^2,\end{aligned}$$

which implies that V_k maps $\mathcal{D}(\Delta_{\varphi_k}^{1/2})$ to $\mathcal{D}(\Delta_\varphi^{1/2})$ and $\|\Delta_\varphi^{1/2} V_k\xi\| \le \|\Delta_{\varphi_k}^{1/2}\xi\|$ for all $\xi \in \mathcal{D}(\Delta_{\varphi_k}^{1/2})$. Hence the result follows from Proposition B.11 of Appendix B.

Step 2. We show that $\lim_k \|\Delta_\varphi^{1/4}(\alpha_k(u_{k,t}) - u_t)\Phi\| = 0$ for all $t \in \mathbb{R}$. Let $x_k := \alpha_k(u_{k,t}) - u_t \in M$. By (7.2) one has $\|x_k\Phi\| \to 0$ and $\|\Delta_\varphi^{1/2} x_k\Phi\| = \|x_k^*\Phi\| \to 0$. With the spectral decomposition $\Delta_\varphi = \int_0^\infty s\, de_s$ one has

$$\begin{aligned}\|\Delta_\varphi^{1/4} x_k\Phi\|^2 &= \int_0^\infty s^{1/2}\, d\|e_s(x_k\Phi)\|^2\\ &= \left(\int_0^\infty d\|e_s(x_k\Phi)\|^2\right)^{1/2}\left(\int_0^\infty s\, d\|e_s(x_k\Phi)\|^2\right)^{1/2}\\ &= \|x_k\Phi\|\, \|\Delta_\varphi^{1/2} x_k\Phi\| \longrightarrow 0.\end{aligned}$$

Step 3. We show that $\lim_k \|\Delta_{\varphi_k}^{1/4} u_{k,t}\Phi_k\| = \|\Delta_\varphi^{1/4} u_t\Phi\|$ for all $t \in \mathbb{R}$. Since $\lambda^{-1}\varphi \le \psi \le \lambda\varphi$, it follows from Lemma A.58 of Sect. A.7 that the function $s \in \mathbb{R} \mapsto [D\psi : D\varphi]_s$ (resp., $[D\varphi : D\psi]_s$) extends to a strongly continuous (M-valued) function $[D\psi : D\varphi]_z$ (resp., $[D\varphi : D\psi]_z$) on the strip $-1/2 \le \operatorname{Im} z \le 0$ which is analytic in the interior. Since $[D\psi : D\varphi]_s = [D\varphi : D\psi]_s^*$, we see that $u_s = [D\psi : D\varphi]_s$ has the (strongly) analytic continuation u_z to the strip $-1/2 < \mathrm{Im}\, z < 1/2$ given by

$$u_z := \begin{cases} [D\psi : D\varphi]_z, & -1/2 < \mathrm{Im}\, z \le 0, \\ [D\varphi : D\psi]_{\bar{z}}^*, & 0 \le \mathrm{Im}\, z < 1/2. \end{cases}$$

Since $\sigma_s^\varphi(u_t) = u_s^* u_{s+t}$ by Theorem A.52 (ii), we further see that the function $s \in \mathbb{R} \mapsto \sigma_s^\varphi(u_t)$, with $t \in \mathbb{R}$ fixed, has the analytic continuation to the strip $-1/2 < \mathrm{Im}\, z < 1/2$ as

$$\sigma_z^\varphi(u_t) = u_{\bar{z}}^* u_{z+t}, \qquad -1/2 < \mathrm{Im}\, z < 1/2.$$

Similarly, for each k the function $s \in \mathbb{R} \mapsto u_{k,s}$ has the analytic continuation $u_{k,z}$ to $-1/2 < \mathrm{Im}\, z < 1/2$ so that $s \in \mathbb{R} \mapsto \sigma_s^{\varphi_k}(u_{k,t})$, with t fixed, has the analytic continuation $\sigma_z^{\varphi_k}(u_{k,t}) = u_{k,\bar{z}}^* u_{k,z+t}$ to $-1/2 < \mathrm{Im}\, z < 1/2$. Since

$$\varphi_k(\sigma_s^{\varphi_k}(u_{k,t}^*)\sigma_{-s}^{\varphi_k}(u_{k,t})) = \varphi(\alpha_k(u_{k,s+t}^* u_{k,s} u_{k,-s}^* u_{k,-s+t}))$$

and the repeated use of Lemma 7.4 based on (7.2) gives

$$\alpha_k(u_{k,s+t}^* u_{k,s} u_{k,-s}^* u_{k,-s+t}) \longrightarrow u_{s+t}^* u_s u_{-s}^* u_{s+t} \quad \text{weakly},$$

it follows that

$$\lim_k \varphi_k(\sigma_s^{\varphi_k}(u_{k,t})^* \sigma_{-s}^{\varphi_k}(u_{k,t})) = \varphi(u_{s+t}^* u_s u_{-s}^* u_{s+t}) = \varphi(\sigma_s^\varphi(u_t)^* \sigma_{-s}^\varphi(u_t)).$$

Vitali's theorem (see, e.g., [115, Sec. 7.3, p. 156]) implies that

$$\lim_k \varphi_k(\sigma_{\bar{z}}^{\varphi_k}(u_{k,t})^* \sigma_{-z}^{\varphi_k}(u_{k,t})) = \varphi(\sigma_{\bar{z}}^\varphi(u_t)^* \sigma_{-s}^\varphi(u_t)), \quad -1/2 < \mathrm{Im}\, z < 1/2.$$

Putting $z = i/4$ yields

$$\lim_k \varphi_k(\sigma_{-i/4}^{\varphi_k}(u_{k,t})^* \sigma_{-i/4}^{\varphi_k}(u_{k,t})) = \varphi(\sigma_{-i/4}^\varphi(u_t)^* \sigma_{-i/4}^\varphi(u_t)). \tag{7.21}$$

Since analytic continuation gives $\sigma_{-i/4}^\varphi(u_t)\Phi = \Delta_\varphi^{1/4} u_t \Phi$ and $\sigma_{-i/4}^{\varphi_k}(u_{k,t})\Phi_k = \Delta_{\varphi_k}^{1/4} u_{k,t} \Phi_k$, one can rewrite (7.21) as

$$\lim_k \|\Delta_{\varphi_k}^{1/4} u_{k,t} \Phi_k\|^2 = \|\Delta_\varphi^{1/4} u_t \Phi\|^2$$

for all $t \in \mathbb{R}$, as required.

Step 4. Let $\mathcal{H}_{1/4}$ be the Hilbert space introduced by completing $\mathcal{D}(\Delta_\varphi^{1/4})$ with respect to the inner product $\langle \xi, \eta \rangle_{1/4} := \langle \Delta_\varphi^{1/4}\xi, \Delta_\varphi^{1/4}\eta \rangle$ for $\xi, \eta \in \mathcal{D}(\Delta_\varphi^{1/4})$. For each k let $\mathcal{H}_{k,1/4}$ be the Hilbert space by completing $\mathcal{D}(\Delta_{\varphi_k}^{1/4})$ similarly. Step 1 says that V_k maps $\mathcal{D}(\Delta_{\varphi_k}^{1/4})$ to $\mathcal{D}(\Delta_\varphi^{1/4})$ contractively with respect to the 1/4-inner product, so V_k extends to a contraction from $\mathcal{H}_{k,1/4}$ to $\mathcal{H}_{1/4}$ (denoted by the same V_k). Let $V_k^\sharp : \mathcal{H}_{1/4} \to \mathcal{H}_{k,1/4}$ denote the adjoint of $V_k : \mathcal{H}_{k,1/4} \to \mathcal{H}_{1/4}$. Then we show that

$$\alpha_{k,\varphi}^*(x)\Phi_k = V_k^\sharp x\Phi, \qquad x \in M.$$

For every $x \in M$ and $y \in M_k$ we have

$$\begin{aligned}\langle V_k y\Phi_k, x\Phi \rangle_{1/4} &= \langle \Delta_\varphi^{1/4} V_k y \Phi_k, \Delta_\varphi^{1/4} x\Phi \rangle = \langle \alpha_k(y)\Phi, \Delta_\varphi^{1/2} x\Phi \rangle \\ &= \langle \alpha_k(y)\Phi, Jx^*\Phi \rangle = \langle \alpha_k(y)\Phi, Jx^*J\Phi \rangle \\ &= \langle \alpha_k(y)JxJ\Phi, \Phi \rangle = \langle Jx\Phi, \alpha_k(y^*)\Phi \rangle\end{aligned}$$

and

$$\begin{aligned}\langle y\Phi_k, V_k^\sharp x\Phi \rangle_{1/4} &= \langle \Delta_{\varphi_k}^{1/4} y\Phi_k, \Delta_{\varphi_k}^{1/4} V_k^\sharp x\Phi \rangle = \langle \Delta_{\varphi_k}^{1/2} y\Phi_k, V_k^\sharp x\Phi \rangle \\ &= \langle J_k y^*\Phi_k, V_k^\sharp x\Phi \rangle = \langle J_k V_k^\sharp x\Phi, y^*\Phi_k \rangle.\end{aligned}$$

Therefore,

$$\langle J_k V_k^\sharp x\Phi, y^*\Phi_k \rangle = \langle Jx\Phi, \alpha_k(y^*)\Phi \rangle, \qquad x \in M,\ y \in M_k.$$

Comparing this with condition (6.1) yields the assertion.

Step 5. We show that $\lim_k \|\Delta_\varphi^{1/4}(\alpha_k \circ \alpha_{k,\varphi}^*(u_t) - u_t)\Phi\| = 0$ for all $t \in \mathbb{R}$. For any $t \in \mathbb{R}$ set $\xi_k := u_{k,t}\Phi_k \in \mathcal{H}_{k,1/4}$ and $\eta := u_t\Phi \in \mathcal{H}_{1/4}$. Then, by Step 3,

$$\|\xi_k\|_{1/4} \left(:= \langle \xi_k, \xi_k \rangle_{1/4}^{1/2} \right) = \|\Delta_{\varphi_k}^{1/4} u_{k,t}\Phi_k\| \longrightarrow \|\Delta_\varphi^{1/4} u_t\Phi\| = \|\eta\|_{1/4},$$

and $\|V_k\xi_k - \eta\|_{1/4} \to 0$ by Step 2. Hence Lemma 7.3 implies that $\|V_k^\sharp\eta - \xi_k\|_{1/4} \to 0$. By Step 4 this means that

$$\|\Delta_{\varphi_k}^{1/4}(\alpha_{k,\varphi}^*(u_t) - u_{k,t})\Phi_k\| \longrightarrow 0.$$

Therefore, by Step 1 we have

$$\|\Delta_\varphi^{1/4}\alpha_k(\alpha_{k,\varphi}^*(u_t) - u_{k,t})\Phi\| \longrightarrow 0,$$

and consequently,

$$\begin{aligned}&\|\Delta_\varphi^{1/4}(\alpha_k\circ\alpha_{k,\varphi}^*(u_t)-u_t)\Phi\|\\&\quad\le\|\Delta_\varphi^{1/4}\alpha_k(\alpha_{k,\varphi}^*(u_t)-u_{k,t})\Phi\|+\|\Delta_\varphi^{1/4}(\alpha_k(u_{k,t})-u_t)\Phi\|\longrightarrow 0\end{aligned}$$

thanks to Step 2.

Step 6. We prove (7.13). Note that

$$\|\alpha_k\circ\alpha_{k,\varphi}^*(u_t)\Phi\|^2\le\varphi(\alpha_k\circ\alpha_{k,\varphi}^*(u_t^*u_t))=\varphi(1)=\|u_t\Phi\|^2,\quad t\in\mathbb{R}. \tag{7.22}$$

Step 5 implies that

$$\langle\Delta_\varphi^{1/4}x\Phi,(\alpha_k\circ\alpha_{k,\varphi}^*(u_t)-u_t)\Phi\rangle\longrightarrow 0,\qquad x\in M.$$

Since $\Delta_\varphi^{1/4}M\Phi$ is dense in $\mathcal{H}$ (and $\{\alpha_k\circ\alpha_{k,\varphi}^*(u_t)\Phi\}$ is bounded), it follows that $\alpha_k\circ\alpha_{k,\varphi}^*(u_t)\Phi\to u_t\Phi$ weakly as $k\to\infty$ for all $t\in\mathbb{R}$. This and (7.22) imply that $\|\alpha_k\circ\alpha_{k,\varphi}^*(u_t)\Phi-u_t\Phi\|\to 0$, showing that $\alpha_k\circ\alpha_{k,\varphi}^*(u_t)\to u_t$ strongly. Furthermore, it is immediate to see that Steps 2, 3 and 5 are also valid when $u_t,u_{k,t}$ are replaced with $u_t^*,u_{k,t}^*$, respectively. Hence we have $\alpha_k\circ\alpha_{k,\varphi}^*(u_t^*)\to u_t^*$ strongly as well.

Step 7. We prove (7.14). Let $\beta_k:=\alpha_k\circ\alpha_{k,\varphi}^*:M\to M$. By Proposition 6.3 (1) note that $\langle\Phi,\beta_k(x)\Phi\rangle=\varphi\circ\alpha_k\circ\alpha_{k,\varphi}^*(x)=\varphi(x)=\langle\Phi,x\Phi\rangle$ for all $x\in M$. From this and (7.13) one can use Lemma 7.6 to have

$$\lim_k\langle u_s\Phi,\beta_k(x)u_{-s}\Phi\rangle=\langle u_s\Phi,xu_{-s}\Phi\rangle,\qquad x\in M,\ s\in\mathbb{R}.$$

From the proof of Step 3 the function $s\in\mathbb{R}\mapsto u_s=[D\psi,D\varphi]_s$ extends to a strongly continuous function u_z on $-1/2\le\operatorname{Im}z\le 1/2$ which is analytic in the interior. Hence Vitali's theorem implies that

$$\lim_k\langle u_{\bar z}\Phi,\beta_k(x)u_{-z}\Phi\rangle=\langle u_{\bar z}\Phi,xu_{-z}\Phi\rangle,\qquad -1/2<\operatorname{Im}z<1/2.$$

For $z=ip$ $(0<p<1/2)$ one has

$$\lim_k\langle u_{-ip}\Phi,\beta_k(x)u_{-ip}\Phi\rangle=\langle u_{-ip}\Phi,xu_{-ip}\Phi\rangle,\qquad 0<p<1/2. \tag{7.23}$$

Note (see Lemma A.58) that $\|u_{-ip}\Phi - u_{-i/2}\Phi\| \to 0$ as $0 < p \nearrow 1/2$ and $u_{-i/2}\Phi = \Psi$. Since

$$\begin{aligned}
&|\langle u_{-i/p}\Phi, \beta_k(x)u_{-ip}\Phi\rangle - \langle\Psi, \beta_k(x)\Psi\rangle| \\
&\quad\le |\langle u_{-ip}\Phi - \Psi, \beta_k(x)u_{-ip}\Phi\rangle| + |\langle\Psi, \beta_k(x)(u_{-ip}\Phi - \Psi)\rangle| \\
&\quad\le \|x\|\,\|u_{-ip}\Phi\|\,\|u_{-ip}\Phi - \Psi\| + \|x\|\,\|\Psi\|\,\|u_{-ip}\Phi - \Psi\| \longrightarrow 0
\end{aligned}$$

as $0 < p \nearrow 1/2$ uniformly for k. Hence $\lim_k\langle\Psi, \beta_k(x)\Psi\rangle$ exists and the LHS of (7.23) converges as $p \nearrow 1/2$ to $\lim_k\langle\Psi, \beta_k(x)\Psi\rangle$, while the RHS converges to $\langle\Psi, x\Psi\rangle$. Therefore, $\lim_k\langle\Psi, \beta_k(x)\Psi\rangle = \langle\Psi, x\Psi\rangle$, i.e., $\lim_k \psi(\beta_k(x)) = \psi(x)$.

Problem 7.7 The assumption $\lambda^{-1}\varphi \le \psi \le \lambda\varphi$ in Lemma 7.2 (also Theorem 7.8) may be rather too strong in the von Neumann algebra setting, so it is desirable to remove that. But the problem does not seem easy because the assumption has been used in an essential way in Steps 3 and 7 of the proof of Lemma 7.2 to appeal to Vitali's theorem.

We now present the approximate reversibility theorem, refining [108, Theorem 3.7] (also [101, Theorem 9.12]). The proof based on Lemmas 7.1 and 7.2 is similar to that of Theorem 6.19 based on Lemmas 6.14 and 6.15.

Theorem 7.8 *Let $\alpha_k : M_k \to M$ ($k \in \mathbb{N}$) be a sequence of unital 2-positive maps with von Neumann algebras M_k. Let $\psi, \varphi \in M_*^+$ and assume that $\lambda^{-1}\varphi \le \psi \le \lambda\varphi$ for some $\lambda > 0$. Then the following conditions are equivalent:*

(i) $\lim_k \psi\circ\alpha_k\circ\alpha_{k,\varphi}^* = \psi$ *in $\sigma(M_*, M)$ ($\varphi\circ\alpha\circ\alpha_{k,\varphi}^* = \varphi$ is automatic), where $\alpha_{k,\varphi}^*$ is Petz' recovery map of α_k with respect to φ in the sense of Proposition 6.6.*
(ii) Approximate reversibility: *There exist unital normal 2-positive maps β_k : $M \to M_k$ ($k \in \mathbb{N}$) such that*

$$\psi\circ\alpha_k\circ\beta_k \longrightarrow \psi, \quad \varphi\circ\alpha_k\circ\beta_k \longrightarrow \varphi \quad \textit{in } \sigma(M_*, M).$$

(iii) $\lim_k S_f(\psi\circ\alpha_k\|\varphi\circ\alpha_k) = S_f(\psi\|\varphi)$ *for every operator convex function f on $(0, \infty)$.*
(iv) $\lim_k S_f(\psi\circ\alpha_k\|\varphi\circ\alpha_k) = S_f(\psi\|\varphi)$ *for some operator convex function f on $(0, \infty)$ such that* $\operatorname{supp}\mu_f$ *has a limit point in $(0, \infty)$.*

Proof Write $\psi_k := \psi\circ\alpha_k$, $\varphi_k := \varphi\circ\alpha_k$, $e := s(\varphi) = s(\psi)$ ($\in M$) and $e_k := s(\varphi_k) = s(\psi_k)$ ($\in M_k$).

(i) $\Longrightarrow$ (ii). For each k let $\beta_k := \alpha_{k,\varphi}^* : M \to e_kM_ke_k$ be given as in Proposition 6.6 and define $\widetilde{\beta}_k(x) := \beta_k(x) + \rho(x)(1 - e_k)$ for $x \in M$, where ρ is a normal state on M. Then $\beta_k : M \to M_k$ is a unital normal 2-positive map. From (i) we have for every $x \in M$,

$$\psi\circ\alpha_k\circ\widetilde{\beta}_k(x) = \psi\circ\alpha_k\circ\beta_k(x) \to \psi(x),$$

as well as $\varphi \circ \alpha_k \circ \widetilde{\beta}_k(x) = \varphi \circ \alpha \circ \beta_k(x) = \varphi(x)$ by Proposition 6.6 (1). Hence (ii) holds.

(ii) $\Longrightarrow$ (iii). Assume (ii). For every f as stated in (iii) we have

$$\begin{aligned} S_f(\psi\|\varphi) &\le \liminf_k S_f(\psi \circ \alpha_k \circ \beta_k \| \varphi \circ \alpha_k \circ \beta_k) \\ &\le \liminf_k S_f(\psi \circ \alpha_k \| \varphi \circ \alpha_k) \\ &\le \limsup_k S_f(\psi \circ \alpha_k \| \varphi \circ \alpha_k) \\ &\le S_f(\psi\|\varphi) \end{aligned}$$

by the lower semicontinuity in $\sigma(M_*, M)$ and the monotonicity of S_f, see Theorem 2.7 (i) and (iv).

(iii) $\Longrightarrow$ (iv) is obvious.

(iv) $\Longrightarrow$ (i). For each k let $\hat{\alpha}_k := e\alpha_k(\cdot)e|_{e_k M_k e_k} : e_k M_k e_k \to eMe$, which is a unital normal 2-positive map (see Remark 6.7). Then as in the proof of (iv) $\Longrightarrow$ (i) of Theorem 6.19 the following hold:

$$\psi_k|_{e_k M_k e_k} = (\psi|_{eMe}) \circ \hat{\alpha}_k, \qquad \varphi_k|_{e_k M_k e_k} = (\varphi|_{eMe}) \circ \hat{\alpha}_k,$$

and for any operator convex function f on $(0, \infty)$,

$$S_f(\psi_k\|\varphi_k) = S_f(\psi|_{eMe} \circ \hat{\alpha}_k \| (\varphi|_{eMe}) \circ \hat{\alpha}_k),$$

as well as $S_f(\psi\|\varphi) = S_f(\psi|_{eMe}\|\varphi|_{eMe})$. Here, note that $S_f(\psi\|\varphi) < +\infty$ always holds by Corollary 4.19. Assume (iv) with f as stated. Then it follows from Lemmas 7.1 and 7.2 that

$$(\psi|_{eMe}) \circ \hat{\alpha}_k \circ \hat{\beta}_k \longrightarrow \psi|_{eMe} \quad \text{in } \sigma(M_*, M),$$

where $\hat{\beta}_k : eMe \to e_k M_k e_k$ is Petz' recovery map of $\hat{\alpha}_k$. For every $x \in M$, since $\alpha_{k,\varphi}^*(x) = \hat{\beta}_k(exe)$ (see Remark 6.7), we have

$$\psi_k \circ \alpha_{k,\varphi}^*(x) = (\psi|_{eMe}) \circ \hat{\alpha}_k \circ \hat{\beta}_k(exe) \longrightarrow \psi(exe) = \psi(x).$$

Hence (i) follows.

7.2 Reversibility via Measurements

In this section we will study the reversibility via measurement procedure, i.e., quantum-classical channels given by unital normal positive maps $\alpha : \mathcal{A} \to M$ with commutative $\mathcal{A}$ and its relation with equalities between S_f, $\widehat{S}_f$ and S_f^{meas}, the standard, the maximal and the measured f-divergences discussed in Chaps. 2, 4 and 5. Recall that $S_f^{\mathrm{meas}} \le S_f \le \widehat{S}_f$, see (5.7).

Let f be an operator convex function on $(0, \infty)$ as before, and for $\psi, \varphi \in M_*^+$ we consider the following conditions:

(i) ψ, φ commute (see Lemma 4.20 and Definition A.57).

(ii) *Existence of a sufficient commutative subalgebra:* There exist a commutative von Neumann subalgebra $\mathcal{A}$ of M and a normal conditional expectation $E : M \to \mathcal{A}$ such that $\psi \circ E = \psi$ and $\varphi \circ E = \varphi$.

(iii) *Reversibility via a measurement channel:* There exist unital normal positive maps $\alpha : \mathcal{A} \to M$ and $\beta : M \to \mathcal{A}$ with a commutative von Neumann algebra $\mathcal{A}$ such that

$$\psi = \psi \circ \alpha \circ \beta, \qquad \varphi = \varphi \circ \alpha \circ \beta.$$

(iv) *Approximate reversibility via measurement channels:* There exist unital normal positive maps $\alpha_k : \mathcal{A}_k \to M$ and $\beta_k : M \to \mathcal{A}_k$ $(k \in \mathbb{N})$ with commutative von Neumann algebras $\mathcal{A}_k$ such that

$$\psi \circ \alpha_k \circ \beta_k \longrightarrow \psi, \quad \varphi \circ \alpha_k \circ \beta_k \longrightarrow \varphi \quad \text{in } \sigma(M_*, M).$$

(v) There exists a unital normal positive map $\alpha : \mathcal{A} \to M$ with a commutative von Neumann algebra $\mathcal{A}$ such that

$$S_f(\psi \circ \alpha \| \varphi \circ \alpha) = S_f(\psi \| \varphi).$$

(vi) $S_f^{\mathrm{pr}}(\psi \| \varphi) = S_f(\psi \| \varphi)$ (see Definition 5.1).

(vii) $S_f^{\mathrm{meas}}(\psi \| \varphi) = S_f(\psi \| \varphi)$, that is, there exists a sequence of unital normal positive maps $\alpha_k : \mathcal{A}_k \to M$ with commutative von Neumann algebras $\mathcal{A}_k$ such that

$$\lim_k S_f(\psi \circ \alpha_k \| \varphi \circ \alpha_k) = S_f(\psi \| \varphi).$$

(viii) $S_f^{\mathrm{meas}}(\psi \| \varphi) = \widehat{S}_f(\psi \| \varphi)$.

(ix) $S_f(\psi \| \varphi) = \widehat{S}_f(\psi \| \varphi)$.

We first summarize in the next lemma an easy part of the proof of Theorem 7.10 below.

Lemma 7.9 *For the above conditions we have*

$$\text{(i)} \Longrightarrow \text{(ii)} \Longrightarrow \text{(iii)} \Longrightarrow \text{(iv)} \Longrightarrow \text{(vii)},$$

$$\text{(iii)} \Longrightarrow \text{(v)} \Longrightarrow \text{(vii)},$$

$$\text{(ii)} \Longrightarrow \text{(vi)} \Longrightarrow \text{(vii)},$$

$$\text{(iii)} \Longrightarrow \text{(viii)} \Longrightarrow \text{(ix)}.$$

Proof (i) $\Longrightarrow$ (ii). Assume (i) and let $e := s(\psi + \varphi)$, $\psi_0 := \psi|_{eMe}$ and $\varphi_0 := \varphi|_{eMe}$. From the proof of [55, Proposition 6.7], there exists a commutative von Neumann subalgebra $\mathcal{A}_0$ of eMe and a normal conditional expectation $E_0 : eMe \to \mathcal{A}_0$ such that $\psi_0 = \psi_0 \circ E_0$ and $\varphi_0 = \varphi_0 \circ E_0$. Hence (ii) holds with $\mathcal{A} = \mathcal{A}_0 + \mathbb{C}(1-e)$ and $E : M \to \mathcal{A}$ given by

$$E(x) := E_0(exe) + \rho((1-e)x(1-e))(1-e), \qquad x \in M,$$

where ρ is a normal state on $(1-e)M(1-e)$.

(ii) $\Longrightarrow$ (iii) is clear with the injection $\alpha : \mathcal{A} \to M$ and $\beta = E$, and (iii) $\Longrightarrow$ (iv) is obvious. (iv) $\Longrightarrow$ (vii) is a special case of (ii) $\Longrightarrow$ (iii) of Theorem 7.8, which holds for arbitrary $\psi, \varphi \in M_*^+$.

(iii) $\Longrightarrow$ (v) follows from Theorem 2.7 (iv), and (v) $\Longrightarrow$ (vii) is obvious (see also Proposition 5.2 for (vii)). (ii) $\Longrightarrow$ (vi) is seen as

$$\begin{aligned} S_f(\psi\|\varphi) &= S_f(\psi|_{\mathcal{A}}\|\varphi|_{\mathcal{A}}) \\ &= \sup\left\{S_f(\psi|_{\mathcal{A}_0}\|\varphi|_{\mathcal{A}_0}) : \mathcal{A}_0 \text{ a finite-dimensional subalgebra of } \mathcal{A}\right\} \\ &\le S_f^{\mathrm{pr}}(\psi\|\varphi), \end{aligned}$$

where the second equality is due to Theorem 2.7 (vii). (vi) $\Longrightarrow$ (vii) is obvious from (5.7). (iii) $\Longrightarrow$ (viii) is seen since (iii) implies that

$$\widehat{S}_f(\psi\|\varphi) = \widehat{S}_f(\psi\circ\alpha\circ\beta\|\varphi\circ\alpha\circ\beta) \le S_f(\psi\circ\alpha\|\varphi\circ\alpha) \le S_f^{\mathrm{meas}}(\psi\|\varphi),$$

where the first inequality above is due to Theorem 4.4 (i) and Proposition 4.21. Finally, (viii) $\Longrightarrow$ (ix) is obvious.

The main aim of the present section is to prove the following theorem. In [108, Theorem 3.3] (also [101, Theorem 9.10]) Petz proved the implication (iv) $\Longrightarrow$ (i), so the theorem below supplements Petz' result in [108] with some other equivalent conditions.

Theorem 7.10 *Assume that* supp μ_f, *the support of the representing measure of f (see (2.5)), has a limit point in $(0, \infty)$. If $s(\psi) = s(\varphi)$ and $S_f(\psi\|\varphi) < +\infty$, then the above conditions (i)–(ix) are all equivalent.*

Once we have shown the implications in Lemma 7.9, it only remains to prove (vii) $\Longrightarrow$ (i) and (ix) $\Longrightarrow$ (v).

Proof (Theorem 7.10) In view of Lemma 7.9 the proof will be completed when we show the following two implications.

(ix) $\Longrightarrow$ (v). By Theorem 4.17 there exist a unital normal positive map $\beta : M \to \mathcal{A}$ with a commutative von Neumann algebra $\mathcal{A}$ and $p, q \in \mathcal{A}_*^+$ such that $\psi = p \circ \beta$, $\varphi = q \circ \beta$ and $S_f(p\|q) = \widehat{S}_f(\psi\|\varphi)$. From (ix) this implies that $S_f(p\|q) = S_f(\psi\|\varphi) < +\infty$. Hence, from Theorem 6.19 it follows that β is reversible for $\{p, q\}$, so there exists a unital normal positive map $\alpha : \mathcal{A} \to M$ such that $p = \psi \circ \alpha$ and $q = \varphi \circ \alpha$. Since

$$S_f(\psi \circ \alpha \| \varphi \circ \alpha) = S_f(p\|q) = S_f(\psi\|\varphi),$$

condition (v) holds. (Note that the proof of this part works when $s(\psi) \le s(\varphi)$.)

(vii) $\Longrightarrow$ (i). Condition (vii) means that there exists a sequence of unital normal positive maps $\alpha_k : \mathcal{A}_k \to M$ with commutative $\mathcal{A}_k$ such that (7.1) holds. Put $e := s(\psi) = s(\varphi)$. For each k, let e_k $(\in \mathcal{A}_k)$ be the support projection of the normal positive map $e\alpha_k(\cdot)e : \mathcal{A}_k \to eMe$, and define $\hat{\alpha}_k := e\alpha_k(\cdot)e|_{e_k\mathcal{A}_ke_k} : e_k\mathcal{A}_ke_k \to eMe$, which is obviously unital (i.e., $\hat{\alpha}_k(e_k) = e$) and faithful. Note that

$$S_f(\psi\|\varphi) = S_f(e\psi e\|e\varphi e).$$

Moreover,

$$(\psi \circ \alpha_k)(a) = (e\psi e) \circ \hat{\alpha}_k(e_kae_k), \qquad a \in \mathcal{A}_k,$$

and similarly for φ, so that

$$S_f(\psi \circ \alpha_k\|\varphi \circ \alpha_k) = S_f((e\psi e) \circ \hat{\alpha}_k\|(e\varphi e) \circ \hat{\alpha}_k).$$

Hence, replacing ψ, φ, α_k with $e\psi e, e\varphi e, \hat{\alpha}_k$, we may assume that $s(\psi) = s(\varphi) = 1$ and all α_k are faithful. Now, let $u_{k,t} := [D\psi \circ \alpha_k : D\varphi \circ \alpha_k]_t$ and $u_t := [D\psi : D\varphi]_t$ for $t \in \mathbb{R}$. Then Lemma 7.1 implies that $\alpha_k(u_{k,t}) \to u_t$ strongly* for all $t \in \mathbb{R}$. By Lemma 7.4 we have

$$\alpha_k(u_{k,s}u_{k,t}) \longrightarrow u_su_t \text{ weakly}, \quad s, t \in \mathbb{R}.$$

Since $\mathcal{A}_k$ is commutative and so $\{u_{k,t}\}_{t\in\mathbb{R}}$ is a one-parameter unitary group by Proposition A.56, we have $u_su_t = u_tu_s$ for all $s, t \in \mathbb{R}$, which means by Proposition A.56 again that ψ and φ are commuting.

Theorem 7.11 *Assume that* $\operatorname{supp}\mu_f$ *has a limit point in* $(0, \infty)$. *Let* $\psi, \varphi \in M_*^+$, *and assume that* $s(\psi) \le s(\varphi)$ *and* $S_f(\psi\|\varphi) < +\infty$. *Let (i)′–(ix)′ denote the respective conditions corresponding to (i)–(ix), where* ψ *is replaced with* $\psi + \varphi$; *for example, (vii)′* $S_f^{\mathrm{meas}}(\psi + \varphi\|\varphi) = S_f(\psi + \varphi\|\varphi)$. *Then the conditions (i)–(v), (viii), (ix) and all (i)′–(ix)′ are equivalent.*

Proof Each condition of (i)–(iv) is unchanged when ψ is replaced with $\psi + \varphi$. Note that $s(\psi + \varphi) = s(\varphi)$ from $s(\psi) \le s(\varphi)$ and $S_f(\psi + \varphi\|\varphi) < +\infty$ by Proposition 2.9. Hence we can apply Theorem 7.10 to $\psi + \varphi$ and φ to see that the conditions (i)–(iv) and (i)′–(ix)′ are all equivalent. Furthermore, we have (iii) $\Longrightarrow$ (v) and (iii) $\Longrightarrow$ (viii) $\Longrightarrow$ (ix) by Lemma 7.9, (ix) $\Longrightarrow$ (v) by the above proof of Theorem 7.10 (this part has been done with $s(\psi) \le s(\varphi)$), and (v) $\Longrightarrow$ (iii) from

Theorem 6.19. Hence (v), (viii) and (ix) are also equivalent to (iii), so the result follows.

Problem 7.12 Conditions (vi) and (vii) are missing in Theorem 7.11. But it is unknown to us how to prove the implication (vii) $\Longrightarrow$ (i) under $s(\psi) \le s(\varphi)$. Here, note that although $[D\psi : D\varphi]_t$ is defined even when $s(\psi) \le s(\varphi) = 1$ (in this case, $[D\psi : D\varphi]_0 = s(\psi)$), Lemma 7.1 does not hold in general. Indeed, one can easily construct an example of ψ, φ and von Neumann subalgebras $M_k \subset M$ such that $\psi|_{M_k}$ and $\varphi|_{M_k}$ are faithful and $S_f(\psi \circ \alpha_k \| \varphi \circ \alpha_k) \to S_f(\psi \| \varphi) < +\infty$, for example, in the case where M is commutative and $f(x) = x^\alpha$, $1 < \alpha < 2$. But in this case, $\alpha_k([D\psi_k : D\varphi_k]_0) = \alpha_k(1) = 1 \ne s(\psi) = [D\psi : D\varphi]_0$, so the strong convergence (7.2) is impossible. This phenomenon is similar to convergence of positive self-adjoint operators. Let A_k, A be positive self-adjoint operators on a Hilbert space such that $A_k \to A$ in the sense that $(sI + A_k)^{-1} \to (sI + A)^{-1}$ strongly for all $s > 0$ (see Proposition B.8). When all A_k are non-singular but A is singular, note that A_k^{it} cannot converge to $s(A)A^{it}$ for $t \in \mathbb{R}$; in fact, $A_k^0 = I \not\to s(A)$ for $t = 0$. Therefore, we need to find another route to prove (vii) $\Longrightarrow$ (i) for the case $s(\psi) \le s(\varphi)$.

Problem 7.13 A significant problem is the attainability question for S_f^{meas}, that is, whether there exists a unital normal positive map $\alpha : \mathcal{A} \to M$ with commutative $\mathcal{A}$ such that $S_f(\psi \circ \alpha \| \varphi \circ \alpha) = S_f^{\mathrm{meas}}(\psi \| \varphi)$. Theorem 7.10 tells us that the attainability holds for f satisfying the assumption in the theorem if $s(\psi) = s(\varphi)$ and $S_f^{\mathrm{meas}}(\psi \| \varphi) = S_f(\psi \| \varphi) < +\infty$. When $M = B(\mathcal{H})$ with $\dim \mathcal{H} < \infty$, the attainability is true for arbitrary $\psi, \varphi \in B(\mathcal{H})_*^+$ [58, Proposition 4.17]. If the question were affirmative in general for $\psi, \varphi \in M_*^+$ with $s(\psi) \le s(\varphi)$, then conditions (v) and (vii) would be equivalent, so that Problem 7.12 becomes meaningless.

Chapter 8
Preservation of Maximal f-Divergences

8.1 Maximal f-Divergences and Operator Connections

In this chapter we will characterize the preservation of $\widehat{S}_f$ under a unital normal positive map γ, i.e., the equality case in the monotonicity inequality $\widehat{S}_f(\psi \circ \gamma \| \varphi \circ \varphi) \le \widehat{S}_f(\psi \| \varphi)$. To do this, we work in the standard form

$$(M, L^2(M), J = {}^*, L^2(M)_+),$$

where $L^p(M)$ is Haagerup's L^p-space (see Sect. A.6). Recall that M_* is identified with $L^1(M)$ by the linear bijection $\psi \leftrightarrow h_\psi$ and the tr-functional on $L^1(M)$ is given by $\mathrm{tr}(h_\psi) = \psi(1)$. To present the main result, we introduce the notion of operator connections of $\psi, \varphi \in M_*^+$, which is a type of extension of the Kubo–Ando operator connections [85]. This is an interesting topic on its own, so we develop the theory in some detail in Appendix D separately.

When $A, B \in B(\mathcal{H})_+$ are invertible, the *operator perspective* $P_\phi(A, B)$, introduced in [35, 36], is

$$P_\phi(A, B) := B^{1/2}\phi(B^{-1/2}AB^{-1/2})B^{1/2}$$

for any continuous function ϕ on $(0, \infty)$. The *operator connection* $A\sigma B$ in the Kubo–Ando sense [85] is defined by

$$A\sigma B = A^{1/2}k(A^{-1/2}BA^{-1/2})A^{1/2} \ (= P_k(B, A))$$

for invertible $A, B \in B(\mathcal{H})_+$, corresponding to a non-negative operator monotone function $k = k_\sigma$ on $[0, \infty)$ (called the representing function of σ). The σ is extended

F. Hiai, *Quantum f-Divergences in von Neumann Algebras*,
Mathematical Physics Studies, https://doi.org/10.1007/978-981-33-4199-9_8

to general $A, B \in B(\mathcal{H})_+$ as

$$A\sigma B = \lim_{\varepsilon \searrow 0} (A + \varepsilon I)\sigma(B + \varepsilon I)$$

in the strong operator topology (decreasingly). See the first paragraph of Appendix D for a more intrinsic (axiomatic) definition of operator connections [85].

In the following we modify the operator perspectives and connections given above into those for functionals $\psi, \varphi \in M_*^+$.

Definition 8.1 Let $\psi, \varphi \in M_*^+$ and assume that $(\psi, \varphi) \in (M_*^+ \times M_*^+)_\le$, see the beginning of Sect. 4.1. By Lemma A.24 we have a unique $A \in s(\varphi)Ms(\varphi)$ such that $h_\psi^{1/2} = Ah_\varphi^{1/2}$ (note that $h_\varphi^{1/2} \in L^2(M)_+$ is the vector representative of φ). Define $T_{\psi/\varphi} := A^*A$. For every continuous function ϕ on $[0, \infty)$, since $h_\varphi^{1/2}\phi(T_{\psi/\varphi})h_\varphi^{1/2} \in L^1(M)$, we define the M_*-valued *perspective* $P_\phi(\psi, \varphi)$ by

$$h_{P_\phi(\psi,\varphi)} = h_\varphi^{1/2}\phi(T_{\psi/\varphi})h_\varphi^{1/2}. \tag{8.1}$$

Definition 8.2 Let σ be an operator connection in the Kubo–Ando sense with the representing function k_σ (a non-negative operator monotone function on $[0, \infty)$). When $(\psi, \varphi) \in (M_*^+ \times M_*^+)_\le$, we define the M_*-valued *connection* $\varphi\sigma\psi$ as

$$\varphi\sigma\psi = P_{k_\sigma}(\psi, \varphi), \quad \text{i.e.,} \quad h_{\varphi\sigma\psi} = h_\varphi^{1/2}k_\sigma(T_{\psi/\varphi})h_\varphi^{1/2}. \tag{8.2}$$

For general $\psi, \varphi \in M_*^+$, choosing an $\omega \in M_*^+$ with $\omega \sim \varphi + \psi$, we define

$$\varphi\sigma\psi := \lim_{\varepsilon \searrow 0} (\varphi + \varepsilon\omega)\sigma(\psi + \varepsilon\omega) \in M_*^+ \quad \text{in the norm.} \tag{8.3}$$

In fact, from Lemma D.4 and Theorem D.6 of Appendix D, the limit in (8.3) as $\varepsilon \searrow 0$ exists independently of the choice of $\omega \sim \varphi + \psi$, and it coincides with definition (8.2) when $(\psi, \varphi) \in (M_*^+ \times M_*^+)_\le$. Indeed, a slightly more intrinsic definition of $\varphi\sigma\psi$ is provided in Appendix D.

It is instructive to note here that the connection $\varphi\sigma\psi$ for $\psi, \varphi \in M_*^+$ is explicitly related to the maximal f-divergence $\widehat{S}_f(\psi\|\varphi)$ as follows:

$$(\varphi\sigma\psi)(1) = -\widehat{S}_{-k_\sigma}(\psi\|\varphi), \tag{8.4}$$

where k_σ is the representing function of σ (so $-k_\sigma$ is an operator convex function on $(0, \infty)$). In fact, identity (8.4) is shown in Proposition D.10, and it will be useful in Sect. 8.2.

Now let $\gamma : N \to M$ be a unital normal positive linear map between von Neumann algebras, and let $\gamma_* : L^1(M) \to L^1(N)$ be the predual map of γ via $M_* \cong L^1(M)$ and $N_* \cong L^1(N)$, i.e., $\gamma_*(h_\psi) = h_{\psi\circ\gamma}$ for $\psi \in M_*$ and the corresponding $h_\psi \in L^1(M)$. For every $\varphi \in M_*^+$ let $e_0 \in N$ be the support projection of $\varphi \circ \gamma$, and $\gamma_\varphi^* : M \to e_0 N e_0$ be the (extended) Petz' recovery map of γ with respect to φ given in Proposition 6.6. The next lemma is the description of γ_φ^* in terms of Haagerup's L^1-elements h_φ and $h_{\varphi\circ\gamma}$. Note that (8.5) has a strong resemblance to (6.4) in the finite-dimensional case.

Lemma 8.3 *For every $\varphi \in M_*^+$ let γ_φ^* be as stated above. Then for every $x \in M$, $\gamma_\varphi^*(x) \in e_0 N e_0$ is determined by*

$$h_{\varphi\circ\gamma}^{1/2} \gamma_\varphi^*(x) h_{\varphi\circ\gamma}^{1/2} = \gamma_*\bigl(h_\varphi^{1/2} x h_\varphi^{1/2}\bigr). \tag{8.5}$$

Proof Write $\beta = \gamma_\varphi^*$. For every $x \in M$ the defining condition of $\beta(x)$ is (6.1) for all $y \in e_0 N e_0$, and it can be rewritten in terms of $h_\varphi, h_{\varphi\circ\gamma}$ as

$$\operatorname{tr} \beta(x) h_{\varphi\circ\gamma}^{1/2} y h_{\varphi\circ\gamma}^{1/2} = \operatorname{tr} x h_\varphi^{1/2} \gamma(y) h_\varphi^{1/2}, \qquad y \in e_0 N e_0,$$

that is,

$$\operatorname{tr} h_{\varphi\circ\gamma}^{1/2} \beta(x) h_{\varphi\circ\gamma}^{1/2} y = \operatorname{tr} \gamma_*\bigl(h_\varphi^{1/2} x h_\varphi^{1/2}\bigr) y, \qquad y \in e_0 N e_0.$$

Since $e_0 h_{\varphi\circ\gamma}^{1/2} e_0 = h_{\varphi\circ\gamma}^{1/2}$ and $e_0 \gamma_*\bigl(h_\varphi^{1/2} x h_\varphi^{1/2}\bigr) e_0 = \gamma_*\bigl(h_\varphi^{1/2} x h_\varphi^{1/2}\bigr)$ as easily verified, the above holds for all $y \in N$, which is equivalent to (8.5).

In this section we present the next theorem on the preservation of the maximal f-divergence, i.e., the equality case in the monotonicity inequality of $\widehat{S}_f(\psi\|\varphi)$, which is the extension of [58, Theorem 3.34] to the von Neumann algebra setting. It is remarkable that we treat general $\psi, \varphi \in M_*^+$ with no condition on their support projections, differently from the reversibility theorems in Chaps. 6 and 7, so that even in the finite-dimensional case, the next theorem improves [58, Theorem 3.34].

Theorem 8.4 *Let γ be as above. Let $\psi, \varphi \in M_*^+$ and set $\omega := \psi + \varphi$. As in Definition 8.1 let $T_{\psi/\omega} := A^*A$ with $A \in s(\omega)Ms(\omega)$ satisfying $h_\psi^{1/2} = A h_\omega^{1/2}$, and $T_{\psi\circ\gamma/\omega\circ\gamma} := A_0^* A_0$ with $A_0 \in s(\omega\circ\gamma)Ns(\omega\circ\gamma)$ satisfying $h_{\psi\circ\gamma}^{1/2} = A_0 h_{\omega\circ\gamma}^{1/2}$. Then the following conditions are equivalent:*

- (i) *$\widehat{S}_f(\psi\circ\gamma\|\varphi\circ\gamma) = \widehat{S}_f(\psi\|\varphi)$ for any operator convex function f on $(0,\infty)$;*
- (ii) *$\widehat{S}_f(\psi\circ\gamma\|\varphi\circ\gamma) = \widehat{S}_f(\psi\|\varphi) < +\infty$ for some nonlinear operator convex function f on $(0,\infty)$ with $\lim_{t\to 0^+} tf(t) = 0$ (in particular, this is the case if $f(0^+) < +\infty$) and $\lim_{t\to\infty} f(t)/t^2 = 0$ (in particular, this is the case if $f'(\infty) < +\infty$);*
- (iii) *$\widehat{S}_f(\psi\circ\gamma\|\omega\circ\gamma) = \widehat{S}_f(\psi\|\omega)$ for any operator convex function f on $[0,\infty)$;*
- (iv) *$\widehat{S}_f(\psi\circ\gamma\|\omega\circ\gamma) = \widehat{S}_f(\psi\|\omega)$ for some nonlinear operator convex function f on $[0,\infty)$;*
- (v) *$S_{t^2}(\psi\circ\gamma\|\omega\circ\gamma) = S_{t^2}(\psi\|\omega)$ (equivalently $D_2(\psi\circ\gamma\|\omega\circ\gamma) = D_2(\psi\|\omega)$ as long as $\psi \ne 0$);*

(vi) $(\varphi \circ \gamma)\sigma(\psi \circ \gamma) = (\varphi\sigma\psi) \circ \gamma$ *for any operator connection* σ*;*
(vii) $(\varphi \circ \gamma)\sigma(\psi \circ \gamma) = (\varphi\sigma\psi) \circ \gamma$ *for some nonlinear operator connection* σ*;*
(viii) $(\omega \circ \gamma)\sigma(\psi \circ \gamma) = (\omega\sigma\psi) \circ \gamma$ *for any operator connection* σ*;*
(ix) $(\omega \circ \gamma)\sigma(\psi \circ \gamma) = (\omega\sigma\psi) \circ \gamma$ *for some nonlinear operator connection* σ*;*
(x) $P_\phi(\psi \circ \gamma, \omega \circ \gamma) = P_\phi(\psi, \omega) \circ \gamma$ *for any continuous function* ϕ *on* $[0, \infty)$*;*
(xi) $h_{\psi\circ\gamma}^{1/2} A_0 A_0^* h_{\psi\circ\gamma}^{1/2} = \gamma_*\big(h_\psi^{1/2} A A^* h_\psi^{1/2}\big)$*;*
(xii) $\gamma_\omega^*((T_{\psi/\omega})^2) = (\gamma_\omega^*(T_{\psi/\omega}))^2$.

Moreover, if γ *is 2-positive and hence so is* γ_ω^**, then the above conditions are also equivalent to*

(xiii) $T_{\psi/\omega} \in \mathcal{M}_{\gamma_\omega^*}$ *(the multiplicative domain of* γ_ω^**).*

For example, the function $f(t) = t \log t$ on $(0, \infty)$ satisfies $f(0^+) < +\infty$ and $\lim_{t\to\infty} f(t)/t^2 = 0$. so that a typical realization of condition (ii) is that $D_{\mathrm{BS}}(\psi \circ \gamma \| \varphi \circ \gamma) = D_{\mathrm{BS}}(\psi \| \varphi) < +\infty$, see Example 4.22.

In the rest of this chapter let $\psi_0 := \psi \circ \gamma$, $\varphi_0 := \varphi \circ \gamma$, $\omega_0 := \omega \circ \gamma = \psi_0 + \varphi_0$, $e := s(\omega) \in M$ and $e_0 := s(\omega_0) \in N$. The following two simple lemmas are included in this section and the proof of Theorem 8.4 will be given in the next section.

Lemma 8.5 *We have* $\gamma_\omega^*(T_{\psi/\omega}) = T_{\psi_0/\omega_0}$.

Proof By (8.5) we have

$$h_{\omega_0}^{1/2} \gamma_\omega^*(T_{\psi/\omega}) h_{\omega_0}^{1/2} = \gamma_*(h_\omega^{1/2} T_{\psi/\omega} h_\omega^{1/2}) = \gamma_*(h_\psi^{1/2} h_\psi^{1/2})$$
$$= \gamma_*(h_\psi) = h_{\psi_0} = h_{\omega_0}^{1/2} T_{\psi_0/\omega_0} h_{\omega_0}^{1/2}.$$

Hence, for every $y, z \in N$,

$$\big\langle h_{\omega_0}^{1/2} y, \big(\gamma_\omega^*(T_{\psi/\omega}) - T_{\psi_0/\omega_0}\big) h_{\omega_0}^{1/2} z\big\rangle = 0.$$

Since $\overline{h_{\omega_0}^{1/2} N} = e_0 L^2(\mathcal{H}_0)$ and $\gamma_\omega^*(T_{\psi/\omega})$, $T_{\psi_0/\omega_0} \in e_0 N e_0$, the assertion follows.

Lemma 8.6 *For every operator convex function* f *on* $[0, \infty)$*,*

$$P_f(\psi \circ \gamma, \omega \circ \gamma) \le P_f(\psi, \omega) \circ \gamma.$$

Proof By definition (8.1) and Lemma 8.5, as well as (8.5) applied to ω, we have

$$h_{P_f(\psi_0,\omega_0)} = h_{\varphi_0}^{1/2} f(T_{\psi_0/\omega_0}) h_{\omega_0}^{1/2} = h_{\omega_0}^{1/2} f(\gamma_\omega^*(T_{\psi/\omega})) h_{\omega_0}^{1/2}$$
$$\le h_{\omega_0}^{1/2} \gamma_\omega^*(f(T_{\psi/\omega})) h_{\omega_0}^{1/2} = \gamma_*\big(h_\omega^{1/2} f(T_{\psi/\omega}) h_\omega^{1/2}\big)$$
$$= \gamma_*(h_{P_f(\psi,\omega)}) = h_{P_f(\psi,\omega)\circ\gamma},$$

where the inequality above is due to the Jensen inequality for operator convex functions (see [25, Theorem 2.1]). Hence we have the asserted inequality.

8.2 Proof of the Theorem

We divide the proof of Theorem 8.4 into two parts. The integral expression of an operator convex function on (0, 1) given in Appendix C will be useful in Part 2.

Proof (Part 1) We prove that (iii)–(v) and (viii)–(xii) are equivalent, and they are equivalent to (xiii) when γ is 2-positive.

In view of (4.5), it is obvious that (iii) $\Longrightarrow$ (v) $\Longrightarrow$ (iv). That (x) $\Longrightarrow$ (viii) $\Longrightarrow$ (ix) is trivial. By (8.4) (Proposition D.10) we have (x) $\Longrightarrow$ (iii) and (ix) $\Longrightarrow$ (iv).

(xi) $\Longleftrightarrow$ (v). Note that

$$\begin{aligned} h_{\psi_0}^{1/2} A_0 A_0^* h_{\psi_0}^{1/2} &= h_{\omega_0}^{1/2} (A_0^* A_0)^2 h_{\omega_0}^{1/2} = h_{P_{t^2}(\psi_0,\omega_0)} \\ &\le h_{P_{t^2}(\psi,\omega)\circ\gamma} \qquad \text{(by Lemma 8.6)} \\ &= \gamma_*\big(h_\omega^{1/2}(A^*A)^2 h_\omega^{1/2}\big) = \gamma_*\big(h_\psi^{1/2} AA^* h_\psi^{1/2}\big). \end{aligned}$$

Hence (xi) is equivalent to

$$\operatorname{tr} h_{\psi_0}^{1/2} A_0 A_0^* h_{\psi_0}^{1/2} = \operatorname{tr} h_\psi^{1/2} AA^* h_\psi^{1/2},$$

or equivalently, $\operatorname{tr} h_{\omega_0}^{1/2}(A_0^*A_0)^2 h_{\omega_0}^{1/2} = \operatorname{tr} h_\omega^{1/2}(A^*A)^2 h_\omega^{1/2}$, which is (v) by (4.5) and Proposition 4.5 (6).

(xi) $\Longleftrightarrow$ (xii). It follows from (8.5) that

$$h_{\omega_0}^{1/2} \gamma_\omega^*((A^*A)^2) h_{\omega_0}^{1/2} = \gamma_*\big(h_\omega^{1/2}(A^*A)^2 h_\omega^{1/2}\big) = \gamma_*\big(h_\psi^{1/2} AA^* h_\psi^{1/2}\big).$$

It also follows from Lemma 8.5 that

$$h_{\omega_0}^{1/2} (\gamma_\omega^*(A^*A))^2 h_{\omega_0}^{1/2} = h_{\omega_0}^{1/2} (A_0^* A_0)^2 h_{\omega_0}^{1/2} = h_{\psi_0}^{1/2} A_0 A_0^* h_{\psi_0}^{1/2}.$$

Hence (xi) is equivalent to $h_{\omega_0}^{1/2}\gamma_\omega^*((A^*A)^2)h_{\omega_0}^{1/2} = h_{\omega_0}^{1/2}(\gamma_\omega^*(A^*A))^2 h_{\omega_0}^{1/2}$, which implies (xii) as in the proof of Lemma 8.5. The converse is obvious.

(xii) $\Longrightarrow$ (x). Let $\widehat{\gamma_\omega^*}$ be the restriction of γ_ω^* to the commutative von Neumann subalgebra of M generated by $T_{\psi/\omega}$ and 1. Then condition (xii) implies that $T_{\psi/\omega}$ is in the multiplicative domain $\mathcal{M}_{\widehat{\gamma_\omega^*}}$ of $\widehat{\gamma_\omega^*}$ (see Definition 6.9). Then, since the restriction of $\widehat{\gamma_\omega^*}$ to $\mathcal{M}_{\widehat{\gamma_\omega^*}}$ is a $*$-homomorphism by Proposition 6.10, we have, for any continuous function ϕ on $[0, \infty)$,

$$\phi(\gamma_\omega^*(T_{\psi/\omega})) = \phi(\widehat{\gamma_\omega^*}(T_{\psi/\omega})) = \widehat{\gamma_\omega^*}(\phi(T_{\psi/\omega})) = \gamma_\omega^*(\phi(T_{\psi/\omega})),$$

which implies by Lemma 8.5 that $\phi(T_{\psi_0/\omega_0}) = \gamma_\omega^*(\phi(T_{\psi/\omega}))$. Multiplying $h_{\omega_0}^{1/2}$ from both sides gives

$$h_{\omega_0}^{1/2}\phi(T_{\psi_0/\omega_0})h_{\omega_0}^{1/2} = h_{\omega_0}^{1/2}\gamma_\omega^*(\phi(T_{\psi/\omega}))h_{\omega_0}^{1/2} = \gamma_*(h_\omega^{1/2}\phi(T_{\psi/\omega})h_\omega^{1/2})$$

by Lemma 8.3. This equality is (x).

(iv) $\Longrightarrow$ (v). Recall [63, Theorem 8.1] that an operator convex function f on $[0,\infty)$ has the integral expression

$$f(t) = f(0^+) + at + bt^2 + \int_{(0,\infty)} \left(\frac{t}{1+s} - \frac{t}{t+s}\right) d\mu(s), \qquad t \in (0,\infty),$$

where $a \in \mathbb{R}$, $b \ge 0$ and μ is a positive measure on $(0,\infty)$ with $\int_{(0,\infty)}(1+s)^{-2}\,d\mu(s) < +\infty$. With $f_s(t) := -t/(t+s)$ ($t \in [0,\infty)$) for $s > 0$, one has

$$\widehat{S}_f(\psi\|\omega) = f(0^+)\omega(1) + a\psi(1) + bS_{t^2}(\psi\|\omega) + \int_{(0,\infty)} \left(\frac{\psi(1)}{1+s} + \widehat{S}_{f_s}(\psi\|\omega)\right) d\mu(s), \tag{8.6}$$

$$\begin{aligned}\widehat{S}_f(\psi_0\|\omega_0) &= f(0^+)\omega_0(1) + a\psi_0(1) + bS_{t^2}(\psi_0\|\omega_0) \\ &\quad + \int_{(0,\infty)} \left(\frac{\psi_0(1)}{1+s} + \widehat{S}_{f_s}(\psi_0\|\omega_0)\right) d\mu(s) \\ &= f(0^+)\omega(1) + a\psi(1) + bS_{t^2}(\psi_0\|\omega_0) \\ &\quad + \int_{(0,\infty)} \left(\frac{\psi(1)}{1+s} + \widehat{S}_{f_s}(\psi_0\|\omega_0)\right) d\mu(s).\end{aligned} \tag{8.7}$$

By comparing (8.6) and (8.7), in view of the monotonicity in Theorem 4.4 (i), one has

$$S_{t^2}(\psi_0\|\omega_0) = S_{t^2}(\psi\|\omega) \quad \text{if } b > 0,$$
$$\widehat{S}_{f_s}(\psi_0\|\omega_0) = \widehat{S}_{f_s}(\psi\|\omega) \quad \text{for } \mu\text{-a.e. } s > 0.$$

Since f is nonlinear, $b > 0$ or $\mu \ne 0$. So it suffices to show that (v) holds if $\widehat{S}_{f_s}(\psi_0\|\omega_0) = \widehat{S}_{f_s}(\psi\|\omega)$ for some $s \in (0,\infty)$. Since $f_s(A^*A) = -1 + s(A^*A + s1)^{-1}$ and similarly for $f_s(A_0^*A_0)$, the assumption means that for some $s \in (0,\infty)$,

$$\operatorname{tr} h_{\omega_0}^{1/2}(A_0^*A_0 + se_0)^{-1}h_{\omega_0}^{1/2} = \operatorname{tr} h_\omega^{1/2}(A^*A + se)^{-1}h_\omega^{1/2}, \tag{8.8}$$

where $(A^*A + se)^{-1}$ is the inverse in eMe and $(A_0^*A_0 + se_0)^{-1}$ is in e_0Ne_0. We have $h_\omega^{1/2} = Bh_{\psi+s\omega}^{1/2}$ with an invertible $B \in eMe$ and $h_{\omega_0}^{1/2} = B_0h_{\psi_0+s\omega_0}^{1/2}$ with an invertible $B_0 \in e_0Ne_0$. Note that

$$h_{\psi+s\omega}^{1/2} = B^{-1}h_\omega^{1/2} = h_\omega^{1/2}B^{*-1}, \quad h_{\psi_0+s\omega_0}^{1/2} = B_0^{-1}h_{\omega_0}^{1/2} = h_{\omega_0}^{1/2}B_0^{*-1}. \tag{8.9}$$

Since

$$h_\omega^{1/2}(A^*A + se)h_\omega^{1/2} = h_\psi + sh_\omega = h_{\psi+s\omega} = h_\omega^{1/2}(BB^*)^{-1}h_\omega^{1/2},$$

we have $A^*A + se = (BB^*)^{-1}$. Similarly, $A_0^*A_0 + se_0 = (B_0B_0^*)^{-1}$. Hence (8.8) is rewritten as

$$\operatorname{tr} h_{\omega_0}^{1/2}B_0B_0^*h_{\omega_0}^{1/2} = \operatorname{tr} h_\omega^{1/2}BB^*h_\omega^{1/2},$$

that is,

$$\operatorname{tr} h_{\psi_0+s\omega_0}^{1/2}(B_0^*B_0)^2h_{\psi_0+s\omega_0}^{1/2} = \operatorname{tr} h_{\psi+s\omega}^{1/2}(B^*B)^2h_{\psi+s\omega}^{1/2},$$

which means that (v) holds for ω, $\psi + s\omega$ in place of ψ, ω. Here note that the above proofs of (v) $\iff$ (xi) $\iff$ (xii) $\implies$ (x) $\implies$ (iii) have been carried out in the setting of arbitrary $(\psi, \omega) \in (M_*^+ \times M_*^+)_\le$. So we find that (iii) holds for ω, $\psi + s\omega$ in place of ψ, ω. Applying this to $f(t) = (t + \varepsilon)^{-1}$ for any $\varepsilon > 0$ gives

$$\operatorname{tr} h_{\psi_0+s\omega_0}^{1/2}(B_0^*B_0 + \varepsilon e_0)^{-1}h_{\psi_0+s\omega_0}^{1/2} = \operatorname{tr} h_{\psi+s\omega}^{1/2}(B^*B + \varepsilon e)^{-1}h_{\psi+s\omega}^{1/2}.$$

Letting $\varepsilon \searrow 0$ gives

$$\operatorname{tr} h_{\psi_0+s\omega_0}^{1/2}(B_0^*B_0)^{-1}h_{\psi_0+s\omega_0}^{1/2} = \operatorname{tr} h_{\psi+s\omega}^{1/2}(B^*B)^{-1}h_{\psi+s\omega}^{1/2}.$$

Therefore, from (8.9) we have

$$\operatorname{tr} h_{\omega_0}^{1/2}(B_0B_0^*)^{-2}h_{\omega_0}^{1/2} = \operatorname{tr} h_\omega^{1/2}(BB^*)^{-2}h_\omega^{1/2},$$

i.e.,

$$\operatorname{tr} h_{\omega_0}^{1/2}(A_0^*A_0 + se_0)^2h_{\omega_0}^{1/2} = \operatorname{tr} h_\omega^{1/2}(A^*A + se)^2h_\omega^{1/2},$$

which means by (4.5) that

$$S_{t^2}(\psi_0\|\omega_0) + 2s\psi_0(1) + s^2\omega_0(1) = S_{t^2}(\psi\|\omega) + 2s\psi(1) + s^2\omega(1),$$

so that (v) holds.

Finally, assume that γ is 2-positive. Then so is γ_ω^* by Proposition 6.6 (2), and (xii) $\Longleftrightarrow$ (xiii) is obvious by Definition 6.9.

Proof (Part 2) We prove that (i), (ii), (vi) and (vii) are equivalent to the conditions treated in Part 1.

(x) $\Longrightarrow$ (i). Let f be any operator convex function on $(0, \infty)$. Set $g(t) := (1-t)f\big(\frac{t}{1-t}\big)$ for $t \in [0, 1]$, where $g(0) := f(0^+)$ and $g(1) := f'(\infty)$ are in $(-\infty, +\infty]$. For each $n \in \mathbb{N}$ let $g_n(t) := g(t) \wedge n$, $t \in [0, 1]$. Then by (x) we have

$$h_{\omega_0}^{1/2} g_n(T_{\psi_0/\omega_0}) h_{\omega_0}^{1/2} = \gamma_*(h_\omega^{1/2} g_n(T_{\psi/\omega}) h_\omega^{1/2}).$$

Taking tr of both sides above yields that

$$\int_0^1 g_n(t)\, d\|E_{\psi_0/\omega_0}(t) h_{\omega_0}^{1/2}\|^2 = \int_0^1 g_n(t)\, d\|E_{\psi/\omega}(t) h_\omega^{1/2}\|^2,$$

where $T_{\psi/\omega} = \int_0^1 t\, dE_{\psi/\omega}(t)$ and $T_{\psi_0/\omega_0} = \int_0^1 t\, dE_{\psi_0/\omega_0}(t)$ are the spectral decompositions. By the monotone convergence theorem letting $n \to \infty$ gives

$$\int_0^1 g(t)\, d\|E_{\psi_0/\omega_0}(t) h_{\omega_0}^{1/2}\|^2 = \int_0^1 g(t)\, d\|E_{\psi/\omega}(t) h_\omega^{1/2}\|^2,$$

which means that $\widehat{S}_f(\psi_0\|\varphi_0) = \widehat{S}_f(\psi\|\varphi)$ by Theorem 4.8.

(i) $\Longrightarrow$ (vi). Let σ be any operator connection with the representing operator monotone function $k_\sigma \geq 0$ on $[0, \infty)$. Then by (i) for $f = -k_\sigma$ and by (8.4) (Proposition D.10) we have

$$\begin{aligned}(\varphi_0 \sigma \psi_0)(1_N) &= -\widehat{S}_{-k_\sigma}(\psi_0\|\varphi_0) = -\widehat{S}_{-k_\sigma}(\psi\|\varphi)\\ &= (\varphi\sigma\psi)(1_M) = ((\varphi\sigma\psi)\circ\gamma)(1_N).\end{aligned}$$

Since $\varphi_0\sigma\psi_0 \geq (\varphi\sigma\psi)\circ\gamma$ by Proposition D.9, (vi) follows.

(vi) $\Longrightarrow$ (vii) is trivial.

(vii) $\Longrightarrow$ (ii). Assume that (vii) holds for a nonlinear operator connection σ with the representation function of k_σ; then $f := -k_\sigma$ is a nonlinear operator convex function on $[0, \infty)$ and $f'(\infty) < +\infty$. Similarly to the proof of (i) $\Longrightarrow$ (vi) we have

$$-\widehat{S}_f(\psi_0\|\varphi_0) = (\varphi_0\sigma\psi_0)(1_N) = (\varphi\sigma\psi)(1_M) = -\widehat{S}_f(\psi\|\varphi).$$

(ii) $\Longrightarrow$ (v). Assume that (ii) holds with f as stated, and let $g(t) := (1-t)f\big(\frac{t}{1-t}\big)$ for $t \in [0, 1]$, where $g(0) := f(0^+)$ and $g(1) := f'(\infty) \in (-\infty, +\infty]$. Then by Theorem 4.8 we have

$$\int_0^1 g(t)\, d\|E_{\psi_0/\omega_0}(t) h_{\omega_0}^{1/2}\|^2 = \int_0^1 g(t)\, d\|E_{\psi/\omega}(t) h_\omega^{1/2}\|^2 \tag{8.10}$$

with the spectral measures $E_{\varphi/\omega}$ of $T_{\varphi/\omega}$ and E_{φ_0/ω_0} of T_{φ_0/ω_0}. Since

$$\lim_{t\to 0^+} tg(t) = \lim_{t\to 0^+} tf(t) = 0, \qquad \lim_{t\to 1^-} (1-t)g(t) = \lim_{t\to\infty} \frac{f(t)}{t^2} = 0,$$

it follows from Corollary C.3 that g has the integral expression

$$g(t) = \alpha + \beta t + \int_{(-\infty,1)} \frac{(2t-1)^2}{1-st}\, d\nu(s), \qquad t \in [0,1], \tag{8.11}$$

where $\alpha, \beta \in \mathbb{R}$ and ν is a positive measure on $(-\infty, 1)$ with $\int_{(-\infty,1)}(2-s)^{-1}\, d\nu(s) < +\infty$. We set

$$\phi_s(t) := \frac{(2t-1)^2}{1-st}, \qquad t \in [0,1],\ s \in (-\infty,1).$$

By Fubini's theorem the LHS of (8.10) is

$$\alpha\omega_0(1) + \beta\psi_0(1) + \int_{(-\infty,1)} \left[\int_0^1 \phi_s(t)\, d\|E_{\psi_0/\omega_0}(t)h_{\omega_0}^{1/2}\|^2\right] d\nu(s)$$
$$= \alpha\omega(1) + \beta\psi(1) + \int_{(-\infty,1)} \operatorname{tr} h_{\omega_0}^{1/2}\phi_s(T_{\psi_0/\omega_0})h_{\omega_0}^{1/2}\, d\nu(s)$$

and the RHS of (8.10) is

$$\alpha\omega(1) + \beta\psi(1) + \int_{(-\infty,1)} \operatorname{tr} h_\omega^{1/2}\phi_s(T_{\psi/\omega})h_\omega^{1/2}\, d\nu(s).$$

Therefore, (8.10) implies that

$$\int_{(-\infty,1)} \operatorname{tr} h_{\omega_0}^{1/2}\phi_s(T_{\psi_0/\omega_0})h_{\omega_0}^{1/2}\, d\nu(s) = \int_{(-\infty,1)} \operatorname{tr} h_\omega^{1/2}\phi_s(T_{\psi/\omega})h_\omega^{1/2}\, d\nu(s). \tag{8.12}$$

For every $s \in (-\infty, 1)$, since ϕ_s is operator convex on $[0, 1]$, note by Lemma 8.5 that

$$h_{\omega_0}^{1/2}\phi_s(T_{\psi_0/\omega_0})h_{\omega_0}^{1/2} = h_{\omega_0}^{1/2}\phi_s(\gamma_\omega^*(T_{\psi/\omega}))h_{\omega_0}^{1/2}$$
$$\le h_{\omega_0}^{1/2}\gamma_\omega^*(\phi_s(T_{\psi/\omega}))h_{\omega_0}^{1/2} = \gamma_*(h_\omega^{1/2}\phi_s(T_{\psi/\omega})h_\omega^{1/2}),$$

where we have used [25, Theorem 2.1] for the inequality above and the last equality is due to (8.5). Therefore,

$$\operatorname{tr} h_{\omega_0}^{1/2}\phi_s(T_{\psi_0/\omega_0})h_{\omega_0}^{1/2}) \le \operatorname{tr} h_\omega^{1/2}\phi_s(T_{\psi/\omega})h_\omega^{1/2}), \qquad s \in (-\infty,1). \tag{8.13}$$

Since f is nonlinear and so is g, note that $\nu((-\infty,1)) > 0$ in expression (8.11). Hence by (8.12) and (8.13) we find that there exists an $s \in (-\infty,1)$ such that

$$\operatorname{tr} h_{\omega_0}^{1/2}\phi_s(T_{\psi_0/\omega_0})h_{\omega_0}^{1/2} = \operatorname{tr} h_\omega^{1/2}\phi_s(T_{\psi/\omega})h_\omega^{1/2}. \tag{8.14}$$

When $s = 0$, since $\phi_0(t) = (2t-1)^2$, equality (8.14) implies that $\operatorname{tr} h_{\omega_0}^{1/2}T_{\psi_0/\omega_0}^2 h_{\omega_0}^{1/2} = \operatorname{tr} h_\omega^{1/2}T_{\psi/\omega}^2 h_\omega^{1/2}$ so that $\widehat{S}_{t^2}(\psi_0\|\omega_0) = \widehat{S}_{t^2}(\psi\|\omega)$, i.e., (v) holds.

Next, when $s \neq 0$, note that

$$\phi_s(t) = \frac{\left(1-\frac{2}{s}+\frac{2}{s}(1-st)\right)^2}{1-st} = \left(1-\frac{2}{s}\right)^2\frac{1}{1-st} - \frac{4(1-s)}{s^2} - \frac{4}{s}t.$$

Hence, (8.14) implies that

$$\operatorname{tr} h_{\omega_0}^{1/2}(e_0 - sA_0^*A_0)^{-1}h_{\omega_0}^{1/2} = \operatorname{tr} h_\omega^{1/2}(e - sA^*A)^{-1}h_\omega^{1/2}. \tag{8.15}$$

Now, since $\omega \sim (1-s)\psi + \varphi$, by Lemma A.24 choose invertible $B \in eMe$ and $B_0 \in e_0Ne_0$ such that

$$h_\omega^{1/2} = Bh_{(1-s)\psi+\varphi}^{1/2}, \qquad h_{\omega_0}^{1/2} = B_0h_{(1-s)\psi_0+\varphi_0}^{1/2}.$$

We then have

$$h_\omega^{1/2}(e - sA^*A)h_\omega^{1/2} = h_\omega - sh_\psi = h_{(1-s)\psi+\varphi} = h_\omega^{1/2}(BB^*)^{-1}h_\omega^{1/2}$$

thanks to $h_{(1-s)\psi+\varphi}^{1/2} = B^{-1}h_\omega^{1/2} = h_\omega^{1/2}B^{*-1}$. Therefore, we have $e - sA^*A = (BB^*)^{-1}$ so that $(e - sA^*A)^{-1} = BB^*$, and similarly $e_0 - sA_0^*A_0 = (B_0B_0^*)^{-1}$ so that $(e_0 - sA_0^*A_0)^{-1} = B_0B_0^*$. Thus, (8.15) is rewritten as

$$\operatorname{tr} h_{\omega_0}^{1/2}B_0B_0^*h_{\omega_0}^{1/2} = \operatorname{tr} h_\omega^{1/2}BB^*h_\omega^{1/2},$$

that is,

$$\operatorname{tr} h_{(1-s)\psi_0+\varphi_0}^{1/2}(B_0^*B_0)^2h_{(1-s)\psi_0+\varphi_0}^{1/2} = \operatorname{tr} h_{(1-s)\psi+\varphi}^{1/2}(B^*B)^2h_{(1-s)\psi+\varphi}^{1/2},$$

which means that (v) holds for ω, $(1-s)\psi+\varphi$ in place of ψ,ω. Furthermore, similarly to an argument in the proof of (iv) $\Longrightarrow$ (v) of Part 1 based on (v) $\Longrightarrow$ (iii), we find that

$$\operatorname{tr} h_{\omega_0}^{1/2}(B_0B_0^*)^{-2}h_{\omega_0}^{1/2} = \operatorname{tr} h_\omega^{1/2}(BB^*)^{-2}h_\omega^{1/2}.$$

Since $(BB^*)^{-2} = (e - sA^*A)^2$ and $(B_0B_0^*)^{-2} = (e_0 - sA_0^*A_0)^2$, we have

$$\omega_0(1) - 2s\psi_0(1) + s^2 S_{t^2}(\psi_0\|\omega_0) = \omega(1) - 2s\psi(1) + s^2 S_{t^2}(\psi\|\omega),$$

so that (v) holds.

Remarks 8.7

(1) When $(\psi, \varphi) \in (M_*^+ \times M_*^+)_\leq$, we can replace ω with φ in conditions (v) and (x)–(xiii) of Theorem 8.4. Indeed, all the proofs of Part 1 are valid for ψ, φ themselves instead of replacing φ with $\omega = \psi + \varphi$.

(2) As remarked in Remark 6.21, even in the finite-dimensional case, the preservation of S_{t^2} $(= \widehat{S}_{t^2})$ under a CP map γ does not imply the reversibility of γ. Therefore, by Theorem 8.4 (and (1) above) we see that a CP map γ is not necessarily reversible for $(\psi, \varphi) \in (M_*^+ \times M_*^+)_\leq$ even if $\widehat{S}_f(\psi \circ \gamma\|\varphi \circ \gamma) = \widehat{S}_f(\psi\|\varphi)$ holds for all operator convex functions f on $(0, \infty)$. Thus, the preservation of $\widehat{S}_f$ is strictly weaker than that of S_f.

(3) Assume that the domain N of γ is abelian (i.e., γ is a quantum-classical channel). By Theorem 4.18 and Proposition 4.21, condition (iii) implies that $S_f(\psi \circ \gamma\|(\psi+\varphi)\circ\gamma) = S_f(\psi\|\psi+\varphi)$ for any operator convex function f on $[0, \infty)$, which implies by Theorem 6.19 that γ is reversible for $\{\psi, \psi + \varphi\}$. Hence from Theorem 7.11 it follows that ψ, $\psi + \varphi$ commute and so ψ, φ commute. Hence, in this case, the conditions of Theorem 8.4 are also equivalent to the commutativity of ψ, φ. Here note that ψ, φ are arbitrary without the assumption $s(\psi) \leq s(\varphi)$, while it is assumed in Theorem 7.11.

Problem 8.8 It is desirable to remove the assumption $\lim_{t\to 0^+} tf(t) = 0$ and $\lim_{t\to\infty} f(t)/t^2 = 0$ on f in condition (ii) of Theorem 8.4. To do this, we need to take care of only the case where $f(t) = t^2$ and $f(t) = t^{-1}$ (see the proof of (ii) $\Longrightarrow$ (v) of Part 2). Since t^{-1} is the transpose of t^2, the question is whether or not $S_{t^2}(\psi \circ \gamma\|\varphi \circ \gamma) = S_{t^2}(\psi\|\varphi) < +\infty$ implies (v). In view of Remark 8.7 (1), this holds true when $(\psi, \varphi) \in (M_*^+ \times M_*^+)_\leq$. By Proposition 4.5 (4) note that $S_{t^2}(\psi\|\varphi) < +\infty$ implies $s(\psi) \leq s(\varphi)$. Therefore, the question is affirmative in the finite-dimensional case, because $s(\psi) \leq s(\varphi) \iff (\psi, \varphi) \in (M_*^+ \times M_*^+)_\leq$ in that case.

Appendix A
Preliminaries on von Neumann Algebras

A.1 Introduction of von Neumann Algebras

Appendix A[1] consists of eight sections, in which we give concise accounts of selected topics of von Neumann algebras used in the main body of this monograph.

The first section is a very brief introduction of von Neumann algebras. The set $B(\mathcal{H})$ of all bounded linear operators on a Hilbert space $\mathcal{H}$ with the inner product $\langle\cdot,\cdot\rangle$ is a vector space with the operator sum $a+b$ and the scalar multiplication λa ($a,b \in B(\mathcal{H})$, $\lambda \in \mathbb{C}$) and is a Banach space with the operator norm

$$\|a\| := \sup\{\|a\xi\| : \xi \in H,\ \|\xi\| \le 1\}.$$

Moreover, $B(\mathcal{H})$ becomes a Banach $*$-algebra with the operator product ab and the adjoint operation $a \mapsto a^*$. A subspace of $B(\mathcal{H})$ is called an algebra if it is closed under the product, and a $*$-subalgebra if it is further closed under the $*$-operation. In general, an *operator algebra* means a $*$-subalgebra of $B(\mathcal{H})$.

The (operator) norm topology, the strong operator topology and the weak operator topology are defined on $B(\mathcal{H})$, which are weaker in this order. For a net $\{a_\alpha\}$ and a in $B(\mathcal{H})$, we say that $\{a_\alpha\}$ converges to a in the strong operator topology (or $a_\alpha \to a$ *strongly*) if $\|(a_\alpha - a)\xi\| \to 0$ for all $\xi \in \mathcal{H}$, and $\{a_\alpha\}$ converges to a in the weak operator topology (or $a_\alpha \to a$ *weakly*) if $\langle\xi,(a_\alpha - a)\eta\rangle \to 0$ for all $\xi,\eta \in \mathcal{H}$. Moreover, the convergence $a_\alpha \to a$ in the strong* operator topology (or *strongly**) means that $\|(a_\alpha - a)\xi\| \to 0$ and $\|(a_\alpha - a)^*\xi\| \to 0$ for all $\xi \in \mathcal{H}$. Since $B(\mathcal{H})$ is the dual Banach space of the Banach space $C_1(\mathcal{H})$ of trace-class operators with the trace-norm, the weak topology $\sigma(B(\mathcal{H}),C_1(\mathcal{H}))$ is also defined

[1] The contents of Appendix A are extracted from the author's lecture notes [56], a corrected and enlarged version of the manuscript for an intensive course (2019) at Budapest University of Technology and Economics. The proofs and more details omitted in this appendix are found in [56].

F. Hiai, *Quantum f-Divergences in von Neumann Algebras*,
Mathematical Physics Studies, https://doi.org/10.1007/978-981-33-4199-9

on $B(\mathcal{H})$, which is called the *σ-weak topology* and is particularly important in the study of von Neumann algebras. In fact, a net $\{a_\alpha\}$ in $B(\mathcal{H})$ converges to $a \in B(\mathcal{H})$ in the σ-weak topology (we write $a_\alpha \to a$ *σ-weakly*) if $\sum_{n=1}^{\infty} \langle \xi_n (a_\alpha - a)\eta_n \rangle \to 0$ for all $\xi_n, \eta_n \in \mathcal{H}$ $(n \in \mathbb{N})$ with $\sum_n \|\xi_n\|^2 < +\infty$, $\sum_n \|\eta_n\|^2 < +\infty$. We write $B(\mathcal{H})_{\rm sa}$ for the set of all self-adjoint (i.e., $a = a^*$) $a \in B(\mathcal{H})$. The order $a \le b$ for $a, b \in B(\mathcal{H})_{\rm sa}$ means that $\langle \xi, a\xi \rangle \le \langle \xi, b\xi \rangle$ for all $\xi \in \mathcal{H}$, which is a partial order on $B(\mathcal{H})_{\rm sa}$. When a net $\{a_\alpha\}$ in $B(\mathcal{H})_{\rm sa}$ is increasing and bounded above, it has the supremum $a \in B(\mathcal{H})_{\rm sa}$ and $a_\alpha \to a$ strongly; in this case, we write $a_\alpha \nearrow a$.

A ∗-subalgebra of $B(\mathcal{H})$ is called a *von Neumann algebra* (also *W^*-algebra*) if it contains the identity operator 1 and is closed in the weak operator topology. For $S \subset B(\mathcal{H})$, define the *commutant* S' of S by

$$S' := \{a \in B(H) : ab = ba \text{ for all } b \in S\},$$

and also $S'' := (S')'$. For a ∗-subalgebra M of $B(\mathcal{H})$ with $1 \in M$, the *double commutation theorem* or *von Neumann's density theorem* says that the following three conditions are equivalent:

- M is a von Neumann algebra (i.e., closed in the weak operator topology);
- M is closed in the strong (equivalently strong*) operator topology;
- $M'' = M$.

From this, the polar decompositions and the spectral decompositions of operators in a von Neumann algebra M are taken inside M itself, so M contains plenty of projections. Starting from the above theorem (1929), J. von Neumann developed basics, including the classification, as we will explain shortly, of the von Neumann algebra theory in a series of joint papers with F.J. Murray.

C^*-algebras are another major subject of operator algebras, which are faithfully represented as norm-closed subalgebras of $B(\mathcal{H})$. Sakai [116] gave the abstract characterization of von Neumann algebras in such a way that a C^*-algebra is isomorphic to a von Neumann algebra if and only if it is the dual Banach space of some Banach space. In this case, the predual space is unique in a strong sense. The term W^*-algebra is often used to stress this abstract (or Hilbert space-free) situation. Although von Neumann algebras are special C^*-algebras, both categories of operator algebras are quite different theoretically and methodologically.

Let M_* be the set of all σ-weakly continuous linear functionals on a von Neumann algebra M, which is a Banach space as a closed subspace of the dual Banach space M^*. Then the dual Banach space of M_* is isometric to M, so M_* is identified with the *predual* of M. A positive linear functional φ on M is in M_* if and only if φ is *normal* (i.e., $a_\alpha \nearrow a \implies \varphi(a_\alpha) \nearrow \varphi(a)$). Let M, N be von Neumann algebras. A ∗-homomorphism $\pi : M \to N$ is normal (i.e., $a_\alpha \nearrow a$ in M implies $\pi(a_\alpha) \nearrow \pi(a)$) if and only if it is continuous with respect to the σ-weak topologies on M, N. A ∗-homomorphism $\pi : M \to B(\mathcal{K})$ ($\mathcal{K}$ is a Hilbert space) is called a *∗-representation* (or simply representation) of M. When π is a normal representation of M, the range $\pi(M)$ is a von Neumann algebra and the kernel of π is represented as $M(1 - e)$ for some central projection e (i.e., a projection in the *center* $Z(M) := M \cap M'$), so π induces a $*$-isomorphism between Me and $\pi(M)$. Note that a faithful representation is normal automatically.

The notion of *Murray–von Neumann equivalence* on the set $\mathrm{Proj}(M)$ of all projections in a von Neumann algebra M is important: $e, f \in \mathrm{Proj}(M)$ is said to be equivalent ($e \sim f$) if there is a $v \in M$ such that $v^*v = e$ and $vv^* = f$. A projection $e \in \mathrm{Proj}(M)$ is called an abelian projection if eMe is an abelian algebra, and a finite projection if, for $f \in \mathrm{Proj}(M)$, $e \sim f \leq e \implies f = e$.

The von Neumann algebra M is said to be *finite* if 1 is a finite projection, and *semifinite* if, for every central projection $e \neq 0$, there is a finite projection $f \in \mathrm{Proj}(M)$ such that $0 \neq f \leq e$. If M has no finite central projection ($\neq 0$), then M is said to be *properly infinite*. If M has no finite projection ($\neq 0$), then M is said to be *purely infinite* or *of type III*. A von Neumann algebra M is properly infinite if and only if M is isomorphic to the tensor product $M \otimes B(\mathcal{H})$ for separable Hilbert spaces $\mathcal{H}$, and M is semifinite if and only if M has a faithful normal and semifinite trace.

A von Neumann algebra M is called a *factor* if the center is trivial (i.e., $Z(M) = \mathbb{C}1$). The factors are classified into one of the following types:

- type I_n ($n \in \mathbb{N}$) if M is isomorphic to the $n \times n$ matrix algebra $\mathbb{M}_n$ (type I_n for some $n \in \mathbb{N}$ if M is finite and has a finite abelian projection $\neq 0$),
- type I_∞ if M is isomorphic to $B(\mathcal{H})$ ($\dim \mathcal{H} = \infty$), or equivalently, M is properly infinite and has an abelian projection $\neq 0$,
- type II_1 if M is finite and has no abelian projection $\neq 0$,
- type II_∞ if M is semifinite and properly infinite, and has no abelian projection $\neq 0$,
- type III if M has no finite projection $\neq 0$.

A finite factor has a faithful finite normal trace (unique up to positive constants). A factor of type II_∞ is represented as (a factor of type II_1) $\otimes B(\mathcal{H})$. Corresponding to the types of factors, the quotient set $\mathrm{Proj}(M)/\sim$ is identified with one of the following: $\{0, 1, \dots, n\}$ for type I_n, $\{0, 1, \dots, \infty\}$ for type I_∞, $[0, 1]$ for type II_1, $[0, \infty]$ for type II_∞, and $\{0, \infty\}$ for type III. Von Neumann algebras of type I (i.e., a direct sum of factors of type I) are said to be *discrete* or *atomic*. Type III factors were further classified by Connes [27] as follows: type III_1, type III_λ ($0 < \lambda < 1$), and type III_0.

Assume that a von Neumann algebra M is on a separable Hilbert space, or equivalently, M has the separable predual M_*. If M is generated by an increasing sequence of finite-dimensional $*$-subalgebras, then M is said to be *hyperfinite* or *AFD* (*approximately finite dimensional*). The uniqueness of type II_1 AFD factors is an old result of Murray and von Neumann. A von Neumann algebra $M \subset B(\mathcal{H})$ is said to be *injective* if there exists a (not necessarily normal) norm one projection from $B(\mathcal{H})$ onto M. In 1976, Connes [29] proved that a von Neumann algebra is AFD if and only if it is injective, and at the same time, he proved that injective factors of types II_1, II_∞, III_λ ($0 < \lambda < 1$) are unique for each type. In this fundamental article on the subject, it was also proved that any injective factor of type III_0 is a *Krieger factor* (i.e., the crossed product of an abelian von Neumann algebra by an ergodic automorphism). Thus, the complete classification of injective factors of type III_0 was built upon Krieger's work [84] in 1976. Later in 1985, Haagerup [47] (also [48]) proved the uniqueness of injective factors of type III_1, thus completing the classification of AFD (= injective) factors. In other words, the conjugacy class of the *flow of weights* due to Connes and Takesaki [30] is a complete invariant for injective factors of type III.

We end the overview with a list [72, 102, 116, 123, 127, 128] of standard textbooks on von Neumann algebras.

A.2 Tomita–Takesaki Modular Theory

Let M be a von Neumann algebra. Throughout the rest of Appendix A, we assume[2] that there is a faithful $\omega \in M_*^+$, or equivalently, M is *σ-finite*, i.e., mutually orthogonal projections in M are at most countable. By making the *GNS (Gelfand–Naimark–Segal) cyclic representation* $\{\pi_\omega, \mathcal{H}_\omega, \Omega\}$ of M with respect to ω, we have a $*$-isomorphism $\pi_\omega : M \to B(\mathcal{H}_\omega)$ with a cyclic and separating vector $\Omega \in \mathcal{H}_\omega$ for $\pi_\omega(M)$ for which

$$\omega(x) = \langle \Omega, \pi_\omega(x)\Omega \rangle, \qquad x \in M.$$

Thus, by identifying M with $\pi_\omega(M)$, we may assume that M itself is a von Neumann algebra on $\mathcal{H}$ with a cyclic and separating vector Ω for M. Here, Ω is *cyclic* for M if $\overline{M\Omega} = \mathcal{H}$ (where $M\Omega := \{x\Omega : x \in M\}$), and Ω is *separating* for M if, for $x \in M$, $x\Omega = 0 \implies x = 0$, equivalently $\overline{M'\Omega} = \mathcal{H}$.

We begin to define two conjugate-linear operators S_0 and F_0 with the dense domains $\mathcal{D}(S_0) = M\Omega$ and $\mathcal{D}(F_0) = M'\Omega$ by

$$S_0 x\Omega := x^*\Omega, \qquad x \in M,$$
$$F_0 x'\Omega := x'^*\Omega, \qquad x' \in M'.$$

For any $x \in M$ and $x' \in M'$ note that

$$\langle x'\Omega, S_0 x\Omega \rangle = \langle x'\Omega, x^*\Omega \rangle = \langle x\Omega, x'^*\Omega \rangle = \langle x\Omega, F_0 x'\Omega \rangle,$$

which implies that S_0 and F_0 are closable, $\overline{F_0} \subset S_0^*$ and $\overline{S_0} \subset F_0^*$. So we set $S := \overline{S_0}$ and $F := S^* = S_0^*$, and take the polar decomposition of S as

$$S = J\Delta^{1/2}, \qquad \Delta := S^*S = FS.$$

[2]When M is not σ-finite, the construction of the modular theory is essentially similar to the σ-finite case (but technically more complicated) with a faithful normal semifinite weight on M based on the left Hilbert algebra theory.

Since the ranges of S and S^* are dense, it follows that J is a conjugate-linear unitary and Δ is a non-singular self-adjoint operator. We then have the following:

(1) $J = J^*$ and $J^2 = 1$.
(2) $\Delta = FS$ and $\Delta^{-1} = SF$.
(3) $S = J\Delta^{1/2} = \Delta^{-1/2}J$ and $F = J\Delta^{-1/2} = \Delta^{1/2}J$.
(4) $\Delta^{-1} = J\Delta J$ and $J\Delta^{it} = \Delta^{-it}J$ for all $t \in \mathbb{R}$.
(5) $J\Omega = \Omega$ and $\Delta\Omega = \Omega$.

The next theorem is *Tomita's fundamental theorem.*

Theorem A.1 (Tomita) *With J and Δ given above, we have*

$$JMJ = M', \tag{A.1}$$

$$\Delta^{it}M\Delta^{-it} = M, \qquad t \in \mathbb{R}. \tag{A.2}$$

Definition A.2 The operator Δ is called the *modular operator* with respect to Ω (or ω), and J is called the *modular conjugation* with respect to Ω (or ω). The one-parameter automorphism group $\sigma_t = \sigma_t^\omega$ of M defined by $\sigma_t^\omega(x) := \Delta^{it}x\Delta^{-it}$ ($x \in M, t \in \mathbb{R}$) is called the *modular automorphism group* with respect to Ω (or ω).

Proposition A.3 *If $x \in M \cap M'$ (the center of M), then*

$$JxJ = x^*, \qquad \sigma_t(x) = \Delta^{it}x\Delta^{-it} = x, \qquad t \in \mathbb{R}.$$

Example A.4 Consider the simple case where $\omega = \tau$ is a faithful normal finite trace of M, so M is a finite von Neumann algebra of type II_1 (or M is finite dimensional). Since $\|x\Omega\|^2 = \tau(x^*x) = \tau(xx^*) = \|x^*\Omega\|^2$ for all $x \in M$, S is a conjugate-linear unitary, which means that $S = J$ and $\Delta = 1$. Hence (A.2) trivially holds. For every $x, y, y_1 \in M$ note that

$$JxJyy_1\Omega = Jxy_1^*y^*\Omega = yy_1x^*\Omega = yJxy_1^*\Omega = yJxJy_1\Omega$$

so that $JxJy = yJxJ$. Hence $JMJ \subset M'$. Moreover, for every $x \in M$ and $x' \in M'$ note that

$$\langle x\Omega, Jx'\Omega\rangle = \langle x'\Omega, Jx\Omega\rangle = \langle x'\Omega, x^*\Omega\rangle = \langle x\Omega, x'^*\Omega\rangle$$

so that $Jx'\Omega = x'^*\Omega$. Hence, similarly to the above, $JM'J \subset M$, so $M' \subset JMJ$. Therefore, (A.1) holds. In this way, Tomita's theorem in this case is quite easy.

Let α_t ($t \in \mathbb{R}$) be a one-parameter weakly continuous (hence automatically strongly* continuous) automorphism group of M. For $\beta \in \mathbb{R}$ with $\beta \neq 0$, if $\beta < 0$,

$$D_\beta := \{z \in \mathbb{C} : 0 < \operatorname{Im} z < -\beta\}, \qquad \overline{D}_\beta := \{z \in \mathbb{C} : 0 \le \operatorname{Im} z \le -\beta\},$$

and if $\beta > 0$, $D_\beta := \{z \in \mathbb{C} : -\beta < \operatorname{Im} z < 0\}$ and $\overline{D}_\beta$ is similar.

Definition A.5 A functional $\varphi \in M_*^+$ is said to satisfy the *KMS (Kubo–Martin–Schwinger) condition* with respect to α_t at β, or the (α_t, β)-KMS condition, if, for every $x, y \in M$, there is a bounded continuous function $f_{x,y}(z)$ on $\overline{D}_\beta$, analytic in D_β, such that

$$f_{x,y}(t) = \varphi(\alpha_t(x)y), \qquad f_{x,y}(t - i\beta) = \varphi(y\alpha_t(x)), \qquad t \in \mathbb{R}.$$

This condition proposed by Haag, Hugenholtz and Winnink serves as a mathematical formulation of equilibrium states in the quantum statistical mechanics, which is also defined and more useful in C^*-algebraic dynamical systems, see [23]. To illustrate that, given a Hamiltonian $H \in B(\mathcal{H})_{\mathrm{sa}}$ in a finite-dimensional $\mathcal{H}$, consider the Gibbs state $\varphi(x) = \mathrm{Tr}\, e^{-\beta H}x/\mathrm{Tr}\, e^{-\beta H}$ and the corresponding dynamics $\alpha_t(x) = e^{itH}xe^{-itH}$, $t \in \mathbb{R}$. (The physical meaning of β is the inverse temperature.) For any $x, y \in B(\mathcal{H})$ the entire function $f_{x,y}(z) := \mathrm{Tr}(e^{-\beta H}e^{izH}xe^{-izH}y)/\mathrm{Tr}\, e^{-\beta H}$ satisfies

$$\begin{aligned} f_{x,y}(t) &= \mathrm{Tr}(e^{-\beta H}e^{itH}xe^{-itH}y)/\mathrm{Tr}\, e^{-\beta H} = \varphi(\alpha_t(x)y), \\ f_{x,y}(t - i\beta) &= \mathrm{Tr}(e^{itH}xe^{-itH}e^{-\beta H}y)/\mathrm{Tr}\, e^{-\beta H} = \varphi(y\alpha_t(x)), \qquad t \in \mathbb{R}, \end{aligned}$$

so that φ satisfies the (α_t, β)-KMS condition. Moreover, the Gibbs state φ is a unique state satisfying the (α_t, β)-KMS condition.

The following proposition is another justification for the KMS condition to describe equilibrium states.

Proposition A.6 *If $\varphi \in M_*^+$ satisfies the (α_t, β)-KMS condition, then $\varphi \circ \alpha_t = \varphi$ for all $t \in \mathbb{R}$.*

The following is Takesaki's theorem in [124], an important ingredient of the Tomita–Takesaki theory in addition to Tomita's theorem.

Theorem A.7 (Takesaki) *In the same situation as in Theorem A.1, the ω satisfies the $(\sigma_t^\omega, -1)$-KMS condition. Furthermore, σ_t^ω is uniquely determined as a weakly continuous one-parameter automorphism group of M for which ω satisfies the KMS condition at $\beta = -1$.*

Definition A.8 The *centralizer* of ω is defined as

$$M_\omega := \{x \in M : \omega(xy) = \omega(yx), \ y \in M\}.$$

It is obvious that M_ω is a subalgebra of M including the center $M \cap M'$.

Proposition A.9 *The centralizer M_ω coincides with the fixed-point algebra of σ_t^ω, i.e.,*

$$M_\omega = \{x \in M : \sigma_t^\omega(x) = x, \ t \in \mathbb{R}\}.$$

Proposition A.9 contains the second assertion of Proposition A.3. Also, it follows that ω is a trace if and only if $\sigma_t^\omega = \mathrm{id}$ for all $t \in \mathbb{R}$.

The notion of conditional expectations in the von Neumann algebra setting was first introduced by Umegaki [132]. If M_0 is a von Neumann subalgebra of M and $E : M \to M_0$ is a norm one projection onto M_0, then we call E a *conditional expectation* onto M_0. Tomiyama [131] showed that even when M is a C^*-algebra and M_0 is a C^*-subalgebra of M, a norm one projection $E : M \to M_0$ satisfies (1) E is positive, (2) $E(y_1 x y_2) = y_1 E(x) y_2$ for all $x \in M$ and $y_1, y_2 \in N$, and (3) $E(x)^* E(x) \le E(x^* x)$ for all $x \in M$. In fact, it is well-known that E is completely positive, see the first paragraph of Sect. 6.1.

Takesaki's theorem [125] characterizes von Neumann subalgebras onto which the normal conditional expectation exists for a given faithful $\omega \in M_*^+$.

Theorem A.10 ([125]) *For a von Neumann subalgebra M_0 of M and a faithful $\omega \in M_*^+$, there exists a (faithful normal) conditional expectation $E : M \to M_0$ satisfying $\omega = \omega \circ E$ on M if and only if M_0 is globally invariant under σ_t^ω, i.e., $\sigma_t^\omega(M_0) = M_0$ for all $t \in \mathbb{R}$.*

Furthermore, in this case, such an E as above is unique and $\sigma_t^{\omega|M_0} = \sigma_t^\omega|_{M_0}$ for all $t \in \mathbb{R}$.

A.3 Standard Forms

The theory of standard forms of von Neumann algebras was developed, independently, by Araki [9] and Connes [28] in the case of σ-finite von Neumann algebras, and by Haagerup [42, 43] in the general case.

Let M be a σ-finite von Neumann algebra, thus represented on a Hilbert space $\mathcal{H}$ with a cyclic and separating vector Ω, for which we have the modular operator Δ and the modular conjugation J, as in Sect. A.2. Let $j : M \to M'$ be the conjugate-linear $*$-isomorphism defined by $j(x) := JxJ$, $x \in M$.

Definition A.11 The *natural positive cone* $\mathcal{P} = \mathcal{P}^\natural$ in $\mathcal{H}$ associated with (M, Ω) is defined by

$$\mathcal{P} := \overline{\{xj(x)\Omega : x \in M\}} = \overline{\{xJx\Omega : x \in M\}}. \tag{A.3}$$

Theorem A.12 *We have:*

(1) $\mathcal{P} = \overline{\Delta^{1/4} M_+ \Omega} = \overline{\Delta^{-1/4} M'_+ \Omega}$. *In particular, $\mathcal{P}$ is a closed cone.*
(2) $J\xi = \xi$ *for all* $\xi \in \mathcal{P}$.
(3) $\Delta^{it}\mathcal{P} = \mathcal{P}$ *for all* $t \in \mathbb{R}$.
(4) $xj(x)\mathcal{P} \subset \mathcal{P}$ *for all* $x \in M$.
(5) $\mathcal{P}$ *is self-dual, i.e.,* $\mathcal{P} = \{\eta \in \mathcal{H} : \langle \xi, \eta \rangle \ge 0,\ \xi \in \mathcal{P}\}$.

We thus conclude that any (σ-finite) von Neumann algebra is faithfully represented on a Hilbert space $\mathcal{H}$ with a conjugate-linear involution J and a self-dual cone $\mathcal{P}$ such that:

(a) $JMJ = M'$ (Theorem A.1),
(b) $JxJ = x^*$ for all $x \in M \cap M'$ (Proposition A.3),
(c) $J\xi = \xi$ for all $\xi \in \mathcal{P}$,
(d) $xj(x)\mathcal{P} \subset \mathcal{P}$ for all $x \in M$, where $j(x) := JxJ$.

Definition A.13 A quadruple $(M, \mathcal{H}, J, \mathcal{P})$ satisfying the above conditions (a)–(d) is called a *standard form* of a von Neumann algebra M. This is the abstract (or axiomatic) definition, and we have shown the existence of a standard form for any σ-finite von Neumann algebra.

Examples A.14

(1) Let $(X, \mathcal{X}, \mu)$ be a σ-finite (or more generally, localizable) measure space. The commutative von Neumann algebra $M = L^\infty(X, \mu)$ is faithfully represented on the Hilbert space $L^2(X, \mu)$ as multiplication operators $\pi(f)\xi := f\xi$ for $f \in L^\infty(X, \mu)$ and $\xi \in L^2(X, \mu)$. The standard form of M is

$$(L^\infty(X, \mu), L^2(X, \mu), J\xi = \overline{\xi}, L^2(X, \mu)_+),$$

where $L^2(X, \mu)_+$ is the cone of non-negative functions $\xi \in L^2(X, \mu)$.

(2) Let $M = B(\mathcal{H})$, a factor of type I. Let $C_2(\mathcal{H})$ be the space of Hilbert–Schmidt operators (i.e., $a \in B(\mathcal{H})$ with $\operatorname{Tr} a^*a < +\infty$), which is a Hilbert space with the Hilbert–Schmidt inner product $\langle a, b\rangle := \operatorname{Tr} a^*b$ for $a, b \in C_2(\mathcal{H})$. Then $M = B(\mathcal{H})$ is faithfully represented on $C_2(\mathcal{H})$ as left multiplication operators $\pi(x)a := xa$ for $x \in B(\mathcal{H})$ and $a \in C_2(\mathcal{H})$. The standard form of M is

$$(B(\mathcal{H}), C_2(\mathcal{H}), J = {}^*, C_2(\mathcal{H})_+),$$

where $J = {}^*$ is the adjoint operation and $C_2(\mathcal{H})_+$ is the cone of positive $a \in C_2(\mathcal{H})$. In this case, note that for any $x \in B(\mathcal{H})$, JxJ is the right multiplication of x^* on $C_2(\mathcal{H})$ and $xj(x)C_2(\mathcal{H})_+ = xC_2(\mathcal{H})_+x^* \subset C_2(\mathcal{H})_+$.

The following gives geometric properties of the cone $\mathcal{P}$.

Proposition A.15 *Let $(M, \mathcal{H}, J, \mathcal{P})$ be a standard form. Then:*

(1) *$\mathcal{P}$ is a pointed cone, i.e., $\mathcal{P} \cap (-\mathcal{P}) = \{0\}$.*
(2) *If $\xi \in \mathcal{H}$ and $J\xi = \xi$, then ξ has unique decomposition $\xi = \xi_1 - \xi_2$ with $\xi_1, \xi_2 \in \mathcal{P}$ and $\xi_1 \perp \xi_2$.*
(3) *$\mathcal{H}$ is linearly spanned by $\mathcal{P}$.*

The next proposition is the description of the standard form of a reduced von Neumann algebra eMe, where e is a projection in M.

Proposition A.16 *Let $(M, \mathcal{H}, J, \mathcal{P})$ be a standard form. Let $e \in M$ be a projection and set $q := ej(e)$. Then:*

(1) *$exe \mapsto qxq$ $(x \in M)$ is a $*$-isomorphism of eMe onto qMq. In particular, $e \neq 0 \iff q \neq 0$.*
(2) *$qJ = Jq$ and $(qMq, q\mathcal{H}, qJq, q\mathcal{P})$ is a standard form of $qMq \cong eMe$.*

The next proposition shows the universality of $(J, \mathcal{P})$ for the choice of a cyclic and separating vector ξ in $\mathcal{P}$.

Proposition A.17 *Let $(M, \mathcal{H}, J, \mathcal{P})$ be a standard form. If $\xi \in \mathcal{P}$ is cyclic and separating for M, then*

$$J_\xi = J, \qquad \mathcal{P}_\xi = \mathcal{P},$$

where J_ξ is the modular conjugation and $\mathcal{P}_\xi$ is the natural positive cone associated with (M, ξ).

For each $\xi \in \mathcal{H}$ we denote by ω_ξ $(\in M_*^+)$ the vector functional $x \mapsto \langle \xi, x\xi \rangle$ on M. The following is the most important property of the standard form, establishing the relation between $\mathcal{P}$ and M_*^+.

Theorem A.18 *Let $(M, \mathcal{H}, J, \mathcal{P})$ be a standard form. For every $\varphi \in M_*^+$ there exists a $\xi \in \mathcal{P}$ such that $\varphi(x) = \langle \xi, x\xi \rangle$ for all $x \in M$. Furthermore, the following estimates hold:*

$$\|\xi - \eta\|^2 \leq \|\omega_\xi - \omega_\eta\| \leq \|\xi - \eta\| \, \|\xi + \eta\|, \qquad \xi, \eta \in \mathcal{P}.$$

Consequently, the map $\xi \mapsto \omega_\xi$ is a homeomorphism from $\mathcal{P}$ onto M_*^+ when $\mathcal{P}$ and M_*^+ are equipped with the norm topology.

The vector $\xi \in \mathcal{P}$ such that $\omega_\xi = \varphi$ is called the *vector representative* of $\varphi \in M_*^+$.

Another essential property of the standard form is the universality (uniqueness) in the following strict sense.

Theorem A.19 *Let $(M, \mathcal{H}, J, \mathcal{P})$ and $(M_1, \mathcal{H}_1, J_1, \mathcal{P}_1)$ be standard forms of von Neumann algebras of M and M_1, respectively. If $\gamma : M \to M_1$ is a $*$-isomorphism, then there exists a unique unitary $U : \mathcal{H} \to \mathcal{H}_1$ such that:*

(1) *$\gamma(x) = UxU^*$ for all $x \in M$,*
(2) *$J_1 = UJU^*$,*
(3) *$\mathcal{P}_1 = U\mathcal{P}$.*

Example A.20 Let $(M, \mathcal{H}, J, \mathcal{P})$ be a standard form. We describe here the standard form of $M^{(2)} := M \otimes \mathbb{M}_2$, the tensor product of M with the 2×2 matrix algebra $\mathbb{M}_2(\mathbb{C})$. To do this, choose a cyclic and separating vector $\Omega \in \mathcal{P}$ and consider a faithful $\omega^{(2)} \in (M^{(2)})_*^+$ defined by $\omega^{(2)}\left(\begin{bmatrix} x_{11} & x_{12} \\ x_{21} & x_{22} \end{bmatrix}\right); = \omega_\Omega(x_{11}) + \omega_\Omega(x_{22})$. Then the GNS cyclic representation $\{\pi^{(2)}, \mathcal{H}^{(2)}, \Omega^{(2)}\}$ of $M^{(2)}$ with respect to $\omega^{(2)}$ is given as follows:

$$\mathcal{H}^{(2)} = \mathcal{H} \oplus \mathcal{H} \oplus \mathcal{H} \oplus \mathcal{H} = \left\{ \begin{bmatrix} \xi_{11} & \xi_{12} \\ \xi_{21} & \xi_{22} \end{bmatrix} : \xi_{ij} \in \mathcal{H} \right\}, \tag{A.4}$$

$$\Omega^{(2)} = \begin{bmatrix} \Omega & 0 \\ 0 & \Omega \end{bmatrix},$$

and $\pi^{(2)}\left(\begin{bmatrix} x_{11} & x_{12} \\ x_{21} & x_{22} \end{bmatrix}\right)$ acts like 2×2 matrix product as $\begin{bmatrix} x_{11} & x_{12} \\ x_{21} & x_{22} \end{bmatrix} \begin{bmatrix} \xi_{11} & \xi_{12} \\ \xi_{21} & \xi_{22} \end{bmatrix}$, whose 4×4 representation is

$$\begin{bmatrix} x_{11} & 0 & x_{12} & 0 \\ 0 & x_{11} & 0 & x_{12} \\ x_{21} & 0 & x_{22} & 0 \\ 0 & x_{21} & 0 & x_{22} \end{bmatrix} \begin{bmatrix} \xi_{11} \\ \xi_{12} \\ \xi_{21} \\ \xi_{22} \end{bmatrix} \quad \text{for} \quad \begin{bmatrix} \xi_{11} \\ \xi_{12} \\ \xi_{21} \\ \xi_{22} \end{bmatrix} \in \mathcal{H}^{(2)}. \tag{A.5}$$

Let $S^{(2)}$ and $\Delta^{(2)}$ be the operators for $(M^{(2)}, \Omega^{(2)})$ corresponding to S and Δ for (M, Ω) (see Sect. A.2). Since

$$S^{(2)}\left(\begin{bmatrix} x_{11} & x_{12} \\ x_{21} & x_{22} \end{bmatrix} \Omega^{(2)}\right) = \begin{bmatrix} x_{11}^*\Omega & x_{21}^*\Omega \\ x_{12}^*\Omega & x_{22}^*\Omega \end{bmatrix},$$

one can write

$$S^{(2)} = \begin{bmatrix} S & 0 & 0 & 0 \\ 0 & 0 & S & 0 \\ 0 & S & 0 & 0 \\ 0 & 0 & 0 & S \end{bmatrix} \quad \text{and} \quad \Delta^{(2)} = \begin{bmatrix} \Delta & 0 & 0 & 0 \\ 0 & \Delta & 0 & 0 \\ 0 & 0 & \Delta & 0 \\ 0 & 0 & 0 & \Delta \end{bmatrix}. \tag{A.6}$$

From this with $S = J\Delta^{1/2}$ one has the polar decomposition $S^{(2)} = J^{(2)}(\Delta^{(2)})^{1/2}$, where

$$J^{(2)} = \begin{bmatrix} J & 0 & 0 & 0 \\ 0 & 0 & J & 0 \\ 0 & J & 0 & 0 \\ 0 & 0 & 0 & J \end{bmatrix}. \tag{A.7}$$

Therefore, the standard form of $M^{(2)}$ is given as $(M^{(2)}, \mathcal{H}^{(2)}, J^{(2)}, \mathcal{P}^{(2)})$ with identifications (A.4), (A.5) and (A.7). Moreover, by (A.3), $\mathcal{P}^{(2)} = \overline{\{xJ^{(2)}x\Omega^{(2)} : x \in M^{(2)}\}}$. In particular, restricting $x \in M^{(2)}$ to $\begin{bmatrix} x_1 & 0 \\ 0 & x_2 \end{bmatrix}$ one has

$$\mathcal{P}^{(2)} \supset \left\{ \begin{bmatrix} \xi & 0 \\ 0 & \eta \end{bmatrix} : \xi, \eta \in \mathcal{P} \right\}.$$

A.4 Relative Modular Operators

The notion of relative modular operators was introduced by Araki [11] to extend the relative entropy to the general von Neumann algebra setting. Let $(M, \mathcal{H}, J, \mathcal{P})$ be a standard form. Let $\psi, \varphi \in M_*^+$, whose vector representatives are $\Psi, \Phi \in \mathcal{P}$, respectively. Define the operators $S^0_{\psi,\varphi}$ and $F^0_{\psi,\varphi}$ by

$$S^0_{\psi,\varphi}(x\Phi + \eta) := s_M(\varphi)x^*\Psi, \qquad x \in M,\ \eta \in (1 - s_{M'}(\varphi))\mathcal{H},$$

$$F^0_{\psi,\varphi}(x'\Phi + \zeta) := s_{M'}(\varphi)x'^*\Psi, \qquad x' \in M',\ \zeta \in (1 - s_M(\varphi))\mathcal{H},$$

where $s_M(\varphi)$ is the orthogonal projection onto $\overline{M'\Phi}$ (the M-support) and $s_{M'}(\varphi)$ is that onto $\overline{M\Phi}$ (the M'-support). Note that $Js_M(\varphi)J = s_{M'}(\varphi)$. In fact, since $J\Phi = \Phi$ and $JM'J = M$ (Theorem A.1),

$$Js_M(\varphi)J\mathcal{H} = Js_M(\varphi)\mathcal{H} = \overline{JM'\Phi} = \overline{JM'J\Phi} = \overline{M\Phi} = s_{M'}(\varphi)\mathcal{H}.$$

Note that $S^0_{\psi,\varphi}$ and $F^0_{\psi,\varphi}$ are well defined. In fact, assume that $x_1\Phi + \eta_1 = x_2\Phi + \eta_2$ for $x_i \in M$ and $\eta_i \in (1 - s_{M'}(\varphi))\mathcal{H}$. Then $(x_1 - x_2)\Phi = \eta_2 - \eta_1 = 0$, so $\eta_1 = \eta_2$ and $x_1 s_M(\varphi) = x_2 s_M(\varphi)$. Hence $s_M(\varphi)x_1^*\Psi = s_M(\varphi)x_2^*\Psi$, showing that $S^0_{\psi,\varphi}$ is well defined and similarly for $F^0_{\psi,\varphi}$. For every $x \in M$, $\eta \in (1 - s_{M'}(\varphi))\mathcal{H}$ and $x' \in M'$, $\zeta \in (1 - s_M(\varphi))\mathcal{H}$, one has

$$\begin{aligned}
\langle S^0_{\psi,\varphi}(x\Phi + \eta), x'\Phi + \zeta\rangle &= \langle s_M(\varphi)x^*\Psi, x'\Phi + \zeta\rangle = \langle x^*\Psi, x'\Phi\rangle \\
&= \langle x'^*\Psi, x\Phi\rangle = \langle s_{M'}(\varphi)x'^*\Psi, x\Phi + \eta\rangle \\
&= \langle F^0_{\psi,\varphi}(x'\Phi + \zeta), x\Phi + \eta\rangle.
\end{aligned}$$

Since $S^0_{\psi,\varphi}$ and $F^0_{\psi,\varphi}$ have the dense domains, we have their closures $S_{\psi,\varphi}$ and $F_{\psi,\varphi}$ satisfying

$$S^*_{\psi,\varphi} = F_{\psi,\varphi}, \qquad F^*_{\psi,\varphi} = S_{\psi,\varphi}. \tag{A.8}$$

Definition A.21 For every $\psi, \varphi \in M_*^+$ define $\Delta_{\psi,\varphi} := S^*_{\psi,\varphi}S_{\psi,\varphi}$, called the *relative modular operator* with respect to ψ, φ.

The notion of the relative modular operator plays a crucial role in this monograph, so we give the next proposition with proof.

Proposition A.22 *Let $\psi, \varphi, \psi_i, \varphi_i \in M_*^+$ $(i = 1, 2)$.*

(1) *The support projection of $\Delta_{\psi,\varphi} := S^*_{\psi,\varphi}S_{\psi,\varphi}$ is $s_M(\psi)s_{M'}(\varphi)$.*
(2) $S_{\psi,\varphi} = J\Delta^{1/2}_{\psi,\varphi}$ *(the polar decomposition).*

(3) $\Delta_{\varphi,\psi}^{-1} = J\Delta_{\psi,\varphi}J$, *where* $\Delta_{\varphi,\psi}^{-1}$ *is defined with restriction to the support of* $\Delta_{\varphi,\psi}$, *i.e., in the sense of the generalized inverse.*

(4) *If* $\psi_1 \le \psi_2$, *then* $\Delta_{\psi_1,\varphi} \le \Delta_{\psi_2,\varphi}$. *If* $s(\varphi_1) = s(\varphi_2)$ *and* $\varphi_1 \ge \varphi_2$, *then* $\Delta_{\psi,\varphi_1} \le \Delta_{\psi,\varphi_2}$.

(5) $\Delta_{\psi_1+\psi_2,\varphi} = \Delta_{\psi_1,\varphi} \dot{+} \Delta_{\psi_2,\varphi}$ *(the form sum, see around* (B.5) *of Appendix B).*

Proof

(1) By (A.8) the support projection of $S_{\psi,\varphi}^* S_{\psi,\varphi}$ is the orthogonal projection onto the closure of the range of $F_{\psi,\varphi}$, which is $\overline{s_{M'}(\varphi)M'\Psi} = s_{M'}(\varphi)s_M(\psi)\mathcal{H}$. Hence the assertion follows.

(2) First, prove the case $\psi = \varphi$. We write $S_\varphi := S_{\varphi,\varphi}$ and $\Delta_\varphi := \Delta_{\varphi,\varphi}$. Let $e := s_M(\varphi)$. $e' := s_{M'}(\varphi)$ and $q := ee' = eJeJ$. By Proposition A.16, $(qMq, q\mathcal{H}, qJq, q\mathcal{P})$ is a standard form of $qMq \cong eMe$. Define $\overline{\varphi} \in (qMq)_*^+$ by $\overline{\varphi}(qxq) := \varphi(exe)$ for $x \in M$, whose vector representative is $\Phi = q\Phi$. Note that Φ is cyclic and separating for qMq on $q\mathcal{H}$. Let $\Delta_{\overline{\varphi}}$ and $J_{\overline{\varphi}}$ be the modular operator and the modular conjugation with respect to $\overline{\varphi}$. Since $S_\varphi((1-e)x\Phi) = ex^*(1-e)\Phi = 0$ for all $x \in M$, $S_\varphi|_{(1-e)e'\mathcal{H}} = 0$ as well as $S_\varphi|_{(1-e')\mathcal{H}} = 0$. Since $(1-e') + (1-e)e' = 1-q$, we have $S_\varphi|_{(1-q)\mathcal{H}} = 0$. Moreover, for every $x \in M$, note that $ex\Phi = exq\Phi = qxq\Phi$ and $S_\varphi(ex\Phi) = ex^*\Phi = (qxq)^*\Phi$, so $S_\varphi|_{q\mathcal{H}}$ coincides with $S_{\overline{\varphi}}$. Therefore,

$$S_\varphi = S_{\overline{\varphi}} \oplus 0, \quad \text{so} \quad \Delta_\varphi = \Delta_{\overline{\varphi}} \oplus 0$$

on the decomposition $\mathcal{H} = q\mathcal{H} \oplus (1-q)\mathcal{H}$. Since $J_{\overline{\varphi}} = qJq$ by Proposition A.17, one has

$$S_\varphi = J_{\overline{\varphi}}\Delta_{\overline{\varphi}}^{1/2} \oplus 0 = (Jq \oplus J(1-q))(\Delta_{\overline{\varphi}}^{1/2} \oplus 0) = J\Delta_\varphi^{1/2}.$$

Next, prove the case of general ψ, φ. Let $M^{(2)} := M \otimes \mathbb{M}_2$, whose standard form $(M^{(2)}, \mathcal{H}^{(2)}, J^{(2)}, \mathcal{P}^{(2)})$ was described in Example A.20. Define $\theta \in (M^{(2)})_*^+$ by $\theta\left(\begin{bmatrix} x_{11} & x_{12} \\ x_{21} & x_{22} \end{bmatrix}\right) := \varphi(x_{11}) + \psi(x_{22})$, whose vector representative in $\mathcal{P}^{(2)}$ is $\Theta = \begin{bmatrix} \Phi & 0 \\ 0 & \Psi \end{bmatrix}$. It is clear that $s_{M^{(2)}}(\theta) = \begin{bmatrix} s_M(\varphi) & 0 \\ 0 & s_M(\psi) \end{bmatrix}$ or in the 4×4 form,

$$s_{M^{(2)}}(\theta) = \begin{bmatrix} s_M(\varphi) & 0 & 0 & 0 \\ 0 & s_M(\varphi) & 0 & 0 \\ 0 & 0 & s_M(\psi) & 0 \\ 0 & 0 & 0 & s_M(\psi) \end{bmatrix}, \tag{A.9}$$

and also by (A.7),

$$s_{(M^{(2)})'}(\theta) = J^{(2)} s_{M^{(2)}}(\theta) J^{(2)} = \begin{bmatrix} s_{M'}(\varphi) & 0 & 0 & 0 \\ 0 & s_{M'}(\psi) & 0 & 0 \\ 0 & 0 & s_{M'}(\varphi) & 0 \\ 0 & 0 & 0 & s_{M'}(\psi) \end{bmatrix}. \tag{A.10}$$

Furthermore, S_θ is defined as

$$S_\theta\left(\begin{bmatrix} x_{11} & x_{12} \\ x_{21} & x_{22} \end{bmatrix} \Theta + \begin{bmatrix} \eta_{11} & \eta_{12} \\ \eta_{21} & \eta_{22} \end{bmatrix}\right) = s_{M^{(2)}}(\theta) \begin{bmatrix} x_{11}^* & x_{21}^* \\ x_{12}^* & x_{22}^* \end{bmatrix} \Theta \tag{A.11}$$

for $\begin{bmatrix} x_{11} & x_{12} \\ x_{21} & x_{22} \end{bmatrix} \in M^{(2)}$ and $\begin{bmatrix} \eta_{11} & \eta_{12} \\ \eta_{21} & \eta_{22} \end{bmatrix} \in \big(1 - s_{(M^{(2)})'}(\theta)\big)\mathcal{H}^{(2)}$. Rewrite (A.11) in the 4×4 form and combine with (A.9) and (A.10). Then extending (A.6) we find that

$$S_\theta = \begin{bmatrix} S_\varphi & 0 & 0 & 0 \\ 0 & 0 & S_{\psi,\varphi} & 0 \\ 0 & S_{\varphi,\psi} & 0 & 0 \\ 0 & 0 & 0 & S_\psi \end{bmatrix} \tag{A.12}$$

and so

$$\Delta_\theta = S_\theta^* S_\theta = \begin{bmatrix} S_\varphi^* S_\varphi & 0 & 0 & 0 \\ 0 & S_{\varphi,\psi}^* S_{\varphi,\psi} & 0 & 0 \\ 0 & 0 & S_{\psi,\varphi}^* S_{\psi,\varphi} & 0 \\ 0 & 0 & 0 & S_\psi^* S_\psi \end{bmatrix} = \begin{bmatrix} \Delta_\varphi & 0 & 0 & 0 \\ 0 & \Delta_{\varphi,\psi} & 0 & 0 \\ 0 & 0 & \Delta_{\psi,\varphi} & 0 \\ 0 & 0 & 0 & \Delta_\psi \end{bmatrix}. \tag{A.13}$$

From the first case (applied to $\theta \in (M^{(2)})_*^+$) and (A.7) it follows that

$$S_\theta = J^{(2)} \Delta_\theta^{1/2} = \begin{bmatrix} J\Delta_\varphi^{1/2} & 0 & 0 & 0 \\ 0 & 0 & J\Delta_{\psi,\varphi}^{1/2} & 0 \\ 0 & J\Delta_{\varphi,\psi}^{1/2} & 0 & 0 \\ 0 & 0 & 0 & J\Delta_\psi^{1/2} \end{bmatrix}. \tag{A.14}$$

Comparing the (2,3)-entries of (A.12) and (A.14) implies that $S_{\psi,\varphi} = J\Delta_{\psi,\varphi}^{1/2}$.

(3) For the case $\psi = \varphi$, from the above proof of (2) and the property (4) of Sect. A.2, we have

$$\Delta_\varphi^{-1} = \Delta_{\overline{\varphi}}^{-1} \oplus 0 = (J_{\overline{\varphi}} \Delta_{\overline{\varphi}} J_{\overline{\varphi}}) \oplus 0 = J\Delta_\varphi J.$$

For general ψ, φ apply the above case to θ in the proof of (2); then by (A.13) and (A.7) we have

$$\begin{bmatrix} \Delta_\varphi^{-1} & 0 & 0 & 0 \\ 0 & \Delta_{\varphi,\psi}^{-1} & 0 & 0 \\ 0 & 0 & \Delta_{\psi,\varphi}^{-1} & 0 \\ 0 & 0 & 0 & \Delta_\psi^{-1} \end{bmatrix} = \begin{bmatrix} J\Delta_\varphi J & 0 & 0 & 0 \\ 0 & J\Delta_{\psi,\varphi}J & 0 & 0 \\ 0 & 0 & J\Delta_{\varphi,\psi}J & 0 \\ 0 & 0 & 0 & J\Delta_\psi J \end{bmatrix},$$

showing that $\Delta_{\psi,\varphi}^{-1} = J\Delta_{\varphi,\psi}J$.

(4) Let Ψ_i be the vector representative of ψ_i. If $\psi_1 \le \psi_2$, then for every $x \in M$ and $\eta \in (1 - s_{M'}(\varphi))\mathcal{H}$ one has

$$\begin{aligned} \|\Delta_{\psi_1,\varphi}^{1/2}(x\Phi + \eta)\|^2 &= \|S_{\psi_1,\varphi}(x\Phi + \eta)\|^2 = \|x^*\Psi_1\|^2 \\ &= \psi_1(xx^*) \le \psi_2(xx^*) = \|\Delta_{\psi_2,\varphi}^{1/2}(x\Phi + \eta)\|^2. \end{aligned}$$

Since $M\Phi + (1 - s_{M'}(\varphi))\mathcal{H}$ is a core of $\Delta_{\psi_i,\varphi}^{1/2}$, it follows from Proposition B.4 of Appendix B that $\Delta_{\psi_1,\varphi} \le \Delta_{\psi_2,\varphi}$. Hence the first assertion holds. For the second, assume that $s(\varphi_1) = s(\varphi_2)$ and $\varphi_1 \ge \varphi_2$. Then $s(\Delta_{\varphi_1,\psi}) = s(\Delta_{\varphi_2,\psi})$ by (1) above. Hence from the first assertion and Proposition B.6 one has $\Delta_{\varphi_1,\psi}^{-1} \le \Delta_{\varphi_2,\psi}^{-1}$. Since $\Delta_{\psi,\varphi_i} = J\Delta_{\varphi_i,\psi}^{-1}J$ by (3) above, the second assertion holds.

(5) Let Ψ_3 be the vector representative of $\psi_1 + \psi_2$ as well as Ψ_i of ψ_i. Note that $M\Phi + (1 - s_{M'}(\varphi))\mathcal{H}$ is the common core of $\Delta_{\psi_i,\varphi}^{1/2}$ $(i = 1, 2)$ and $\Delta_{\psi_1+\psi_2,\varphi}^{1/2}$. For every $x \in M$ and $\eta \in (1 - s_{M'}(\varphi))\mathcal{H}$ one has

$$\begin{aligned} &\|\Delta_{\psi_1,\varphi}^{1/2}(x\Phi + \eta)\|^2 + \|\Delta_{\psi_2,\varphi}^{1/2}(x\Phi + \eta)\|^2 \\ &= \|s_M(\varphi)x^*\Psi_1\|^2 + \|s_M(\varphi)x^*\Psi_2\|^2 = \psi_1(xs_M(\varphi)x^*) + \psi_2(xs_M(\varphi)x^*) \\ &= (\psi_1 + \psi_2)(xs_M(\varphi)x^*) = \|s_M(\varphi)x^*\Psi_3\|^2 = \|\Delta_{\psi_1+\psi_2,\varphi}^{1/2}(x\Phi + \eta)\|^2. \end{aligned}$$

This immediately implies that

$$\|\Delta_{\psi_1+\psi_2}^{1/2}\xi\|^2 = \|\Delta_{\psi_1,\varphi}^{1/2}\xi\|^2 + \|\Delta_{\psi_2,\varphi}^{1/2}\xi\|^2, \qquad \xi \in \mathcal{D}(\Delta_{\psi_1,\varphi}^{1/2}) \cap \mathcal{D}(\Delta_{\psi_2,\varphi}^{1/2}),$$

which means that $\Delta_{\psi_1+\psi_2,\varphi} = \Delta_{\psi_1,\varphi} \dot{+} \Delta_{\psi_2,\varphi}$.

Examples A.23

(1) Let $M = L^\infty(X, \mu)$ as in Example A.14 (1). For every $\psi, \varphi \in L^1(X, \mu)_+ \cong M_*^+$, it is easy to verify that $\Delta_{\psi,\varphi}$ is the multiplication of $1_{\{\varphi>0\}}(\psi/\varphi)$, which is the Radon–Nikodym derivative of $\psi\, d\mu$ with respect to $\varphi\, d\mu$ (restricted to the support of φ) in the classical sense.

(2) Let $M = B(\mathcal{H})$ as in Example A.14 (2). For every $\psi, \varphi \in B(\mathcal{H})_*^+$ we have the density operators (positive trace-class operators) D_ψ, D_φ such that $\psi(X) = \operatorname{Tr} D_\psi X$ for $X \in B(\mathcal{H})$ and similarly for D_φ. Let $D_\psi = \sum_{a>0} a P_a$ and $D_\varphi = \sum_{b>0} b Q_b$ be the spectral decompositions of D_ψ, D_φ, where P_a and Q_b are finite-dimensional orthogonal projections. Then the relative modular operator $\Delta_{\psi,\varphi}$ on $C_2(\mathcal{H})$ is given as

$$\Delta_{\psi,\varphi} = L_{D_\psi} R_{D_\varphi^{-1}} = \sum_{a>0,\, b>0} ab^{-1} L_{P_a} R_{Q_b}, \tag{A.15}$$

where $L_{[-]}$ and $R_{[-]}$ denote the left and the right multiplications and D_φ^{-1} is the generalized inverse of D_φ.

The relative modular operator $\Delta_{\psi,\varphi}$ is a type of division ψ/φ regarded as the "non-commutative Radon–Nikodym derivative" of ψ with respect to φ. When ψ is dominated by φ, i.e., $\psi \le \lambda\varphi$, one can consider another type (a slightly more primitive) of division ψ/φ as defined in the next lemma, which is a variant of the factorization technique due to [34]. The lemma is often used in the main body of this monograph, so we give a proof for the convenience of the reader.

Lemma A.24 *Assume that $\psi \le \lambda\varphi$ for some $\lambda > 0$. Let $\Phi, \Psi \in \mathcal{P}$ be the vector representatives of φ, ψ, respectively. Then there exists a unique $A \in s(\varphi)Ms(\varphi)$ such that $\Psi = A\Phi$. The A satisfies $\|A\| \le \lambda^{1/2}$ and $s(AA^*) = s(\psi)$.*

Moreover, if $\nu\varphi \le \psi \le \lambda\varphi$ for some $\lambda, \nu > 0$, then the above A satisfies $\nu s(\varphi) \le A^*A \le \lambda s(\varphi)$.

Proof Since

$$\begin{aligned} \|x'\Psi\|^2 &= \|Jx'J\Psi\|^2 = \psi((Jx'J)^*(Jx'J)) \\ &\le \lambda\varphi((Jx'J)^*(Jx'J)) = \lambda\|x'\Phi\|^2, \qquad x' \in M', \end{aligned}$$

there is a unique $A \in B(\mathcal{H})$ such that $A(1 - s(\varphi)) = 0$ and $A(x'\Phi) = x'\Psi$ for all $x' \in M$. Moreover, $\|A\| \le \lambda^{1/2}$, and for every unitary u' in $M's(\varphi)$,

$$u'^*Au'(x'\Phi) = u'^*u'x'\Psi = x'\Psi = A(x'\Phi), \qquad x' \in M,$$

which implies that $u'^*Au' = A$, so $A \in (M's(\varphi))' = s(\varphi)Ms(\varphi)$. Since the closure of the range of A is $\overline{M'\Psi}$, we have $s(AA^*) = s(\psi)$.

Next, assume that $\nu\varphi \le \psi$ in addition to $\psi \le \lambda\varphi$. Then $s(\psi) = s(\varphi)$, and there is a unique $B \in s(\varphi)Ms(\varphi)$ such that $\Phi = B\Psi$. It is easy to see that $AB = BA = s(\varphi)$, hence $B = A^{-1}$ in $s(\varphi)Ms(\varphi)$. Since $BB^* \le \nu^{-1}s(\varphi)$, we have $\nu s(\varphi) \le A^*A \le \lambda s(\varphi)$.

Consider Example A.23 (2) with $\dim \mathcal{H} < \infty$. When $\psi \le \lambda\varphi$ the operator A given in Lemma A.24 may be written as $A = D_\psi^{1/2} D_\varphi^{-1/2}$, so we have $A^*A =$

$D_\varphi^{-1/2} D_\psi D_\varphi^{-1/2}$. Both of the relative modular operator $\Delta_{\psi,\varphi}$ and the operator $D_\varphi^{-1/2} D_\psi D_\varphi^{-1/2}$ play an equally important role in quantum information, but in separate situations (see, e.g., [58, Sec. 3.1]).

A.5 τ-Measurable Operators

The non-commutative integration theory was initiated by Segal [118] and further developed in, e.g., [32, 121, 137], where measurable operators affiliated with a von Neumann algebra with a trace were discussed. Later in [99], Nelson introduced the notion of τ-measurable operators in a stricter connection with a given trace τ. This section is a brief survey of the theory of τ-measurable operators. Assume that M is a semifinite von Neumann algebra on a Hilbert space $\mathcal{H}$ and τ is a faithful normal semifinite trace on M.

Let $a : \mathcal{D}(a) \to \mathcal{H}$ be a linear operator, where $\mathcal{D}(a)$ is a linear subspace of $\mathcal{H}$. We say that a is *affiliated with* M, denoted by $a\,\eta M$, if $x'a \subset ax'$ for all $x' \in M'$, or equivalently, if $u'au'^* = a$ for all unitaries $u' \in M'$. The following facts are easy to verify:

(a) If a, b are linear operators affiliated with M, then $a + b$ with $\mathcal{D}(a + b) = \mathcal{D}(a) \cap \mathcal{D}(b)$ and ab with $\mathcal{D}(ab) = \{\xi \in \mathcal{D}(b) : b\xi \in D(a)\}$ are affiliated with M.
(b) If a is densely defined and $a\,\eta M$, then $a^*\,\eta M$.
(c) If a is closable and $a\,\eta M$, then $\overline{a}\,\eta M$.
(d) Assume that a is densely defined and closed, so we have the polar decomposition $a = w|a|$ and the spectral decomposition $|a| = \int_0^\infty \lambda\, de_\lambda$. Then $a\,\eta M$ if and only if $w, e_\lambda \in M$ for all $\lambda \geq 0$.

Definition A.25 Let a be a densely defined closed operator such that $a\,\eta M$. We say that a is *τ-measurable* if, for any $\delta > 0$, there exists an $e \in \mathrm{Proj}(M)$ such that $e\mathcal{H} \subset \mathcal{D}(a)$ and $\tau(e^\perp) \leq \delta$ (note that $e\mathcal{H} \subset \mathcal{D}(a) \iff \|ae\| < \infty$ due to the closed graph theorem). We denote by $\widetilde{M}$ the set of such τ-measurable operators.

Proposition A.26 *Let a be a densely defined closed operator affiliated with M with $a = w|a|$ and $|a| = \int_0^\infty \lambda\, de_\lambda$ as above. Then the following conditions are equivalent:*

(i) $a \in \widetilde{M}$;
(ii) $|a| \in \widetilde{M}$;
(iii) $\tau(e_\lambda^\perp) \to 0$ *as* $\lambda \to \infty$;
(iv) $\tau(e_\lambda^\perp) < +\infty$ *for some* $\lambda > 0$.

For each $\varepsilon, \delta > 0$ define

$$\mathcal{N}(\varepsilon, \delta) := \{a \in \widetilde{M} : \|ae\| \leq \varepsilon \text{ and } \tau(e^\perp) \leq \delta \text{ for some } e \in \mathrm{Proj}(M)\}.$$

Then the following assertions hold:

(1) If $a \in \widetilde{M}$, then $a^* \in \widetilde{M}$. Moreover, $a \in \mathcal{N}(\varepsilon, \delta) \iff a^* \in \mathcal{N}(\varepsilon, \delta)$.
(2) If $a, b \in \widetilde{M}$, then $a + b$ and ab are densely defined and closable, and $\overline{a+b}$, $\overline{ab} \in \widetilde{M}$. Moreover, if $a \in \mathcal{N}(\varepsilon_1, \delta_1)$ and $b \in \mathcal{N}(\varepsilon_2, \delta_2)$, then $\overline{a+b} \in \mathcal{N}(\varepsilon_1 + \varepsilon_2, \delta_1 + \delta_2)$ and $\overline{ab} \in \mathcal{N}(\varepsilon_1\varepsilon_2, \delta_1 + \delta_2)$.

In view of the above assertions, for every $a, b \in \widetilde{M}$ we may use the convention that $a+b$ and ab mean the strong sum $\overline{a+b}$ and the strong product $\overline{ab}$, respectively. A big advantage of τ-measurable operators is that we can freely take adjoint, sum and product in $\widetilde{M}$. The following is the main result on τ-measurable operators.

Theorem A.27 *$\widetilde{M}$ is a complete metrizable Hausdorff topological $*$-algebra with the strong sum, the strong product, and $\{\mathcal{N}(\varepsilon, \delta) : \varepsilon, \delta > 0\}$ as a neighborhood basis of* 0. *Moreover, M is dense in $\widetilde{M}$.*

Let $\widetilde{M}_+$ be the set of positive self-adjoint $a \in \widetilde{M}$ (denoted by $a \geq 0$). Note that $\widetilde{M}_+$ is a closed convex cone in $\widetilde{M}$ and $\widetilde{M}$ is an ordered topological space with the order $a \geq b$ defined as $a - b \geq 0$ for self-adjoint $a, b \in \widetilde{M}$. The trace τ on M_+ extends to all $a \in \widetilde{M}_+$ as $\tau(a) := \int_0^\infty \lambda \, d\tau(e_\lambda)$, where $a = \int_0^\infty \lambda \, de_\lambda$ is the spectral decomposition.

The topology on $\widetilde{M}$ given in Theorem A.27 is called the *measure topology*. This topology is not necessarily locally convex. For instance, if M is a finite non-atomic von Neumann algebra with a faithful normal finite trace τ, then a non-empty open convex set in $\widetilde{M}$ is only the whole $\widetilde{M}$ and there is no non-zero continuous linear functional on $\widetilde{M}$.

Examples A.28

(1) When $M = B(\mathcal{H})$ and τ is the usual trace Tr, $\widetilde{M} = B(\mathcal{H})$ and the measure topology is the operator norm topology.
(2) Let M be finite with a faithful normal finite trace τ. Then $\widetilde{M}$ is the set of all densely defined closed operators $x \, \eta M$.
(3) Let $(X, \mathcal{X}, \mu)$ be a localizable measure space, where $(X, \mathcal{X}, \mu)$ is *localizable* if, for every $A \in \mathcal{X}$, there is a $B \in \mathcal{X}$ such that $B \subset A$ and $\mu(B) < +\infty$. For an abelian von Neumann algebra $\mathcal{A} = L^\infty(X, \mu) = L^1(X, \mu)^*$ with $\tau(f) := \int_X f \, d\mu$ for $f \in L^\infty(X, \mu)_+$, $\widetilde{\mathcal{A}}$ is the space of measurable functions f on X such that there is an $A \in \mathcal{X}$ such that $\mu(A) < +\infty$ and f is bounded on $X \setminus A$, where $f = g$ in $\widetilde{\mathcal{A}} \iff f(x) = g(x)$ μ-a.e.

We end the section with the following two definitions.

Definition A.29 For $a \in \widetilde{M}$ and $t > 0$ the (tth) *generalized s-number* $\mu_t(a)$ is defined to be

$$\mu_t(a) := \inf\{s \geq 0 : \tau(e_{(s,\infty)}(|a|) \leq t\},$$

where $e_{(s,\infty)}(|a|)$ denotes the spectral projection of $|a|$ corresponding to the interval (s, ∞). Note that $\mu_t(a) < +\infty$ for all $t > 0$ since $\tau(e_{(s,\infty)}(|a|)) \to 0$ as $s \to \infty$ by Proposition A.26.

Definition A.30 For each $a \in \widetilde{M}$ define

$$\|a\|_p := \tau(|a|^p)^{1/p} \in [0, +\infty], \qquad 0 < p < \infty.$$

Furthermore, $\|a\|_\infty := \|a\| \in [0, +\infty]$, the operator norm with the convention that $\|a\| = +\infty$ unless $a \in M$. The *non-commutative L^p-space* on (M, τ) is defined as

$$L^p(M, \tau) := \{a \in \widetilde{M} : \|a\|_p < +\infty\}, \qquad 0 < p \le \infty.$$

(In particular, $L^\infty(M, \tau) = M$.)

The trace τ uniquely extends to a positive linear functional on $L^1(M, \tau)$ and it satisfies $|\tau(a)| \le \|a\|_1$ for all $a \in L^1(M, \tau)$. When $1 \le p \le \infty$ and $1/p+1/q = 1$, for every $a \in L^p(M, \tau)$ and $b \in L^q(M, \tau)$ we have $ab \in L^1(M, \tau)$ and

$$|\tau(ab)| \le \|ab\|_1 \le \|a\|_p \|b\|_q \quad \textit{(Hölder's inequality)}.$$

More generally, when $p, q, r > 0$ and $1/p + 1/q = 1/r$, Hölder's inequality holds as

$$\|ab\|_r \le \|a\|_p \|b\|_q, \qquad a, b \in \widetilde{M}.$$

If $1 \le p < \infty$ and $1/p + 1/q = 1$, then $L^p(M, \tau)$ is a Banach space with respect to the norm $\|\cdot\|_p$ and its dual Banach space is $L^q(M, \tau)$ under the duality pairing

$$(a, b) \in L^p(M, \tau) \times L^q(M, \tau) \longmapsto \tau(ab) \in \mathbb{C}.$$

In particular, we have $M_* = L^1(M, \tau)$ via the correspondence $\psi \in M_* \leftrightarrow a \in L^1(M, \tau)$ given by $\psi(x) = \tau(ax)$, $x \in M$, and $\psi \in M_*^+ \iff a \in L^1(M, \tau)_+$ $(= L^1(M, \tau) \cap \widetilde{M}_+)$. This a is often called the *Radon–Nikodym derivative* of ψ with respect to τ and denoted by $d\psi/d\tau$.

More details on the generalized s-numbers and the non-commutative $L^p(M, \tau)$ are found in [37], which is the best literature on those. More general Haagerup's L^p-spaces will be explained in the next section.

A.6 Haagerup's L^p-Spaces

We begin with the crossed product of M by the modular automorphism group, based on which Haagerup's L^p-spaces are constructed. Let $\sigma_t = \sigma_t^\omega$ $(t \in \mathbb{R})$ be the modular automorphism group with respect to a faithful $\omega \in M_*^+$ (see

Definition A.2). (We may take a faithful normal semifinite weight ω when M is not σ-finite.) Define a faithful normal representation π_σ of M and a one-parameter strongly continuous unitary representation λ on $L^2(\mathbb{R}, \mathcal{H}) \cong \mathcal{H} \otimes L^2(\mathbb{R}, dt)$ by

$$(\pi_\sigma(x)\xi)(s) := \sigma_{-s}(x)\xi(s), \quad (\lambda(t)\xi)(s) := \xi(s-t), \quad s \in \mathbb{R},\ \xi \in L^2(\mathbb{R}, \mathcal{H}),$$

satisfying the covariance property $\pi_\sigma(\sigma_t(x)) = \lambda(t)\pi_\sigma(x)\lambda(t)^*$ for all $x \in M$ and $t \in \mathbb{R}$. Then the *crossed product*

$$N := M \rtimes_\sigma \mathbb{R} = \{\pi_\sigma(M) \cup \lambda(\mathbb{R})\}''$$

is the von Neumann algebra generated by $\pi_\sigma(x)$ $(x \in M)$ and $\lambda(t)$ $(t \in \mathbb{R})$. The crossed product construction is more generally performed for a continuous (locally compact abelian) group action on M, for which Takesaki's duality theorem [126] was established, where the dual action plays a role. In the case of σ^ω the *dual action* $\theta = \widehat{\sigma^\omega}$ is determined by $\theta_s(\pi_\sigma(x)) = \pi_\sigma(x)$ and $\theta_s(\lambda(t)) = e^{ist}\lambda(t)$ for $x \in M$ and $s, t \in \mathbb{R}$. Then $N^\theta = \pi_\sigma(M)$ holds, where N^θ is the fixed-point algebra of θ. We may assume that $M \subset N$ under identifying $x \in M$ with $\pi_\sigma(x)$.

An important fact is that N is a *semifinite* von Neumann algebra with a faithful normal semifinite trace τ (called the *canonical trace* on N) satisfying the trace scaling property

$$\tau \circ \theta_s = e^{-s}\tau, \qquad s \in \mathbb{R}.$$

This fact is shown by making use of the *dual weight* $\widetilde{\omega}$ on N, while we omit its details here (see [44, 45] for theory of dual weights). Thus, let $\widetilde{N}$ be the space of τ-measurable operators affiliated with $N = M \rtimes_\sigma \mathbb{R}$. It is easy to see that θ_s $(s \in \mathbb{R})$ on N uniquely extends to $\widetilde{N}$ as a one-parameter group of homeomorphic $*$-isomorphisms with respect to the measure topology. It is also worth noting that the triplet

$$(N := M \rtimes_{\sigma^\omega} \mathbb{R},\ \theta := \widehat{\sigma^\omega},\ \tau)$$

is canonical in the sense that for another faithful $\omega_1 \in M_*^+$ there is a unitary U on $L^2(\mathbb{R}, \mathcal{H})$ such that

$$UNU^* = N_1, \quad \mathrm{Ad}(U) \circ \theta_s = \theta_{1s} \circ \mathrm{Ad}(U)\ \ (s \in \mathbb{R}), \quad \tau = \tau_1 \circ \mathrm{Ad}(U),$$

where (N_1, θ_1, τ_1) is the triplet associated with ω_1. When M is a factor of type III, the *flow of weights* of M mentioned at the end of Sect. A.1 is defined by

$$(X, F_t^M) := (Z(N), \theta_t|_{Z(N)}),$$

which is an ergodic flow and can be used for the type III$_\lambda$ $(0 \le \lambda \le 1)$ classification, see [30].

Definition A.31 ([46]) For $0 < p \le \infty$, *Haagerup's L^p-space* $L^p(M)$ is defined by

$$L^p(M) := \{x \in \widetilde{N} : \theta_s(x) = e^{-s/p}x,\ s \in \mathbb{R}\}.$$

Let $L^p(M)_+ = L^p(M) \cap \widetilde{N}_+$ where $\widetilde{N}_+$ is the positive part of $\widetilde{N}$.

Clearly, $L^p(M)$'s are closed linear subspaces of $\widetilde{N}$, which are M-bimodules and closed under $a \mapsto a^*$ and $a \mapsto |a|$. Moreover, they are linearly spanned by their positive part $L^p(M)_+$. It follows that every $a \in L^\infty(M)$ is bounded, so we have $L^\infty(M) = M$ since $N^\theta = M$. The following is the starting point of the theory of Haagerup's L^p-spaces developed in [129].

Lemma A.32 *There is a bijection $\psi \in M_*^+ \mapsto h_\psi \in L^1(M)_+$ such that for every $\psi, \varphi \in M_*^+$ and $x \in M$,*

$$h_{\psi+\varphi} = h_\psi + h_\varphi, \qquad h_{x\psi x^*} = x h_\psi x^*,$$

where $(x\psi x^*)(y) := \psi(x^* y x)$, $y \in M$.

Although we omit the details here, note that the bijection $\psi \in M_*^+ \mapsto h_\psi \in L^1(M)_+$ is determined in such a way that the dual weight $\widetilde{\psi}$ on N is equal to $\tau(h_\psi \cdot)$, and that $s(\psi) = s(h_\psi)$ for the support projections. By linearly extending the above bijection we have the next theorem.

Theorem A.33 *There is a linear bijection $\psi \in M_* \mapsto h_\psi \in L^1(M)$ such that for every $\psi \in M_*$ and $x, y \in M$,*

$$h_{x\psi y^*} = x h_\psi y^*, \qquad h_{\psi^*} = h_\psi^*.$$

Moreover, if $\psi = u|\psi|$ is the polar decomposition of $\psi \in M_*$ (see [127, Sec. III.4]), then $h_\psi = u h_{|\psi|}$ is the polar decomposition of h_ψ. Hence

$$|h_\psi| = h_{|\psi|}$$

and the partial isometry part of h_ψ is $u \in M$.

Due to the linear bijection in Theorem A.33, define a linear functional tr on $L^1(M)$ by

$$\mathrm{tr}(h_\psi) := \psi(1), \qquad \psi \in M_*.$$

Then we have

$$\mathrm{tr}(|h_\psi|) = \mathrm{tr}(h_{|\psi|}) = |\psi|(1) = \|\psi\|, \qquad \psi \in M_*. \tag{A.16}$$

Lemma A.34 *Let $a \in \widetilde{N}$ with the polar decomposition $a = u|a|$. Then for every $p \in [1, \infty)$,*

$$a \in L^p(M) \iff u \in M \text{ and } |a|^p \in L^1(M).$$

Definition A.35 In view of Lemma A.34, for every $a \in L^p(M)$ define $\|a\|_p \in [0, +\infty)$ by

$$\|a\|_p := \mathrm{tr}(|a|^p)^{1/p} \quad \text{if } 0 < p < \infty,$$
$$\|a\|_\infty := \|a\| \quad \text{if } p = \infty.$$

In the case $p = 1$, by (A.16), $\|a\|_1 := \mathrm{tr}(|a|)$ for $a \in L^1(M)$ is the norm on $L^1(M)$ copied from the norm on M_* by the linear bijection $\psi \mapsto h_\psi$. In this way,

$$(L^1(M), \|\cdot\|_1)$$

becomes a Banach space identified with M_*.

Example A.36 Assume that M is semifinite with a faithful normal semifinite trace τ_0. Then (N, θ_s, τ) is identified with

$$\left(M \otimes L^\infty(\mathbb{R}),\ \mathrm{id} \otimes (f \mapsto f(\cdot + s)),\ \tau_0 \otimes \int_{\mathbb{R}} \cdot\, e^t\, dt\right).$$

In this case, for each $\psi \in M_*^+$ we have $h_\psi = (d\psi/d\tau_0) \otimes e^{-t}$, where $d\psi/d\tau_0$ is the Radon–Nikodym derivative of ψ with respect to τ_0 (see the end of Sect. A.5). Hence, for each $p \in (0, \infty]$,

$$L^p(M) = L^p(M, \tau_0) \otimes e^{-t/p}$$

and $\|a \otimes e^{-t/p}\|_{L^p(M)} = \|a\|_{L^p(M,\tau_0)}$ for all $a \in L^p(M, \tau_0)$. With neglecting the superfluous tensor factor $e^{-t/p}$, we may identify $L^p(M)$ with $L^p(M, \tau_0)$.

In the rest of the section we present properties of Haagerup's L^p-spaces. See [129] for their proofs and more details.

Proposition A.37 *Let $1 \le p, q \le \infty$ with $1/p + 1/q = 1$. If $a \in L^p(M)$ and $b \in L^q(M)$, then $ab, ba \in L^1(M)$ and*

$$\mathrm{tr}(ab) = \mathrm{tr}(ba).$$

Theorem A.38 (Hölder's Inequality) *Let $1 \le p, q \le \infty$ with $1/p + 1/q = 1$. If $a \in L^p(M)$ and $b \in L^q(M)$, then*

$$|\mathrm{tr}(ab)| \le \|ab\|_1 \le \|a\|_p \|b\|_q.$$

Moreover, for every $p, q, r \in (0, \infty]$ with $1/p + 1/q = 1/r$, if $a \in L^p(M)$ and $b \in L^q(M)$, then $ab \in L^r(M)$ and

$$\|ab\|_r \leq \|a\|_p \|b\|_q.$$

The reference [129] contains only the first assertion of the above theorem, while the second assertion and other inequalities for Haagerup's $\|\cdot\|_p$ are found in [37].

Proposition A.39 *Let $1 \leq p, q \leq \infty$ with $1/p + 1/q = 1$. Then for every $a \in L^p(M)$,*

$$\|a\|_p = \sup\{|\mathrm{tr}(ab)| : b \in L^q(M),\ \|b\|_q \leq 1\}. \tag{A.17}$$

Theorem A.40

(1) *For every $p \in [1, \infty]$, $(L^p(M), \|\cdot\|_p)$ is a Banach space.*
(2) *In particular, $L^2(M)$ is a Hilbert space with respect to the inner product $\langle a, b\rangle := \mathrm{tr}(a^*b)$ $(= \mathrm{tr}(ba^*))$ for $a, b \in L^2(M)$.*
(3) *For any $p \in [1, \infty)$, the norm topology on $L^p(M)$ coincides with the relative topology induced from the measure topology on $\widetilde{N}$. More precisely, the uniform structure on $L^p(M)$ by $\|\cdot\|_p$ coincides with that induced from $\widetilde{N}$.*

Theorem A.41 *Let $1 \leq p < \infty$ and $1/p + 1/q = 1$. Then the dual Banach space of $L^p(M)$ is $L^q(M)$ under the duality pairing*

$$(a, b) \in L^p(M) \times L^q(M) \longmapsto \mathrm{tr}(ab) \in \mathbb{C}.$$

Proposition A.42 *Let $1 \leq p, q \leq \infty$ with $1/p + 1/q = 1$. Let $a \in L^q(M)$. Then*

$$a \geq 0 \iff \mathrm{tr}(ab) \geq 0 \ \text{for all } b \in L^p(M)_+.$$

Theorem A.43 *For each $x \in M$ we define the left action $\lambda(x)$ and the right action $\rho(x)$ on the Hilbert space $L^2(M)$ by*

$$\lambda(x)a := xa, \quad \rho(x)a := ax, \qquad a \in L^2(M),$$

and the involution J on $L^2(M)$ by $Ja := a^*$. Then:

(1) *λ (resp., ρ) is a normal faithful representation (resp., anti-representation) of M on $L^2(M)$.*
(2) *The von Neumann algebras $\lambda(M)$ and $\rho(M)$ are the commutants of each other and*

$$\rho(M) = J\lambda(M)J.$$

(3) *$(\lambda(M), L^2(M), J, L^2(M)_+)$ is a standard form of M.*

Remark A.44 For any projection $e \in M$, Haagerup's L^p-space $L^p(eMe)$ is identified with $eL^p(M)e$. Furthermore, since $ej(e)L^2(M) = eL^2(M)e$, we see from Proposition A.16 that

$$(eMe, eL^2(M)e, J = {}^*, eL^2(M)_+e)$$

is a standard form of eMe, where eMe acts on $eL^2(M)e$ as the left multiplication.

Finally in this section we give a lemma (due to Kosaki [78]), which will be useful in the next section.

Lemma A.45 *For every $\varphi \in M_*^+$ let $s(\varphi)$ be the M-support of φ and σ_t^φ be the modular automorphism group with respect to $\varphi|_{s(\varphi)Ms(\varphi)}$. Then*

$$\sigma_t^\varphi(x) = h_\varphi^{it} x h_\varphi^{-it}, \qquad x \in s(\varphi)Ms(\varphi),\ t \in \mathbb{R}.$$

Proof First, assume that φ is faithful. Then $\alpha_t(x) := h_\varphi^{it} x h_\varphi^{-it}$ ($x \in M, t \in \mathbb{R}$) defines a strongly continuous one-parameter automorphism group. Let $x, y \in M$ and assume that x is entire α-analytic with the analytic extension $\alpha_z(x)$ ($z \in \mathbb{C}$), see, e.g., [22, Sec. 2.5.3]. By analytic continuation it follows that $h_\varphi^{is}\alpha_z(x) = \alpha_{z+s}(x)h_\varphi^{is}$ for all $s \in \mathbb{R}$ and $z \in \mathbb{C}$. For every $\xi \in \mathcal{D}(h_\varphi)$ ($\subset \mathcal{K} := L^2(\mathbb{R}, \mathcal{H})$, the representing Hilbert space for N), by Theorem B.1 of Appendix B there exists a $\mathcal{K}$-valued bounded strongly continuous function $f(\zeta)$ on $-1 \le \operatorname{Im}\zeta \le 0$, analytic in $-1 < \operatorname{Im}\zeta < 0$, such that $f(s) = h_\varphi^{is}\xi$ ($s \in \mathbb{R}$). Then, for each $z \in \mathbb{C}$, $\alpha_{z+\zeta}(x)f(\zeta)$ is a bounded strongly continuous function on $-1 \le \operatorname{Im}\zeta \le 0$, analytic in $-1 < \operatorname{Im}\zeta < 0$, such that $\alpha_{z+s}(x)f(s) = h_\varphi^{is}\alpha_z(x)\xi$ ($s \in \mathbb{R}$). Hence by Theorem B.1 again, $\alpha_z(x)\xi \in \mathcal{D}(h_\varphi)$ and $h_\varphi\alpha_z(x)\xi = \alpha_{z-i}(x)h_\varphi\xi$, from which we find that $h_\varphi\alpha_z(x) = \alpha_{z-i}(x)h_\varphi$ ($z \in \mathbb{C}$), see [129, Chap. I, Proposition 12]. Since

$$\varphi(\alpha_t(x)y) = \operatorname{tr}(h_\varphi\alpha_t(x)y) = \operatorname{tr}(\alpha_{t-i}(x)h_\varphi y) = \varphi(y\alpha_{t-i}(x)), \qquad t \in \mathbb{R},$$

it follows that φ satisfies the $(\alpha_t, -1)$-KMS condition, see [23, Definition 5.3.1 and Proposition 5.3.7]. Hence Theorem A.7 implies that $\sigma_t^\varphi = \alpha_t$. For general $\varphi \in M_*^+$ let $e := s(\varphi)$. Since $h_\varphi \in eL^1(M)e$ corresponds to $\varphi|_{eMe}$ (see Remark A.44), the result follows from the above case.

The above lemma says that $\Delta_\varphi^{it}(xh_\varphi^{1/2}) = h_\varphi^{it}(xh_\varphi^{1/2})h_\varphi^{-it}$ for all $x \in M$ and $t \in \mathbb{R}$. This shows that $\Delta_\varphi^{it}\xi = h_\varphi^{it}\xi h_\varphi^{-it}$ for all $\xi \in L^2(M)$ and $t \in \mathbb{R}$. Also, from the uniqueness of analytic continuation it follows that $\Delta_\varphi^{p/2}(xh_\varphi^{1/2}) = h_\varphi^{p/2}xh_\varphi^{(1-p)/2}$ for $0 \le p \le 1$. Furthermore, in a way similar to the proof of Proposition A.22 (2), we have for any $\psi, \varphi \in M_*^+$,

$$\Delta_{\psi,\varphi}^{it}\xi = h_\psi^{it}\xi h_\varphi^{-it}, \qquad \xi \in L^2(M),\ t \in \mathbb{R}, \tag{A.18}$$

$$\Delta_{\psi,\varphi}^{p/2}(xh_\varphi^{1/2}) = h_\psi^{p/2}xh_\varphi^{(1-p)/2}, \quad x \in M,\ 0 \le p \le 1, \tag{A.19}$$

with the convention that $h_\psi^0 = s(\psi)$, $h_\varphi^0 = s(\varphi)$ and $\Delta_{\psi,\varphi}^0 = s(\psi)Js(\varphi)J$.

A.7 Connes' Cocycle Derivatives

Consider the tensor product $M^{(2)} := M \otimes \mathbb{M}_2$ of M with the 2×2 matrix algebra $\mathbb{M}_2(\mathbb{C})$. For each $\varphi, \psi \in M_*^+$ the *balanced functional* $\theta = \theta(\varphi, \psi)$ on $M^{(2)}$ is defined by

$$\theta\left(\sum_{i,j=1}^{2} x_{ij} \otimes e_{ij}\right) := \varphi(x_{11}) + \psi(x_{22}), \qquad x_{ij} \in M,$$

where e_{ij} $(i, j = 1, 2)$ are the matrix units of $\mathbb{M}_2$. The functional θ was already used in the proof of Proposition A.22 (2). The support projection of θ is $s(\theta) = s(\varphi) \otimes e_{11} + s(\psi) \otimes e_{22}$. Concerning the modular automorphism group σ_t^θ on $s(\theta)M^{(2)}s(\theta)$, the following hold:

(1) $s(\varphi) \otimes e_{11}, s(\psi) \otimes e_{22} \in \big(s(\theta)M^{(2)}s(\theta)\big)_\theta$ (the centralizer of $\theta|_{s(\theta)M^{(2)}s(\theta)}$),
(2) $\sigma_t^\theta(x \otimes e_{11}) = \sigma_t^\varphi(x) \otimes e_{11}$ for all $t \in \mathbb{R}$ and $x \in s(\varphi)Ms(\varphi)$,
(3) $\sigma_t^\theta(x \otimes e_{22}) = \sigma_t^\psi(x) \otimes e_{22}$ for all $t \in \mathbb{R}$ and $x \in s(\psi)Ms(\psi)$,
(4) $\sigma_t^\theta(s(\psi)Ms(\varphi) \otimes e_{21}) \subset s(\psi)Ms(\varphi) \otimes e_{21}$.

Definition A.46 ([26]) Let $\varphi, \psi \in M_*^+$ and $\theta = \theta(\varphi, \psi)$. By the above (4) there exists a strongly* continuous map $t \in \mathbb{R} \mapsto u_t \in s(\psi)Ms(\varphi)$ such that

$$\sigma_t^\theta(s(\psi)s(\varphi) \otimes e_{21}) = u_t \otimes e_{21}, \qquad t \in \mathbb{R}.$$

The map $t \mapsto u_t$ is called *Connes' cocycle (Radon–Nikodym) derivative* of ψ with respect to φ, and denoted by $u_t = [D\psi : D\varphi]_t$, $t \in \mathbb{R}$.

The next proposition specifies the relation between Connes' cocycle derivative $[D\psi : D\varphi]_t$ and Araki's relative modular operator $\Delta_{\psi,\varphi}$.

Proposition A.47 *For every $\varphi, \psi \in M_*^+$ we have*

$$[D\psi : D\varphi]_t Js(\varphi)J = \Delta_{\psi,\varphi}^{it}\Delta_\varphi^{-it}, \quad [D\psi : D\varphi]_t Js(\psi)J = \Delta_\psi^{it}\Delta_{\varphi,\psi}^{-it}, \qquad t \in \mathbb{R}. \tag{A.20}$$

Remark A.48 For any projection $e \in M$, recall that $x \in eMe \mapsto xe' \in eMee'$ $(e' := JeJ)$ is a $*$-isomorphism (see Proposition A.16 (1)). This may justify writing (A.20) in a simpler way as follows:

$$[D\psi : D\varphi]_t = \Delta_{\psi,\varphi}^{it}\Delta_\varphi^{-it} \qquad \text{if } s(\psi) \le s(\varphi),$$

$$[D\psi : D\varphi]_t = \Delta_\psi^{it}\Delta_{\varphi,\psi}^{-it} \qquad \text{if } s(\varphi) \le s(\psi).$$

The expression of $[D\psi : D\varphi]_t$ in terms of Haagerup's L^1-elements is quite convenient to derive properties of Connes' cocycle derivative $[D\psi : D\varphi]_t$. To prove this, we first briefly examine Haagerup's L^p-spaces for $M^{(2)} = M \otimes \mathbb{M}_2$. Take the tensor product $\omega \otimes \mathrm{Tr}$ of a faithful $\omega \in M_*^+$ and the trace functional Tr on $\mathbb{M}_2$. Then $\sigma_t^{\omega\otimes\mathrm{Tr}} = \sigma_t^\omega \otimes \mathrm{id}_2$, where id_2 is the identity map on $\mathbb{M}_2$. From the construction of the crossed products $N := M \rtimes_{\sigma^\omega} \mathbb{R}$ and $N^{(2)} := M^{(2)} \rtimes_{\sigma^\omega\otimes\mathrm{id}_2} \mathbb{R}$ (see the first paragraph of Sect. A.6), it is easy to see that:

(a) $N^{(2)} = N \otimes \mathbb{M}_2$,
(b) the canonical trace on $N^{(2)}$ is $\tau \otimes \mathrm{Tr}$, where τ is the canonical trace on N,
(c) the dual action on $N^{(2)}$ is $\theta_s \otimes \mathrm{id}_2$ ($s \in \mathbb{R}$), where θ_s is the dual action on N.

Based on these facts we have $\widetilde{N^{(2)}} = \widetilde{N} \otimes \mathbb{M}_2$, where $\widetilde{N}$ and $\widetilde{N^{(2)}}$ are the spaces of τ-measurable and $\tau \otimes \mathrm{Tr}$-measurable operators affiliated with N and $N^{(2)}$, respectively. Therefore, for $0 < p \le \infty$, Haagerup's L^p-space

$$L^p(M^{(2)}) := \{a \in \widetilde{N^{(2)}} = \widetilde{N} \otimes \mathbb{M}_2 : (\theta_s \otimes \mathrm{id}_2)(a) = e^{-s/p}a,\ s \in \mathbb{R}\}$$

is given as

$$L^p(M^{(2)}) = L^p(M) \otimes \mathbb{M}_2 = \left\{a = \begin{bmatrix} a_{11} & a_{12} \\ a_{21} & a_{22} \end{bmatrix} : a_{ij} \in L^p(M),\ i, j = 1, 2\right\},$$

and its positive part is $(L^p(M) \otimes \mathbb{M}_2) \cap \big(\widetilde{N^{(2)}}\big)_+$. Moreover, by closely looking the construction of the functional tr, we notice that

(d) the tr-functional on $L^1(M^{(2)}) = L^1(M) \otimes \mathbb{M}_2$ is $\mathrm{tr} \otimes \mathrm{Tr}$, where tr is the tr-functional on $L^1(M)$.

In this way, the standard form of $M^{(2)} = M \otimes \mathbb{M}_2$ in terms of Haagerup's L^2-space is canonically given as

$$(M \otimes \mathbb{M}_2, L^2(M) \otimes \mathbb{M}_2, J = {}^*, (L^2(M) \otimes \mathbb{M}_2)_+),$$

where $[x_{ij}]_{i,j=1}^2 \in M \otimes \mathbb{M}_2$ acts on $L^2(M) \otimes \mathbb{M}_2$ as the 2×2 matrix left multiplication $\begin{bmatrix} x_{11} & x_{12} \\ x_{21} & x_{22} \end{bmatrix} \begin{bmatrix} \xi_{11} & \xi_{12} \\ \xi_{21} & \xi_{22} \end{bmatrix}$, and $J = {}^*$ is the matrix $*$-operation $\begin{bmatrix} \xi_{11} & \xi_{12} \\ \xi_{21} & \xi_{22} \end{bmatrix}^* = \begin{bmatrix} \xi_{11}^* & \xi_{21}^* \\ \xi_{12}^* & \xi_{22}^* \end{bmatrix}$ for $[\xi_{ij}]_{i,j=1}^2 \in L^2(M) \otimes \mathbb{M}_2$.

Lemma A.49 *For every $\varphi, \psi \in M_*^+$ we have*

$$[D\psi : D\varphi]_t = h_\psi^{it} h_\varphi^{-it}, \qquad t \in \mathbb{R}, \tag{A.21}$$

where h_φ, h_ψ are the elements of Haagerup's L^1-space $L^1(M)$ corresponding to φ, ψ.

Proof In view of the above description of $L^p(M \otimes \mathbb{M}_2)$, in particular, the fact in (d), the element of $L^1(N)$ corresponding to $\theta(\varphi, \psi)$ is $h_\theta = \begin{bmatrix} h_\varphi & 0 \\ 0 & h_\psi \end{bmatrix}$. Hence it follows from Lemma A.45 that for $[x_{ij}] \in s(\theta)Ns(\theta)$,

$$\sigma_t^\theta\left(\begin{bmatrix} x_{11} & x_{12} \\ x_{21} & x_{22} \end{bmatrix}\right) = \begin{bmatrix} h_\varphi & 0 \\ 0 & h_\psi \end{bmatrix}^{it} \begin{bmatrix} x_{11} & x_{12} \\ x_{21} & x_{22} \end{bmatrix} \begin{bmatrix} h_\varphi & 0 \\ 0 & h_\psi \end{bmatrix}^{-it}$$

$$= \begin{bmatrix} h_\varphi^{it} x_{11} h_\varphi^{-it} & h_\varphi^{it} x_{12} h_\psi^{-it} \\ h_\psi^{it} x_{21} h_\varphi^{-it} & h_\psi^{it} x_{22} h_\psi^{-it} \end{bmatrix}, \qquad t \in \mathbb{R}. \tag{A.22}$$

Therefore,

$$\begin{bmatrix} 0 & 0 \\ [D\psi : D\varphi]_t & 0 \end{bmatrix} = \sigma_t^\theta\left(\begin{bmatrix} 0 & 0 \\ s(\psi)s(\varphi) & 0 \end{bmatrix}\right) = \begin{bmatrix} 0 & 0 \\ h_\psi^{it} h_\varphi^{-it} & 0 \end{bmatrix}$$

so that (A.21) follows.

Remark A.50 In fact, the assertions stated just before Definition A.46 are immediately seen from expression (A.22) and Lemma A.45.

Example A.51 Assume that M is a semifinite von Neumann algebra with a faithful normal semifinite trace τ. As explained in Example A.36, Haagerup's L^1-space $L^1(M)$ in this case is identified with the conventional L^1-space $L^1(M, \tau)$ with respect to τ. More precisely, for each $\psi \in M_*$, h_ψ in $L^1(M)$ and the Radon–Nikodym derivative $d\psi/d\tau \in L^1(M, \tau)$ are in the relation that $h_\psi = (d\psi/d\tau) \otimes e^{-t}$. Hence, for every $\varphi, \psi \in M_*^+$ we have

$$[D\psi : D\varphi]_t = \left(\frac{d\psi}{d\tau}\right)^{it} \left(\frac{d\varphi}{d\tau}\right)^{-it}.$$

In particular, when $B = B(\mathcal{H})$ with the usual trace Tr, we have $[D\psi : D\varphi]_t = D_\psi^{it} D_\varphi^{-it}$, where D_φ, D_ψ are the density (trace-class) operators representing $\varphi, \psi \in B(\mathcal{H})_*^+$.

Below we present important properties of Connes' cocycle derivative $[D\psi : D\varphi]_t$, which were first given by Connes [26, 27] for the case of faithful normal semifinite weights. Their proofs and more details are found in [122, 128].

Theorem A.52 *Let $\varphi, \psi \in M_*^+$ and assume that $s(\psi) \le s(\varphi)$. Then $u_t := [D\psi : D\varphi]_t$ satisfies the following properties:*

(i) $u_t u_t^* = s(\psi) = u_0$ *and* $u_t^* u_t = \sigma_t^\varphi(s(\psi))$ *for all* $t \in \mathbb{R}$. *In particular,* u_t*'s are partial isometries with the final projection* $s(\psi)$.
(ii) $u_{s+t} = u_s \sigma_s^\varphi(u_t)$ *for all* $s, t \in \mathbb{R}$ *(the* cocycle identity*).*
(iii) $u_{-t} = \sigma_{-t}^\varphi(u_t^*)$ *for all* $t \in \mathbb{R}$.

(iv) $\sigma_t^{\psi}(x) = u_t\sigma_t^{\varphi}(x)u_t^*$ *for all* $t \in \mathbb{R}$ *and* $x \in s(\psi)Ms(\psi)$.

(v) *For every* $x \in s(\psi)Ms(\varphi)$ *and* $y \in s(\varphi)Ms(\psi)$, *there exists a continuous bounded function* F *on* $0 \le \operatorname{Im} z \le 1$, *analytic in* $0 < \operatorname{Im} z < 1$, *such that*

$$F(t) = \psi(u_t\sigma_t^{\varphi}(x)y), \qquad F(t+i) = \varphi(yu_t\sigma_t^{\varphi}(x)), \qquad t \in \mathbb{R}.$$

Furthermore, u_t ($t \in \mathbb{R}$) *is uniquely determined by a strongly* continuous map* $t \in \mathbb{R} \mapsto u_t \in M$ *satisfying the above (i), (ii), (iv) and (v). (Note that (iii) follows from (i) and (ii).)*

Proposition A.53 *Let* $\varphi, \psi, \omega \in M_*^+$.

(1) $[D\psi : D\varphi]_t^* = [D\varphi : D\psi]_t$ *for all* $t \in \mathbb{R}$.
(2) *If either* $s(\psi) \le s(\omega)$ *or* $s(\varphi) \le s(\omega)$, *then* $[D\psi : D\omega]_t[D\omega : D\varphi]_t = [D\psi : D\varphi]_t$ *for all* $t \in \mathbb{R}$ *(the* chain rule*)*.
(3) *If* $s(\psi) \le s(\omega)$ *and* $s(\varphi) \le s(\omega)$, *then* $[D\psi; D\omega]_t = [D\varphi : D\omega]_t$ *for all* $t \in \mathbb{R}$ *if and only if* $\psi = \varphi$.
(4) *For every* $\alpha \in \operatorname{Aut}(M)$ *(the automorphisms of* M*)*, $[D(\psi \circ \alpha) : D(\varphi \circ \alpha)]_t = \alpha^{-1}([D\psi : D\varphi]_t)$ *for all* $t \in \mathbb{R}$.

The following is *Connes' inverse theorem* in [27, Theorem 1.2.4].

Theorem A.54 ([27]) *Let* φ *is a faithful normal semifinite weight on* M. *Let* $t \in \mathbb{R} \mapsto u_t \in M$ *is a strongly* continuous map satisfying*

$$\begin{aligned} u_{s+t} &= u_s\sigma_s^{\varphi}(u_t), && t \in \mathbb{R}, \\ u_{-t} &= \sigma_{-t}^{\varphi}(u_t^*), && t \in \mathbb{R}. \end{aligned} \tag{A.23}$$

Then there exists a unique normal semifinite weight ψ *on* M *such that* $u_t = [D\psi : D\varphi]_t$ *for all* $t \in \mathbb{R}$.

In [27] u_t's are assumed to be unitaries in M. In this case, (A.23) is redundant and ψ given in the theorem is faithful as well. The above version without u_t's being unitaries is taken from [122, Theorem 51]. Note that even when $\varphi \in M_*^+$, ψ in the theorem cannot be in M_*^+ in general. This fact suggests that the von Neumann algebra theory cannot be self-completed when we stick to functionals in M_*, so the normal semifinite weight theory is unavoidable.

We state the next proposition in terms of weights, which is used in Sect. 6.4 together with Theorem A.54. The proposition was originally given in [27, Lemma 1.4.4] for faithful normal semifinite weights ψ, φ, but it is not difficult to remove the assumption of ψ being faithful.[3]

Proposition A.55 *Let* M_0 *be a von Neumann subalgebra of* M *and* $E : M \to M_0$ *be a faithful normal conditional expectation onto* M_0. *Let* φ, ψ *be normal semifinite*

[3] The author is indebted to A. Jenčová for this observation.

weights on M_0 *and assume that* φ *is faithful. Then*

$$[D(\psi \circ E) : D(\varphi \circ E)]_t = [D\psi : D\varphi]_t, \qquad t \in \mathbb{R}.$$

The next proposition supplements Lemma 4.20 with additional characterizations in terms of $[D\psi : D\varphi]_t$.

Proposition A.56 *Let* $\varphi, \psi \in M_*^+$ *with* $s(\psi) \le s(\varphi)$*. The following conditions are equivalent:*

(i) $\psi \circ \sigma_t^\varphi = \psi$ *for all* $t \in \mathbb{R}$*;*
(ii) $[D\psi : D\varphi]_t \in (s(\varphi)Ms(\varphi))_\varphi$ *(the centralizer of* $\varphi|_{s(\varphi)Ms(\varphi)}$*) for all* $t \in \mathbb{R}$*;*
(iii) $[D\psi : D\varphi]_t \in (s(\psi)Ms(\psi))_\psi$ *for all* $t \in \mathbb{R}$*;*
(iv) $t \in \mathbb{R} \mapsto [D\psi : D\varphi]_t$ *is a one-parameter group of unitaries in* $s(\psi)Ms(\psi)$*;*
(v) $h_\psi^{is} h_\varphi^{it} = h_\varphi^{it} h_\psi^{is}$ *for all* $s, t \in \mathbb{R}$*;*
(vi) $h_\psi h_\varphi = h_\varphi h_\psi$ *as elements of* $\widetilde{N}$ *(the* τ*-measurable operators affiliated with* N*).*

Definition A.57 We say that ψ *commutes* with φ if the equivalent conditions of Proposition A.56 hold. Condition (i) is often used to define the commutativity for normal functionals (also normal semifinite weights), but (v) and (vi) are quite natural definitions of commutativity that are available for any $\varphi, \psi \in M_*^+$ without $s(\psi) \le s(\varphi)$, as stated in Lemma 4.20.

For use in Sects. 3.3 and 5.3 we state the following lemmas. The first lemma was also given in [54, Lemma A.1], which generalizes [26, Theorem 3] and [28, Lemma 3.13]. In fact, the lemma can be shown by (A.21) and [123, Sec. 9.24] together with a common argument in analytic function theory. See also [81]. In the notation of [30, Definition 4.1] condition (ii) can be written as $\psi \le \lambda^{1/\delta}\varphi\ (\delta/2)$.

Lemma A.58 *For every* $\psi, \varphi \in M_*^+$ *and* $\delta > 0$*, the following conditions are equivalent:*

(i) $h_\psi^\delta \le \lambda h_\varphi^\delta$*, i.e.,* $\lambda h_\varphi^\delta - h_\psi^\delta \in L^{1/\delta}(M)_+$ *for some* $\lambda > 0$*;*
(ii) $s(\psi) \le s(\varphi)$ *and* $[D\psi : D\varphi]_t$ *extends to a* σ*-weakly continuous (*M*-valued) function* $[D\psi : D\varphi]_z$ *on* $-\delta/2 \le \operatorname{Im} z \le 0$ *which is analytic in the interior.*

If the above conditions hold, then $\|[D\psi : D\varphi]_z\| \le \lambda^{1/2}$ *and* $[D\psi : D\varphi]_z$ *is strongly continuous on* $-\delta/2 \le \operatorname{Im} z \le 0$*, and*

$$h_\psi^{p/2} = [D\psi : D\varphi]_{-ip/2} h_\varphi^{p/2}, \qquad 0 < p \le \delta. \tag{A.24}$$

The assumption of Lemma A.24 is the $\delta = 1$ case of the above lemma, so A in Lemma A.24 is equal to $[D\psi : D\varphi]_{-i/2}$.

Kosaki [81] proved the following, which contains Sakai's quadratic Radon–Nikodym theorem [116, 1.24.3] as the special case when $\delta = p = 1$.

Lemma A.59 *If $\psi, \varphi \in M_*^+$ and $h_\psi^\delta \le \lambda h_\varphi^\delta$ for some $\delta, \lambda > 0$, then for every $p > 0$ there exists a unique $k_p \in M_+$ with $0 \le k_p \le \lambda^{p/2\delta}$ such that $s(k_p) \le s(\varphi)$ and $h_\psi^p = k_p h_\varphi^p k_p$, where $s(k_p)$ is the support projection of k_p. Moreover, $h_\psi^p = k_p h_\varphi^p k_p$ implies that $(h_\varphi^{p/2} h_\psi^p h_\varphi^{p/2})^{1/2} = h_\varphi^{p/2} k_p h_\varphi^{p/2}$.*

The construction of k_p in [81] is as follows. By virtue of Furuta's inequality [41] (extended to τ-measurable operators) the above assumption implies that $(h_\varphi^{p/2} h_\psi^p h_\varphi^{p/2})^{1/2} \le \lambda^{p/2\delta} h_\varphi^p$ for any $p > 0$. Hence there exists a unique $b_p \in M$ with $\|b_p\| \le \lambda^{p/4\delta}$ such that $s(b_p^* b_p) \le s(\varphi)$ and $(h_\varphi^{p/2} h_\psi^p h_\varphi^{p/2})^{1/4} = b_p h_\varphi^{p/2}$. In fact, $b_p = [D\chi(p) : D\varphi]_{-ip/2}$, where $\chi(p) \in M_*^+$ is given by $h_{\chi(p)} = (h_\varphi^{p/2} h_\psi^p h_\varphi^{p/2})^{1/2p}$ (see (A.24)). Let $k_p := b_p^* b_p$. Then $h_\varphi^{p/2} h_\psi^p h_\varphi^{p/2} = h_\varphi^{p/2} k_p h_\varphi^p k_p h_\varphi^{p/2}$ and so $h_\psi^p = k_p h_\varphi^p k_p$ and $(h_\varphi^{p/2} h_\psi^p h_\varphi^{p/2})^{1/2} = h_\varphi^{p/2} k_p h_\varphi^{p/2}$.

A.8 Kosaki's L^p-Spaces

In this section, in connection with Haagerup's L^p-spaces in Sect. A.6, we give a brief survey of Kosaki's L^p-spaces [78]. A merit of Kosaki's L^p-spaces is that the construction itself allows us to apply complex interpolation techniques of the Riesz–Thorin type to them.

Let $\varphi_0 \in M_*^+$ be a distinguished faithful state and $h_0 := h_{\varphi_0} \in L^1(M)$. For each $\eta \in [0, 1]$, M is embedded into $L^1(M)$ by $x \in M \mapsto h_0^\eta x h_0^{1-\eta} \in L^1(M)$. Define the norm $\|h_0^\eta x h_0^{1-\eta}\| := \|x\|$ on $h_0^\eta M h_0^{1-\eta}$ $(\subset L^1(M))$, so that $h_0^\eta M h_0^{1-\eta} \cong M$. Then $(h_0^\eta M h_0^{1-\eta}, L^1(M))$ becomes a pair of compatible Banach spaces.

Definition A.60 Let $1 < p < \infty$ and $0 \le \eta \le 1$. *Kosaki's L^p-space $L^p(M, \varphi_0)_\eta$* with respect to φ_0 is defined to be the *complex interpolation space*

$$C_{1/p}(h_0^\eta M h_0^{1-\eta}, L^1(M))$$

equipped with the complex interpolation norm $\|\cdot\|_{p,\eta}$ $(= \|\cdot\|_{C_{1/p}})$, see [18] for general theory on the complex interpolation method. Moreover, we write $L^1(M, \varphi_0)_\eta := L^1(M)$ with $\|\cdot\|_{1,\eta} = \|\cdot\|_1$ and $L^\infty(M, \varphi_0)_\eta := h_0^\eta M h_0^{1-\eta}$ (identified with M) with $\|h_0^\eta x h_0^{1-\eta}\|_{\infty,\eta} = \|x\|$.

In particular, $L^p(M, \varphi_0)_\eta$ for $\eta = 0, 1$ and $1/2$ are respectively called the *left*, the *right* and the *symmetric L^p-spaces*. We may write the norm $\|\cdot\|_{p,\eta}$ as $\|\cdot\|_{p,\varphi_0,\eta}$ if we need to specify its dependence on φ_0.

Note that a more general embedding $x \in M \mapsto h_0^\eta x k_0^{1-\eta} \in L^1(M)$ was treated in [78], where $k_0 := h_{\psi_0}$ for another faithful state $\psi_0 \in M_*^+$.

Since Hölder's inequality (Theorem A.38) gives

$$\|h_0^\eta x h_0^{1-\eta}\|_1 \le \|h_0\|_1^\eta \|x\| \, \|h_0\|_1^{1-\eta} = \|x\|, \qquad x \in M,$$

we find from general properties of the complex interpolation method (see [18]) that for $1 < p' < p < \infty$,

$$\begin{aligned}(M =)\, L^\infty(M, \varphi_0)_\eta &\subset L^p(M, \varphi_0)_\eta \\ &\subset L^{p'}(M, \varphi_0)_\eta \subset L^1(M, \varphi_0)_\eta \,(= L^1(M)), \\ (\|x\| =)\, \|h_0^\eta x h_0^{1-\eta}\|_{\infty,\eta} &\ge \|h_0^\eta x h_0^{1-\eta}\|_{p,\eta} \\ &\ge \|h_0^\eta x h_0^{1-\eta}\|_{p',\eta} \ge \|h_0^\eta x h_0^{1-\eta}\|_1, \qquad x \in M.\end{aligned}$$

Note that $h_0^\eta M h_0^{1-\eta}$ is dense in $L^p(M, \varphi_0)_\eta$ for every $p \in [1, \infty)$.

The following is a main theorem [78, Theorem 9.1].

Theorem A.61 *Let $1 \le p \le \infty$ and $1/p + 1/q = 1$. Then*

$$L^p(M, \varphi_0)_\eta = h_0^{\eta/q} L^p(M) h_0^{(1-\eta)/q} \quad (\subset L^1(M)),$$

$$\|h_0^{\eta/q} a h_0^{(1-\eta)/q}\|_{p,\eta} = \|a\|_p, \qquad a \in L^p(M).$$

That is, $L^p(M) \cong L^p(M, \varphi_0)_\eta$ by the isometry $a \mapsto h_0^{\eta/q} a h_0^{(1-\eta)/q}$.

In particular, when $\eta = 0$, 1 and $1/2$, we have Kosaki's left, right and symmetric L^p-spaces as

$$L^p(M, \varphi_0)_L := L^p(M) h_0^{1/q}, \qquad L^p(M, \psi_0)_R := h_0^{1/q} L^p(M),$$

$$L^p(M, \varphi_0) := L^p(M, \varphi_0)_{\eta=1/2} = h_0^{1/2q} L^p(M) h_0^{1/2q}$$

with the norms

$$\|a h_0^{1/q}\|_{p,0} = \|h_0^{1/q} a\|_{p,1} = \|h_0^{1/2q} a h_0^{1/2q}\|_{p,1/2} = \|a\|_p, \qquad a \in L^p(M).$$

When $1 \le p < \infty$ and $1/p + 1/q = 1$, the L^p-L^q-duality of Kosaki's L^p-spaces can be given by transforming that of Haagerup's L^p-spaces in view of Theorem A.61. For instance, for $\eta, \eta' \in [0, 1]$ the duality pairing between $L^p(M, \varphi_0)_\eta$ and $L^q(M, \varphi_0)_{\eta'}$ is written as

$$\langle h_0^{\eta/q} a h_0^{(1-\eta)/q}, h_0^{\eta'/p} b h_0^{(1-\eta')/p} \rangle_{p,q} = \mathrm{tr}(ab), \qquad a \in L^p(M),\ b \in L^q(M).$$

When $\eta' = 1 - \eta$, this duality is somewhat convenient in the sense that for every $x, y \in M$,

$$\begin{aligned}
&\langle h_0^\eta x h_0^{1-\eta}, h_0^{1-\eta} y h_0^\eta \rangle_{p,q} \\
&\quad = \langle h_0^{\eta/q} (h_0^{\eta/p} x h_0^{(1-\eta)/p}) h_0^{(1-\eta)/q}, h_0^{(1-\eta)/p} (h_0^{(1-\eta)/q} y h_0^{\eta/q}) h_0^{\eta/p} \rangle_{p,q} \\
&\quad = \mathrm{tr}(h_0^{\eta/p} x h_0^{(1-\eta)/p} h_0^{(1-\eta)/q} y h_0^{\eta/q}) = \mathrm{tr}(h_0^\eta x h_0^{1-\eta} y),
\end{aligned}$$

whose last expression is independent of $p \in [1, \infty)$.

Remark A.62 When $\varphi \in M_*^+$ is not necessarily faithful with the support $e := s(\varphi) \in M$, Kosaki's L^p-space with respect to φ is defined with restriction to eMe. More specifically, for $0 \le \eta \le 1$ and $1 < p < \infty$, with the embedding $x \in eMe \mapsto h_\varphi^\eta x h_\varphi^{1-\eta} \in eL^1(M)e \; (= L^1(eMe))$ (see Remark A.44), Theorem A.61 says that

$$L^p(M, \varphi)_\eta = h_\varphi^{\eta/q} e L^p(M) e h_\varphi^{(1-\eta)/q} = h_\varphi^{\eta/q} L^p(M) h_\varphi^{(1-\eta)/q} \quad (\subset eL^1(M)e)$$

with the norm $\|h_\varphi^{\eta/q} a h_\varphi^{(1-\eta)/q}\|_{p,\eta} = \|a\|_p$ for $a \in eL^p(M)e$, where $1/p+1/q = 1$. In particular, the symmetric case

$$L^p(M, \varphi) := L^p(M, \varphi)_{\eta=1/2} = h_\varphi^{1/2q} L^p(M) h_\varphi^{1/2q}$$

plays a role in Sect. 3.3.

Appendix B
Preliminaries on Positive Self-Adjoint Operators

Let A be a positive self-adjoint operator on a Hilbert space $\mathcal{H}$ having the spectral decomposition

$$A = \int_0^\infty t\, dE_t$$

with the spectral resolution E_t $(0 \le t < \infty)$. It is well-known that $\xi \in \mathcal{H}$ belongs to $\mathcal{D}(A)$, the domain of A, if and only if $\int_0^\infty t^2\, d\|E_t\xi\|^2 < +\infty$, and in this case, $\|A\xi\|^2 = \int_0^\infty t^2\, d\|E_t\xi\|^2$. For a complex-valued Borel function f on $[0, \infty)$, the *Borel functional calculus* $f(A)$ is defined by

$$f(A) := \int_0^\infty f(t)\, dE_t,$$

whose domain is $\mathcal{D}(f(A)) = \left\{\xi \in \mathcal{H} : \int_0^\infty |f(t)|^2\, d\|E_t\xi\|^2 < +\infty\right\}$. If $\xi \in \mathcal{D}(f(A))$, then we have $\|f(A)\xi\|^2 = \int_0^\infty |f(t)|^2\, d\|E_t\xi\|^2$. The operator $f(A)$ is self-adjoint, positive self-adjoint, and bounded if f is real-valued, non-negative, and bounded, respectively. For instance,

$$A^p = \int_0^\infty t^p\, dE_t \qquad (p > 0),$$

$$(s1 + A)^{-1} = \int_0^\infty \frac{1}{s+t}\, dE_t, \quad A(1 + sA)^{-1} = \int_0^\infty \frac{t}{1+st}\, dE_t \qquad (s > 0).$$

Moreover, when A is non-singular (i.e., the support $s(A) = 1$),

$$A^{-1} = \int_0^\infty t^{-1}\, dE_t, \qquad A^{is} = \int_0^\infty t^{is}\, dE_t \quad (s \in \mathbb{R}).$$

F. Hiai, *Quantum f-Divergences in von Neumann Algebras*,
Mathematical Physics Studies, https://doi.org/10.1007/978-981-33-4199-9

In fact, A^{-1} and A^{is} are defined for singular A as well with restriction to $s(A)\mathcal{H}$ (or in the sense of the generalized inverse), that is, $A^{-1} = \int_{(0,\infty)} t^{-1}\,dE_t$ and similarly for A^{is}. (Note that A^0 is used to mean either $A^0 = 1$ or $A^0 = s(A)$ according to the situation.)

A fundamental property of positive self-adjoint operators is stated in the next theorem, whose details are found in [123, Chap. 9] (also [56, Appendix A]).

Theorem B.1 *Let A be a self-adjoint operator on $\mathcal{H}$ with the spectral decomposition $A = \int_0^\infty t\,dE_t$. Let $\alpha > 0$. Then, for every $\xi \in \mathcal{H}$ the following conditions are equivalent:*

(i) $\xi \in \mathcal{D}(A^\alpha)$;
(ii) $\int_0^\infty t^{2\alpha}\,d\|E_t\xi\|^2 < +\infty$;
(iii) *there exists an $\mathcal{H}$-valued bounded weakly (equivalently, strongly) continuous function f on $-\alpha \le \operatorname{Im} z \le 0$, weakly (equivalently, strongly) analytic in $-\alpha < \operatorname{Im} z < 0$, such that $f(s) = A^{is}\xi$ for all $s \in \mathbb{R}$.*

Furthermore, the function $f(z)$ in (iii) is unique and $A^\alpha\xi = f(-i\alpha)$ holds.

The above theorem may be a reformulation of the famous *Stone's representation theorem* from the viewpoint of the analytic generator. For a strongly continuous one-parameter unitary group $U_s = e^{isH}$ on $\mathcal{H}$, where H is a self-adjoint generator, the H is usually obtained by the real analytic method as $H = \lim_{t\to 0}(U_t - 1)/it$, while the above theorem says that the analytic generator $A = e^H$ can be obtained by a complex analytic method.

From the analytic continuation characterization in (iii) above, one can show the following:

Proposition B.2 *Let A be a positive self-adjoint operator on $\mathcal{H}$. Let $\alpha, \beta \ge 0$. Then we have $A^{\alpha+\beta} = A^\alpha A^\beta$; more precisely, for $\xi \in \mathcal{H}$, $\xi \in \mathcal{D}(A^{\alpha+\beta})$ if and only if $\xi \in \mathcal{D}(A^\beta)$ and $A^\beta\xi \in \mathcal{D}(A^\alpha)$, and in this case, $A^{\alpha+\beta}\xi = A^\alpha(A^\beta\xi)$. (Either convention $A^0 = 1$ or $A^0 = s(A)$ is available here.)*

Moreover, we have $(A^{-1})^\alpha = (A^\alpha)^{-1}$ (simply denoted by $A^{-\alpha}$) and $A^{-(\alpha+\beta)} = A^{-\alpha}A^{-\beta}$.

For Borel functional calculus we have general formulas like

$$(f+g)(A) = \overline{f(A) + g(A)}, \qquad (fg)(A) = \overline{f(A)g(A)}.$$

A point of the above proposition is that $A^{\alpha+\beta} = A^\alpha B^\beta$ holds without closure for $\alpha, \beta \in \mathbb{R}$ with $\alpha\beta \ge 0$. The equality further extends to all $\alpha, \beta \in \mathbb{C}$ with $\operatorname{Re}\alpha \cdot \operatorname{Re}\beta \ge 0$.

Another fundamental aspect of positive self-adjoint operators on $\mathcal{H}$ is their correspondence to closed (densely defined) positive quadratic forms on $\mathcal{H}$. A function

$q : \mathcal{D}(q) \to [0, +\infty)$, where $\mathcal{D}(q)$ is a linear subspace of $\mathcal{H}$, is called a *positive quadratic form* if:

- $q(\lambda\xi) = |\lambda|^2 q(\xi)$ for all $\xi \in \mathcal{D}(q)$, $\lambda \in \mathbb{C}$,
- $q(\xi + \eta) + q(\xi - \eta) = 2q(\xi) + 2q(\eta)$ for all $\xi, \eta \in \mathcal{D}(q)$.

Such a q is said to be *closed* if $\{\xi_n\} \subset \mathcal{D}(q)$, $\xi \in \mathcal{H}$, $\|\xi_n - \xi\| \to 0$ and $q(\xi_n - \xi_m) \to 0$ as $n, m \to \infty$, then $\xi \in \mathcal{D}(q)$ and $q(\xi_n - \xi) \to 0$.

Now we state the next important theorem. For details see [74, Chap. 6, Sec. 2.6], [114, Theorem VIII.15], [117, Chap. 10], [119, Theorem 2] or [120, Sec. 7.5] (also [56, Appendix A]).

Theorem B.3 *Let q be a densely defined (i.e., $\mathcal{D}(q)$ is dense in $\mathcal{H}$) positive quadratic form, and define*

$$\overline{q}(\xi) := \begin{cases} q(\xi) & \text{if } \xi \in \mathcal{D}(q), \\ +\infty & \text{if } \xi \in \mathcal{H} \setminus \mathcal{D}(q). \end{cases} \tag{B.1}$$

Then the following conditions are equivalent:

(a) *q is closed;*
(b) *$\overline{q}$ is lower semicontinuous on $\mathcal{H}$;*
(c) *there exists a (unique) positive self-adjoint operator A on $\mathcal{H}$ such that $\mathcal{D}(A^{1/2}) = \mathcal{D}(q)$ and*

$$q(\xi) = \|A^{1/2}\xi\|^2, \qquad \xi \in \mathcal{D}(A^{1/2}). \tag{B.2}$$

In this way, a positive self-adjoint operator A on $\mathcal{H}$ corresponds one-to-one to a closed densely defined positive quadratic form q given in (B.2). Below we will sometimes use the notation $\|A^{1/2}\xi\|^2$ for all $\xi \in \mathcal{H}$ to mean $\overline{q}(\xi)$ in (B.1), that is, we set

$$\|A^{1/2}\xi\|^2 := \int_0^\infty t\, d\|E_t\xi\|^2 = \begin{cases} \|A^{1/2}\xi\|^2 < +\infty & \text{if } \xi \in \mathcal{D}(A^{1/2}), \\ +\infty & \text{otherwise.} \end{cases} \tag{B.3}$$

The following fact is well-known, see, e.g., [122, Sec. A.4].

Proposition B.4 *Let A, B be positive self-adjoint operators on $\mathcal{H}$. Then the following conditions are equivalent:*

(i) *$\mathcal{D}(B^{1/2}) \subset \mathcal{D}(A^{1/2})$ and $\|A^{1/2}\xi\|^2 \le \|B^{1/2}\xi\|^2$ for all $\xi \in \mathcal{D}(B^{1/2})$;*
(ii) *there exists a core $\mathcal{D}$ of $B^{1/2}$ such that $\mathcal{D} \subset \mathcal{D}(A^{1/2})$ and $\|A^{1/2}\xi\|^2 \le \|B^{1/2}\xi\|^2$ for all $\xi \in \mathcal{D}$;*
(iii) *$(s1 + B)^{-1} \le (s1 + A)^{-1}$ for some (equivalently, any) $s > 0$;*
(iv) *$A(1 + sA)^{-1} \le B(1 + sB)^{-1}$ for some (equivalently, any) $s > 0$.*

For positive self-adjoint operators A, B we write $A \le B$ (*in the form sense*) if the equivalent conditions of Proposition B.4 hold. From Proposition B.4 we have the next result.

Proposition B.5 *For densely defined closed operators A, B on $\mathcal{H}$ the following are equivalent:*

(i) $A^*A \le B^*B$;
(ii) *there exists a core $\mathcal{D}$ of B such that $\mathcal{D} \subset \mathcal{D}(A)$ and $\|A\xi\| \le \|B\xi\|$ for all $\xi \in \mathcal{D}$.*

Proposition B.6 *Let A, B be positive self-adjoint operators with $s(A) = s(B)$, where $s(A)$ is the support projection of A. If $A \le B$, then $A^{-1} \ge B^{-1}$, where A^{-1} is defined with restriction to $s(A)\mathcal{H}$ (i.e., in the sense of the generalized inverse).*

Proof Let $e := s(A) = s(B)$. By Proposition B.4 it is clear that $A \le B$ if and only if $(e+A)^{-1} \ge (e+B)^{-1}$, which is equivalent to $(e+A^{-1})^{-1} \le (e+B^{-1})^{-1}$ since $(e+A^{-1})^{-1} = e - (e+A)^{-1}$. Hence the assertion holds.

Lemma B.7 *Let A, B be positive self-adjoint operators on $\mathcal{H}$ such that $A \le B$. Then, for every operator monotone function $f \ge 0$ on $[0, \infty)$ we have $f(A) \le f(B)$, that is,*

$$\|f(A)^{1/2}\xi\|^2 \le \|f(B)^{1/2}\xi\|^2$$

for all $\xi \in \mathcal{H}$ (in the sense of (B.3)*).*

Proof Note that $\|f(A)^{1/2}\xi\|^2 = \int_0^\infty f(t)\,d\|E_t\xi\|^2$ for every $\xi \in \mathcal{H}$ with the spectral decomposition $A = \int_0^\infty t\,dE_t$. Now we recall that f has the integral expression

$$f(t) = a + bt + \int_{[0,\infty)} \frac{t}{t+s}\,d\mu(s), \qquad t \in [0,\infty),$$

where $a, b \ge 0$ and μ is a positive measure on $[0,\infty)$ with $\int_{[0,\infty)} (1+s)^{-1}\,d\mu(s) < +\infty$. By Fubini's theorem we have

$$\begin{aligned}
&\int_0^\infty f(t)\,d\|E_t\xi\|^2 \\
&= a\|\xi\|^2 + b\|A^{1/2}\xi\|^2 + \int_{[0,\infty)} \left[\int_0^\infty \left(1 - \frac{s}{t+s}\right) d\|E_t\xi\|^2\right] d\mu(s) \\
&= a\|\xi\|^2 + b\|A^{1/2}\xi\|^2 + \int_{[0,\infty)} \left[\|\xi\|^2 - s\langle \xi, (s1+A)^{-1}\xi\rangle\right] d\mu(s). \qquad \text{(B.4)}
\end{aligned}$$

The same expression as above holds for B as well. Condition (iii) of Proposition B.4 implies that for every $\xi \in \mathcal{H}$,

$$\int_{[0,\infty)} \big[\|\xi\|^2 - s\langle \xi, (sI + A)^{-1}\xi\rangle\big]\, d\mu(s)$$
$$\le \int_{[0,\infty)} \big[\|\xi\|^2 - s\langle \xi, (sI + B)^{-1}\xi\rangle\big]\, d\mu(s),$$

as well as $\|A^{1/2}\xi\|^2 \le \|B^{1/2}\xi\|^2$. Therefore, we see that $\|f(A)^{1/2}\xi\|^2 \le \|f(B)^{1/2}\xi\|^2$ for all $\xi \in \mathcal{H}$. (In particular, this implies that $\mathcal{D}(f(B)^{1/2}) \subset \mathcal{D}(f(A)^{1/2})$.)

The following fact on convergence of positive self-adjoint operators is well-known, see [114, Sec. VIII.7], [122, Sec. A.3] (also [56, Appendix A]).

Proposition B.8 *Let A_n ($n \in \mathbb{N}$) and A be positive self-adjoint operators on $\mathcal{H}$. Then the following conditions are equivalent:*

(i) $(s1 + A_n)^{-1} \to (s1 + A)^{-1}$ *strongly for some (equivalently, any) $s > 0$;*
(ii) $(i + A_n)^{-1} \to (i + A)^{-1}$ *strongly;*
(iii) $A_n(1 + sA_n)^{-1} \to A(1 + sA)^{-1}$ *strongly for some (equivalently, any) $s > 0$.*

Moreover, assume that all A_n and A are non-singular, so we write $A_n = e^{H_n}$ and $A = e^H$, where $H_n := \log A_n$ and $H := \log A$. Then the above conditions are also equivalent to the following:

(iv) $(i + H_n)^{-1} \to (i + H)^{-1}$ *strongly;*
(v) $A_n^{it} \to A^{it}$ *strongly for all $t \in \mathbb{R}$.*

If the equivalent conditions (i)–(iii) of Proposition B.8 hold, then we say that A_n converges to A *in the strong resolvent sense*. In particular, when $A_1 \le A_2 \le \cdots$ (in the form sense), it is also well-known ([74, Chap. 8, Sec. 3.4], [120, Sec. 7.5]) that $A_n \to A$ in the strong resolvent sense (or $(1 + A_n)^{-1} \searrow (1 + A)^{-1}$ strongly) if and only if $\|A^{1/2}\xi\|^2 = \sup_n \|A_n^{1/2}\xi\|^2$ for all $\xi \in \mathcal{H}$. On the other hand, when $A_1 \ge A_2 \ge \cdots$, the convergence $A_n \to A$ in the strong resolvent sense (or $(1+A_n)^{-1} \nearrow (1+A)^{-1}$ strongly) does not imply that $\|A^{1/2}\xi\|^2 = \inf_n \|A_n^{1/2}\xi\|^2$, because $q_0(\xi) = \inf_n \|A_n^{1/2}\xi\|^2$ is not necessarily lower semicontinuous on $\mathcal{H}$, while $\|A^{1/2}\xi\|^2$ is lower semicontinuous on $\mathcal{H}$ (from condition (b) of Theorem B.3). For more in the latter case, see [120, Sec. 7.5], [82].

The next proposition is used in Sect. 3.3.

Proposition B.9 *Let A_n ($n \in \mathbb{N}$) and A be positive self-adjoint operators on $\mathcal{H}$ such that*

$$(1 + A_n)^{-1} \searrow (1 + A)^{-1} \quad \textit{strongly},$$

that is, A_n increases to A in the strong resolvent sense. Then, for every operator monotone function $f \ge 0$ on $[0, \infty)$, we have

$$(1 + f(A_n))^{-1} \searrow (1 + f(A))^{-1} \quad \textit{strongly},$$

that is, $f(A_n)$ increases to $f(A)$ in the strong resolvent sense.

Proof Since $f(A_1) \le f(A_2) \le \cdots$ by Lemma B.7, in view of the fact mentioned before the proposition, it suffices to show that $\|f(A_n)^{1/2}\xi\|^2 \nearrow \|f(A)^{1/2}\xi\|^2$ for all $\xi \in \mathcal{H}$. Note that $\|f(A)^{1/2}\xi\|^2$ is expressed as in (B.4) and similarly for $\|f(A_n)^{1/2}\xi\|^2$. Since $(s1 + A_n)^{-1} \searrow (s1 + A)^{-1}$ strongly for any $s > 0$ by Proposition B.8, we can apply the monotone convergence theorem to see that

$$\int_{[0,\infty)} \left[\|\xi\|^2 - s\langle \xi, (s1 + A_n)^{-1}\xi\rangle\right] d\mu(s)$$
$$\nearrow \int_{[0,\infty)} \left[\|\xi\|^2 - s\langle \xi, (s1 + A)^{-1}\xi\rangle\right] d\mu(s).$$

Together with $\|A_n^{1/2}\xi\|^2 \nearrow \|A^{1/2}\xi\|^2$ this yields the desired conclusion.

Here we recall the notion of form sums for two positive self-adjoint operators on $\mathcal{H}$, which was introduced in [74]. Let A, B be positive self-adjoint operators on $\mathcal{H}$. Then $q(\xi) := \|A^{1/2}\xi\|^2 + \|B^{1/2}\xi\|^2$ for $\xi \in \mathcal{D}(q) := \mathcal{D}(A^{1/2}) \cap \mathcal{D}(B^{1/2})$ is a closed positive quadratic form on $\mathcal{K} := \overline{\mathcal{D}(q)}$, so there exists a unique self-adjoint operator C on $\mathcal{K}$ such that $\mathcal{D}(C^{1/2}) = \mathcal{D}(q)$ and

$$\|C^{1/2}\xi\|^2 = \|A^{1/2}\xi\|^2 + \|B^{1/2}\xi\|^2, \qquad \xi \in \mathcal{D}(A^{1/2}) \cap \mathcal{D}(B^{1/2}). \tag{B.5}$$

The C is called the *form sum* of A, B and denoted by $A \dotplus B$. The form sum $A \dotplus B$ is a positive self-adjoint operator on $\mathcal{H}$ whenever $\mathcal{D}(A^{1/2}) \cap \mathcal{D}(B^{1/2})$ is dense in $\mathcal{H}$. For instance, this is the case of Proposition A.22 (5).

The next proposition is given here to use it in Sect. 3.1.

Proposition B.10 *Let f be a positive operator monotone decreasing function on $(0, \infty)$ (for example, $f(t) = t^\alpha$ $(t > 0)$ with $-1 \le \alpha \le 0$). Assume that A, B are positive self-adjoint operators on $\mathcal{H}$ such that $\mathcal{D}(A^{1/2}) \cap \mathcal{D}(B^{1/2})$ is dense in $\mathcal{H}$.*[4]

[4] In fact, without this assumption let $\mathcal{K} := \overline{\mathcal{D}(A^{1/2}) \cap \mathcal{D}(B^{1/2})}$. Then $C := \lambda A \dotplus (1 - \lambda)B$ $(0 < \lambda < 1)$ is a positive self-adjoint operator on $\mathcal{K}$ (considered ∞ on $\mathcal{K}^\perp$). The conclusion remains true if $\|f(C)^{1/2}\xi\|$ is understood as

$$\|f(C)^{1/2}\xi\|^2 = \int_{[0,\infty)} f(t)\, d\|G_t\xi\|^2 + f(\infty)\|P_{\mathcal{K}^\perp}\xi\|^2,$$

where $C = \int_0^\infty t\, dG_t$ is the spectral decomposition of C on $\mathcal{K}$, $P_{\mathcal{K}^\perp}$ is the projection onto $\mathcal{K}^\perp$, and $f(\infty) := \lim_{t\to\infty} f(t)$.

Then for every $\xi \in \mathcal{H}$ *we have*

$$\|f(\lambda A \dot{+} (1-\lambda)B)^{1/2}\xi\| \le \|f(A)^{1/2}\xi\|^{\lambda}\|f(B)^{1/2}\xi\|^{1-\lambda}, \qquad 0 \le \lambda \le 1,$$

where $\|f(A)^{1/2}\xi\|$ *etc. are understood in the sense of* (B.3), *i.e.*,

$$\|f(A)^{1/2}\xi\|^2 = \int_{[0,\infty)} f(t)\, d\|E_t\xi\|^2$$

with the spectral decomposition $A = \int_0^\infty t\, dE_t$ *and* $f(0) := f(0^+) \in (0, +\infty]$.

Proof Let $B = \int_0^\infty t\, dF_t$ be the spectral decomposition as well as that of A above. We may assume that $0 < \lambda < 1$. Let $C := \lambda A \dot{+} (1-\lambda)B$. First, when $A, B \in B(\mathcal{H})_+$ and they are invertible, the result is a consequence of [6, Theorem 3.1] saying that $A \mapsto \log\langle \xi, f(A)\xi\rangle$ is convex on $\{A \in B(\mathcal{H})_+ : \text{invertible}\}$ for every $\xi \in \mathcal{H}$. Next, assume that $A, B \ge \varepsilon I$ (hence $C \ge \varepsilon I$ as well) for some $\varepsilon > 0$. For each $n \in \mathbb{N}$ set

$$A_n := \int_{[0,n]} t\, dE_t + n(1 - E_n), \qquad B_n := \int_{[0,n]} t\, dF_t + n(1 - F_n),$$
$$C_n := \lambda A_n + (1-\lambda)B_n.$$

Note that A_n, B_n and C_n are all bounded and invertible. Then from the first case above one has

$$\|f(C_n)^{1/2}\xi\| \le \|f(A_n)^{1/2}\xi\|^{\lambda}\|f(B_n)^{1/2}\xi\|^{1-\lambda}. \tag{B.6}$$

Since $C^{-1} \le C_n^{-1}$ in $B(\mathcal{H})_+$ and $f(t^{-1})$ is operator monotone on $(0, \infty)$, it follows that $f(C) \le f(C_n)$ in $B(\mathcal{H})_+$ and hence

$$\|f(C)^{1/2}\xi\| \le \|f(C_n)^{1/2}\xi\|. \tag{B.7}$$

Moreover, one sees that

$$\|f(A_n)^{1/2}\xi\|^2 = \int_{[0,n]} f(t)\, d\|E_t\xi\|^2 + f(n)\|(1-E_n)\xi\|^2$$
$$\longrightarrow \int_0^\infty f(t)\, d\|E_t\xi\|^2 = \|f(A)^{1/2}\|^2$$

as $n \to \infty$, and similarly $\|f(B_n)^{1/2}\xi\|^2 \to \|f(B)^{1/2}\xi\|^2$. Combining these with (B.6) and (B.7) gives the result in this case.

Finally, let A, B be general as in the proposition. For every $\varepsilon > 0$, since $\lambda(A + \varepsilon 1) \dot{+} (1 - \lambda)(B + \varepsilon 1) = C + \varepsilon 1$, the above second case implies that

$$\|f(C + \varepsilon 1)^{1/2}\xi\| \le \|f(A + \varepsilon 1)^{1/2}\xi\|^{\lambda} \|f(B + \varepsilon 1)^{1/2}\xi\|^{1-\lambda}. \tag{B.8}$$

By the monotone convergence theorem we have

$$\begin{aligned}\|f(A + \varepsilon 1)^{1/2}\xi\|^2 &= \int_{[0,\infty)} f(t + \varepsilon)\, d\|E_t\xi\|^2 \\ &\nearrow \int_{[0,\infty)} f(t)\, d\|E_t\xi\|^2 = \|f(A)^{1/2}\xi\|^2,\end{aligned}$$

and similarly for $\|f(B + \varepsilon 1)^{1/2}\xi\|^2$ and $\|f(C + \varepsilon 1)^{1/2}\xi\|^2$. Therefore, we have the result by taking the limit of (B.8) as $\varepsilon \searrow 0$.

Finally, we state an interpolation result related to positive self-adjoint operators. See [18] and [113, Appendix to IX.4] for general theory of interpolation Banach spaces. Let A be a positive self-adjoint operator on $\mathcal{H}$. For each $\theta \in (0, 1]$ consider the inner product on $\mathcal{D}(A^{\theta/2})$ defined by

$$\langle \xi, \eta \rangle_\theta := \langle A^{\theta/2}\xi, A^{\theta/2}\eta \rangle, \qquad \xi, \eta \in \mathcal{D}(A^{\theta/2}),$$

and let $\mathcal{H}_\theta$ be the Hilbert space by completing $\mathcal{D}(A^{\theta/2})$ with respect to $\langle \cdot, \cdot \rangle_\theta$. (Alternatively, $\mathcal{H}_\theta$ is also obtained by completing $\mathcal{D}(A^\theta)$ with respect to $\langle \xi, \eta \rangle_\theta := \langle \xi, A^\theta \eta \rangle$ for $\xi, \eta \in \mathcal{D}(A^\theta)$.) It follows from [113, p. 35] that $(\mathcal{H}_0 = \mathcal{H}, \mathcal{H}_1)$ becomes a compatible pair of Hilbert spaces. Moreover, it is known [93] that $\mathcal{H}_\theta$ $(0 < \theta < 1)$ becomes an *exact interpolation space* of exponent θ between $\mathcal{H}_0 = \mathcal{H}$ and $\mathcal{H}_1$. In fact, the $\mathcal{H}_\theta$ is an interpolation space in the complex interpolation method and is also called the *geometric interpolation space* of exponent θ. More specifically, we state the next proposition from [93, Theorem 1.1], whose statement is slightly different from (but equivalent to) that in [93] and is convenient to use in Sect. 7.1.

Proposition B.11 *Let A, B be positive self-adjoint operators on Hilbert spaces $\mathcal{H}, \mathcal{K}$, respectively. For each $\theta \in (0, 1]$ let $\mathcal{K}_\theta$ be the interpolation Hilbert space induced from $(\mathcal{K}, B)$ similarly to $\mathcal{H}_\theta$ from $(\mathcal{H}, A)$ stated as above. If $V : \mathcal{H} \to \mathcal{K}$ is a bounded linear operator with norm $\|V\|_0 = \|V\|_{\mathcal{H}\to\mathcal{K}}$ such that $V\mathcal{D}(A^{1/2}) \subset \mathcal{D}(B^{1/2})$ and*

$$\|B^{1/2}V\xi\| \le \|V\|_1 \|A^{1/2}\xi\|, \qquad \xi \in \mathcal{D}(A^{1/2}),$$

with $\|V\|_1 < +\infty$, then for every $\theta \in (0, 1]$ we have $V\mathcal{D}(A^{\theta/2}) \subset \mathcal{D}(B^{\theta/2})$ and

$$\|B^{\theta/2}V\xi\| \le \|V\|_0^{1-\theta} \|V\|_1^{\theta} \|A^{\theta/2}\xi\|, \qquad \xi \in \mathcal{D}(A^{\theta/2}).$$

Appendix C
Operator Convex Functions on (0, 1)

As is well-known, e.g., [51, Theorem 2.7.6], a real function h on $(-1, 1)$ is operator convex if and only if it has the integral expression

$$h(t) = a + bt + \int_{[-1,1]} \frac{t^2}{1 - ut} \, d\lambda(u), \qquad t \in (-1, 1), \tag{C.1}$$

where $a, b \in \mathbb{R}$ and a finite positive measure λ on $[-1, 1]$ are uniquely determined.

On the other hand, a real function f on $(0, \infty)$ is operator convex if and only if it has the integral expression in (2.5) of Sect. 2.2, where $a, b \in \mathbb{R}$, $c \geq 0$ and a positive measure μ on $[0, \infty)$ with $\int_{[0,\infty)} (1 + s)^{-1} \, d\mu(s) < +\infty$ are uniquely determined.

In Theorem 4.8 of Sect. 4.2 the function $g(t) := (1 - t) f\big(\frac{t}{1-t}\big)$ on $(0, 1)$ induced from an operator convex function f on $(0, \infty)$ plays a role. In this appendix we prove the next theorem, which says that the correspondence $f \leftrightarrow g$ is bijective between the operator convex functions on $(0, \infty)$ and those on $(0, 1)$.

Theorem C.1 *Let f be a real function on $(0, \infty)$ and let $g(t) := (1 - t) f\big(\frac{t}{1-t}\big)$ for $t \in (0, 1)$. Then the following conditions are equivalent:*

(a) *f is operator convex on $(0, \infty)$.*
(b) *g is operator convex on $(0, 1)$.*
(c) *There exist $\alpha, \beta \in \mathbb{R}$, $\gamma \geq 0$ and a positive measure ν on $(-\infty, 1]$ satisfying $\int_{(-\infty,1]} (2 - w)^{-1} \, d\nu(w) < +\infty$ such that*

$$g(t) = \alpha + \beta t + \gamma \frac{(2t - 1)^2}{t} + \int_{(-\infty,1]} \frac{(2t - 1)^2}{1 - wt} \, d\nu(w), \quad t \in (0, 1). \tag{C.2}$$

F. Hiai, *Quantum f-Divergences in von Neumann Algebras*,
Mathematical Physics Studies, https://doi.org/10.1007/978-981-33-4199-9

Moreover, in this case,

$$g(0^+) = f(0^+), \qquad g(1^-) = f'(\infty), \tag{C.3}$$

and α, β, γ and the measure ν in (c) are uniquely determined by f (hence by g).

Proof (b) $\iff$ (c). Obviously, g is operator convex on $(0, 1)$ if and only if so is $g\bigl(\frac{t+1}{2}\bigr)$ on $(-1, 1)$. So we may show that expression (C.1) of $h(t) = g\bigl(\frac{t+1}{2}\bigr)$ is transformed into (C.2) of g and vice versa. From (C.1) we write for every $t \in (0, 1)$,

$$\begin{aligned} g(t) &= a + b(2t-1) + \int_{[-1,1]} \frac{(2t-1)^2}{1-u(2t-1)}\, d\lambda(u) \\ &= (a-b) + 2bt + \frac{\lambda(\{-1\})}{2}\frac{(2t-1)^2}{t} + \int_{(-1,1]} \frac{(2t-1)^2}{1-\frac{2u}{1+u}t}\frac{d\lambda(u)}{1+u}. \end{aligned}$$

By transformation $w = \frac{2u}{1+u} : (-1, 1] \to (-\infty, 1]$, define a measure ν on $(-\infty, 1]$ by $d\nu(w) = (1+u)^{-1}\, d\lambda(u)$. Since $\frac{1}{1+u} = \frac{2-w}{2}$, we have

$$\int_{(-\infty,1]} (2-w)^{-1}\, d\nu(w) = \int_{(-1,1]} d\lambda(u) < +\infty.$$

Hence we have expression (C.2) of g. The converse transformation from (C.2) into (C.1) can similarly be shown, so (b) $\iff$ (c) follows.

(a) $\iff$ (b). For this we may show that expression (2.5) of f is transformed to (C.2) of g and vice versa. From (2.5) we write for every $t \in (0, 1)$,

$$\begin{aligned} g(t) &= (1-t)f\left(\frac{t}{1-t}\right) \\ &= a(1-t) + b(2t-1) + c\frac{(2t-1)^2}{1-t} + \int_{[0,\infty)} \frac{(2t-1)^2}{t+s(1-t)}\, d\mu(s) \\ &= (a-b) + (2b-a)t + c\frac{(2t-1)^2}{1-t} + \mu(\{0\})\frac{(2t-1)^2}{t} \\ &\quad + \int_{(0,\infty)} \frac{(2t-1)^2}{1-\frac{s-1}{s}t}\frac{d\mu(s)}{s}. \end{aligned}$$

By transformation $w = \frac{s-1}{s} : (0, \infty) \to (-\infty, 1)$, define a measure ν on $(-\infty, 1)$ by $d\nu(w) = s^{-1}\, d\mu(s)$. Since $\frac{1}{2-w} = \frac{s}{1+s}$, we have

$$\int_{(0,\infty)} (2-w)^{-1}\, d\nu(w) = \int_{(0,\infty)} (1+s)^{-1}\, d\mu(s) < +\infty.$$

Hence, setting $\nu(\{1\}) := c$, we find that g has an expression in the form (C.2).

Conversely, from expression (C.2) of g we have for every $t \in (0, \infty)$,

$$\begin{aligned} f(t) &= (t+1)g\left(\frac{t}{t+1}\right) \\ &= \alpha(t+1) + \beta t + \gamma\frac{(t-1)^2}{t} + \int_{(-\infty,1]} \frac{(t-1)^2}{(t+1)-wt}\,d\nu(w) \\ &= \alpha + (\alpha+\beta)t + \gamma\frac{(t-1)^2}{t} + \nu(\{1\})(t-1)^2 + \int_{(-\infty,1)} \frac{(t-1)^2}{t+\frac{1}{1-w}}\,\frac{d\nu(w)}{1-w}. \end{aligned}$$

By transformation $s = \frac{1}{1-w} : (-\infty, 1) \to (0, \infty)$, define a measure μ on $(0, \infty)$ by $d\mu(s) = (1-w)^{-1}\,d\nu(w)$. Since $\frac{1}{1+s} = \frac{1-w}{2-w}$, we have $\int_{(0,\infty)}(1+s)^{-1}\,d\mu(s) = \int_{(-\infty,1)}(2-w)^{-1}\,d\nu(w) < +\infty$. Hence, setting $\mu(\{0\}) := \gamma$, we find that f has an expression in the form (2.5). So (a) $\Longleftrightarrow$ (b) has been shown.

Finally, the equalities in (C.3) are immediate and the uniqueness in expression (C.2) follows from that in (C.1) (or in (2.5)).

Remark C.2 The above proof indeed presents a proof of the integral expression in (2.5) of an operator convex function f on $(0, \infty)$ based on the well-known integral expression in (C.1) of an operator convex function on $(-1, 1)$.

From $\gamma = \mu(\{0\})$ and $\nu(\{1\}) = c$ given in the proof of Theorem C.1, one can easily find that

$$\gamma = \lim_{t\to 0^+} tf(t) = \lim_{t\to 0^+} tg(t), \quad \nu(\{1\}) = \lim_{t\to\infty} \frac{f(t)}{t^2} = \lim_{t\to 1^-} (1-t)g(t).$$

Thus, we have the following corollary, which is used in the proof of Theorem 8.4 in Sect. 8.2.

Corollary C.3 *Let g be an operator convex function on* $(0, 1)$ *such that*

$$\lim_{t\to 0^+} tg(t) = \lim_{t\to 1^-} (1-t)g(t) = 0.$$

Then g has the integral expression

$$g(t) = \alpha + \beta t + \int_{(-\infty,1)} \frac{(2t-1)^2}{1-wt}\,d\nu(w), \qquad t \in (0, 1),$$

where $\alpha, \beta \in \mathbb{R}$ and ν is a positive measure on $(-\infty, 1)$ with $\int_{(-\infty,1)}(2-w)^{-1}\,d\nu(w) < +\infty$.

Corollary C.4 *Let g be a continuous real function on $[0, 1]$. Then the following conditions are equivalent:*

(i) *g is operator convex on $[0, 1]$.*
(ii) *There exist $\alpha, \beta \in \mathbb{R}$ and an operator monotone function k on $[0, \infty)$ such that*

$$g(t) = \alpha + \beta t - (1-t)k\left(\frac{t}{1-t}\right), \qquad t \in [0, 1).$$

(iii) *There exist $\alpha, \beta \in \mathbb{R}$ and a positive measure μ on $(0, \infty)$ satisfying $\int_{(0,\infty)} (1+s)^{-1}\, d\mu(s) < +\infty$ such that*

$$g(t) = \alpha(1-t) + \beta t - \int_{(0,\infty)} \frac{t(1-t)}{t + s(1-t)}\, d\mu(s), \qquad t \in [0, 1].$$

Proof By Theorem C.1, g is operator convex on $[0, 1]$ if and only if there exists an operator convex function f on $(0, \infty)$ with $f(0^+), f'(\infty) < +\infty$ such that $g(t) = (1-t)f\left(\frac{t}{1-t}\right)$, $t \in (0, 1)$. Hence the corollary is easily seen from [63, Proposition 8.4] and we omit the details.

Appendix D
Operator Connections of Normal Positive Functionals

Let $\mathcal{H}$ be an infinite-dimensional Hilbert space. An *operator connection* σ in the Kubo–Ando sense [85] is a binary operation $(A, B) \in B(\mathcal{H})_+ \times B(\mathcal{H})_+ \mapsto A\sigma B \in B(\mathcal{H})_+$ satisfying the following properties:

(I) *Joint monotonicity:* $A_1 \le A_2$ and $B_1 \le B_2$ imply $A_1\sigma B_1 \le A_2\sigma B_2$.
(II) *Transformer inequality:* $C(A\sigma B)C \le (CAC)\sigma(CBC)$ for any $C \in B(\mathcal{H})_+$.
(III) *Downward continuity:* $A_n \searrow A$ and $B_n \searrow B$ imply $A_n\sigma B_n \searrow A\sigma B$, where $A_n \searrow A$ means that $A_1 \ge A_2 \ge \cdots$ and $A_n \to A$ in the strong operator topology.

It is well-known [85, Theorem 3.3] that there is a one-to-one correspondence between the operator connections σ and the operator monotone functions $k \ge 0$ on $[0, \infty)$, in such a way that k and σ are determined from each other by

$$k(t)I = I\,\sigma\,(tI), \qquad t > 0,$$

$$A\sigma B = A^{1/2}k(A^{-1/2}BA^{-1/2})A^{1/2}$$

for every $A, B \in B(\mathcal{H})_+$ with A invertible. In this case, write $k = k_\sigma$ and call it the *representing function* of σ. The transpose $\widetilde{\sigma}$ of σ is the operator connection corresponding to the operator monotone function $k_{\widetilde{\sigma}}(t) := tk_\sigma(t^{-1})$, $t > 0$, the transpose of k_σ. We have $A\widetilde{\sigma}B = B\sigma A$ for $A, B \in B(\mathcal{H})_+$. The *parallel sum* $A : B$ is the operator connection of A, B corresponding to the operator monotone function $(1 + t^{-1})^{-1} = t/(1 + t)$ on $[0, \infty)$.

The theory of operator connections has recently been extended in [57, 82, 83] to unbounded positive operators (or positive forms), including positive τ-measurable operators affiliated with a semifinite von Neumann algebra and positive elements of Haagerup's L^p-spaces over a general von Neumann algebra.

F. Hiai, *Quantum f-Divergences in von Neumann Algebras*,
Mathematical Physics Studies, https://doi.org/10.1007/978-981-33-4199-9

In this section we will discuss operator connections restricted to functionals in M_*^+ via the identification with Haagerup's $L^1(M)$ under the correspondence $\psi \in M_* \leftrightarrow h_\psi \in L^1(M)$ (Theorem A.33 of Sect. A.6).

We begin with the following definition, which is along the same lines as [111] and also makes sense in view of Lemma D.3 below.

Definition D.1 For each $\varphi, \psi \in M_*^+$ we have positive operators $T_{\varphi/(\varphi+\psi)}$, $T_{\psi/(\varphi+\psi)}$ in $s(\varphi+\psi)Ms(\varphi+\psi)$ (see Lemma A.24 and Definition 8.1). Let σ be an operator connection (in the Kubo–Ando sense). Note that $T_{\varphi/(\varphi+\psi)}\sigma T_{\psi/(\varphi+\psi)} \in M_+$, since

$$\begin{aligned}u'^*(T_{\varphi/(\varphi+\psi)}\sigma T_{\psi/(\varphi+\psi)})u' &= (u'^*T_{\varphi/(\varphi+\psi)}u')\sigma(u'^*T_{\psi/(\varphi+\psi)}u') \\ &= T_{\varphi/(\varphi+\psi)}\sigma T_{\psi/(\varphi+\psi)}\end{aligned}$$

for any unitary $u' \in M'$ (the commutant of M). Hence we can define the *connection* $\varphi\sigma\psi \in M_*^+$ of φ, ψ by

$$h_{\varphi\sigma\psi} = h_{\varphi+\psi}^{1/2}(T_{\varphi/(\varphi+\psi)}\sigma T_{\psi/(\varphi+\psi)})h_{\varphi+\psi}^{1/2}.$$

Proposition D.2 *For every $\varphi, \psi \in M_*^+$,*

$$\varphi\tilde{\sigma}\psi = \psi\sigma\varphi.$$

Proof Since

$$\begin{aligned}h_{\varphi\tilde{\sigma}\psi} &= h_{\varphi+\psi}^{1/2}(T_{\varphi/(\varphi+\psi)}\tilde{\sigma}T_{\psi/(\varphi+\psi)})h_{\varphi+\psi}^{1/2} \\ &= h_{\varphi+\psi}^{1/2}(T_{\psi/(\varphi+\psi)}\sigma T_{\varphi/(\varphi+\psi)})h_{\varphi+\psi}^{1/2} = h_{\psi\sigma\varphi},\end{aligned}$$

the assertion follows.

As for the Kubo–Ando operator connections, the transformer inequality in the above (II) extends to $C^*(A\sigma B)C \le (C^*AC)\sigma(C^*BC)$ for any $C \in B(\mathcal{H})$. Moreover, concerning the equality case of the transformer inequality, the next lemma was shown in [40, Theorem 3], whose proof is included here for the convenience of the reader. The lemma will be useful in showing the main properties of $\varphi\sigma\psi$ below. First recall the integral expression of the representing function k_σ:

$$k_\sigma(t) = \alpha + \beta t + \int_{(0,\infty)} \frac{t(1+s)}{t+s}\,d\mu(s), \qquad t \in [0,\infty), \tag{D.1}$$

where $\alpha, \beta \ge 0$ and μ is a finite positive measure on $(0,\infty)$.

Lemma D.3 *Let $A, B \in B(\mathcal{H})_+$ and $C \in B(\mathcal{H})$ be such that $s(A+B) \le s(CC^*)$, where $s(\cdot)$ denotes the support projection. Then for any connection σ,*

$$C^*(A\sigma B)C = (C^*AC)\sigma(C^*BC).$$

Proof Recall the integral expression [85, Theorem 3.4] corresponding to (D.1):

$$A\sigma B = \alpha A + \beta B + \int_{(0,\infty)} \frac{1+s}{s}\{(sA) : B\}\, d\mu(s). \tag{D.2}$$

Hence it suffices to prove that

$$C^*(A : B)C = (C^*AC) : (C^*BC). \tag{D.3}$$

Assume $C \in B(\mathcal{H})$ with $s(A+B) \le E := s(CC^*)$. Note that $C = EC$ and $\overline{C\mathcal{H}} = E\mathcal{H}$. Recall the well-known variational expression of $A : B$ (see, e.g., [51, Lemma 3.1.5]): for every $\xi \in \mathcal{H}$,

$$\langle \xi, (A : B)\xi \rangle = \inf\{\langle \eta, A\eta \rangle + \langle \zeta, B\zeta \rangle : \eta, \zeta \in \mathcal{H},\ \eta + \zeta = \xi\}. \tag{D.4}$$

This gives

$$\langle \xi, C(A : B)C\xi \rangle = \inf\{\langle \eta, A\eta \rangle + \langle \zeta, B\zeta \rangle : \eta, \zeta \in \mathcal{H},\ \eta + \zeta = C\xi\}.$$

Let $\eta, \zeta \in \mathcal{H}$ be such that $\eta + \zeta = C\xi$ and hence $E\eta + E\zeta = C\xi$. Choose $\eta_n \in \mathcal{H}$ such that $C\eta_n \to E\eta$. Set $\zeta_n := \xi - \eta_n$; then $C\eta_n + C\zeta_n = C\xi$ and $C\zeta_n \to C\xi - E\eta = E\zeta$. By (D.4) we have

$$\langle \eta_n, CAC\eta_n \rangle + \langle \zeta_n, CBC\zeta_n \rangle \ge \langle \xi, \{(CAC) : (CBC)\}\xi \rangle,$$

whose limit as $n \to \infty$ is

$$\langle \eta, A\eta \rangle + \langle \zeta, B\zeta \rangle \ge \langle \xi, \{(CAC) : (CBC)\}\xi \rangle$$

thanks to $s(A+B) \le E$. Therefore, $C(A : B)C \ge (CAC) : (CBC)$, and the reverse inequality is the transformer inequality. So the desired equality follows.

Lemma D.4 *Let $\varphi, \psi, \omega \in M_*^+$ be such that $\varphi + \psi \le \lambda\omega$ for some $\lambda > 0$. Then*

$$h_{\varphi\sigma\psi} = h_\omega^{1/2}(T_{\varphi/\omega}\sigma T_{\psi/\omega})h_\omega^{1/2}.$$

In particular, if $\psi \le \lambda\varphi$ for some $\lambda > 0$, then $h_{\varphi\sigma\psi} = h_\varphi^{1/2} k_\sigma(T_{\psi/\varphi})h_\varphi^{1/2}$, where k_σ is the representing function of σ.

Proof By Lemma A.24 there are $A, B \in s(\varphi+\psi)Ms(\varphi+\psi)$ and $C \in s(\omega)Ms(\omega)$ such that

$$h_\varphi^{1/2} = Ah_{\varphi+\psi}^{1/2}, \qquad h_\psi^{1/2} = Bh_{\varphi+\psi}^{1/2}, \qquad h_{\varphi+\psi}^{1/2} = Ch_\omega^{1/2}.$$

Since $h_\varphi^{1/2} = ACh_\omega^{1/2}$ and $h_\psi^{1/2} = BCh_{\varphi+\psi}^{1/2}$, we have

$$T_{\varphi/\omega} = (AC)^*(AC) = C^*T_{\varphi/(\varphi+\psi)}C, \quad T_{\psi/\omega} = (BC)^*(BC) = C^*T_{\psi/(\varphi+\psi)}C.$$

Moreover, note that $s(T_{\varphi/(\varphi+\psi)}+T_{\psi/(\varphi+\psi)}) \le s(\varphi+\psi) = s(CC^*)$ by Lemma A.24 (in fact, $T_{\varphi/(\varphi+\psi)} + T_{\psi/(\varphi+\psi)} = s(\varphi+\psi)$ holds). Therefore, by Lemma D.3,

$$\begin{aligned} h_\omega^{1/2}(T_{\varphi/\omega}\sigma T_{\psi/\omega})h_\omega^{1/2} &= h_\omega^{1/2}((C^*T_{\varphi/(\varphi+\psi)}C)\sigma(C^*T_{\psi/(\varphi+\psi)}C))h_\omega^{1/2} \\ &= h_\omega^{1/2}C^*(T_{\varphi/(\varphi+\psi)}\sigma T_{\psi/(\varphi+\psi)})Ch_\omega^{1/2} \\ &= h_{\varphi+\psi}^{1/2}(T_{\varphi/(\varphi+\psi)}\sigma T_{\psi/(\varphi+\psi)})h_{\varphi+\psi}^{1/2} = h_{\varphi\sigma\psi}, \end{aligned}$$

showing the first result. The latter result follows from the first by taking $\omega = \varphi$.

Example D.5 Consider the case when $M = B(\mathcal{H})$ on an arbitrary Hilbert space $\mathcal{H}$, which is standardly represented by the left multiplication on the Hilbert–Schmidt class $C_2(\mathcal{H})$, a Hilbert space with the Hilbert–Schmidt inner product. The trace-class $C_1(\mathcal{H})$ is identified with $B(\mathcal{H})_*$ as usual. Let $A, B \in C_1(\mathcal{H})_+$, corresponding to $\varphi = \mathrm{Tr}(A\,\cdot)$, $\psi = \mathrm{Tr}(B\,\cdot) \in B(\mathcal{H})_*^+$. We have $\varphi\sigma\psi$ in Definition D.1 as well as $A\sigma B$ in the Kubo–Ando sense [85]. We write $A = (A+B)^{1/2}T_{A/(A+B)}(A+B)^{1/2}$ and $B = (A+B)^{1/2}T_{B/(A+B)}(A+B)^{1/2}$ with positive operators $T_{A/(A+B)}$, $T_{B/(A+B)}$ in $s(A+B)B(\mathcal{H})s(A+B)$. Note that these are indeed $T_{\varphi/(\varphi+\psi)}$, $T_{\psi/(\varphi+\psi)}$, respectively. By Lemma D.3,

$$A\sigma B = (A+B)^{1/2}(T_{A/(A+B)}\sigma T_{B/(A+B)})(A+B)^{1/2},$$

which means that $\varphi\sigma\psi$ corresponds to $A\sigma B$, or $\varphi\sigma\psi$ coincides with $A\sigma B$.

The next theorem summarizes basic properties of the connection $\varphi\sigma\psi$.

Theorem D.6 *Let $\varphi, \psi, \varphi_i, \psi_i \in M_*^+$.*

(i) Joint monotonicity: *If $\varphi_1 \le \varphi_2$ and $\psi_1 \le \psi_2$, then $\varphi_1\sigma\psi_1 \le \varphi_2\sigma\psi_2$.*
(ii) Transformer inequality: *For every $a \in M$,*

$$a^*(\varphi\sigma\psi)a \le (a^*\varphi a)\sigma(a^*\psi a), \tag{D.5}$$

where $(a^*\varphi a)(x) := \varphi(axa^*)$, $x \in M$. Moreover, if $s(\varphi+\psi) \le s(aa^*)$ (in particular, if a is invertible), then equality holds in inequality (D.5).
(iii) Downward continuity: *If $\varphi_n \searrow \varphi$ and $\psi_n \searrow \psi$ in M_*^+, then $\varphi_n\sigma\psi_n \searrow \varphi\sigma\psi$, where $\varphi_n \searrow \varphi$ means that $\varphi_1 \ge \varphi_2 \ge \cdots$ and $\|\varphi_n - \varphi\| \to 0$. In particular,*

for any $\omega \in M_*^+$,

$$\varphi\sigma\psi = \lim_{\varepsilon\searrow 0}(\varphi + \varepsilon\omega)\sigma(\psi + \varepsilon\omega) \quad \textit{in the norm (decreasingly).} \tag{D.6}$$

(iv) Joint concavity:

$$(\varphi_1 + \varphi_2)\sigma(\psi_1 + \psi_2) \geq (\varphi_1\sigma\psi_1) + (\varphi_2\sigma\psi_2).$$

Proof

(i) Let $\omega := \varphi_2 + \psi_2$. It is easy to see that $T_{\varphi_1/\omega} \leq T_{\varphi_2/\omega}$ and $T_{\psi_1/\omega} \leq T_{\psi_2/\omega}$. Hence by Lemma D.4 we have

$$h_{\varphi_1\sigma\psi_1} = h_\omega^{1/2}(T_{\varphi_1/\omega}\sigma T_{\psi_1/\omega})h_\omega^{1/2} \leq h_\omega^{1/2}(T_{\varphi_2/\omega}\sigma T_{\psi_2/\omega})h_\omega^{1/2} = h_{\varphi_2\sigma\psi_2}.$$

(ii) This will be proved below after we give the integral expression of $\varphi\sigma\psi$.

(iii) The joint monotonicity in (i) says that $\varphi_n\sigma\psi_n$ is monotone decreasing as $n \to \infty$. Hence a positive linear functional ϕ on M is defined by $\phi(x) := \lim_{n\to\infty}(\varphi_n\sigma\psi_n)(x)$, $x \in M$. Since $\varphi_n\sigma\psi_n \geq \phi$, it is clear that ϕ is normal, so $\phi \in M_*^+$ and $\|\varphi_n\sigma\psi_n - \phi\| = (\varphi_n\sigma\psi_n - \phi)(1) \to 0$ as $n \to \infty$. To prove that $\phi = \varphi\sigma\psi$, let $\omega := \varphi_1 + \psi_1$. It follows that $T_{\varphi_n/\omega} \searrow T_1$ and $T_{\psi_n/\omega} \searrow T_2$ for some $T_1, T_2 \in M_+$. For every $x \in M_+$ one has

$$\varphi_n(x) = \mathrm{tr}(h_\omega^{1/2}xh_\omega^{1/2})T_{\varphi_n/\omega} \searrow \mathrm{tr}(h_\omega^{1/2}xh_\omega^{1/2})T_1$$

as $n \to \infty$, so that $\varphi(x) = \mathrm{tr}\, xh_\omega^{1/2}T_1h_\omega^{1/2}$, showing $h_\varphi = h_\omega^{1/2}T_1h_\omega^{1/2}$. This implies that $T_1 = T_{\varphi/\omega}$ and similarly $T_2 = T_{\psi/\omega}$. Hence by Lemma D.4 one has

$$\begin{aligned}(\varphi_n\sigma\psi_n)(x) &= \mathrm{tr}(h_\omega^{1/2}xh_\omega^{1/2})(T_{\varphi_n/\omega}\sigma T_{\psi_n/\omega}) \\ &\searrow \mathrm{tr}(h_\omega^{1/2}xh_\omega^{1/2})(T_{\varphi/\omega}\sigma T_{\psi/\omega}) = (\varphi\sigma\psi)(x)\end{aligned}$$

as $n \to \infty$. Therefore, $\phi = \varphi\sigma\psi$.

(iv) With $\omega := \varphi_1+\varphi_2+\psi_1+\psi_2$, it is easy to see that $T_{(\varphi_1+\varphi_2)/\omega} = T_{\varphi_1/\omega} + T_{\varphi_2/\omega}$ and $T_{(\psi_1+\psi_2)/\omega} = T_{\psi_1/\omega} + T_{\psi_2/\omega}$. Hence by Lemma D.4 one has

$$\begin{aligned}h_{(\varphi_1+\varphi_2)\sigma(\psi_1+\psi_2)} &= h_\omega^{1/2}((T_{\varphi_1/\omega} + T_{\varphi_2/\omega})\sigma(T_{\psi_1/\omega} + T_{\psi_2/\omega}))h_\omega^{1/2} \\ &\geq h_\omega^{1/2}((T_{\varphi_1/\omega}\sigma T_{\psi_1/\omega}) + (T_{\varphi_2/\omega}\sigma T_{\psi_2/\omega}))h_\omega^{1/2} \\ &= h_{\varphi_1\sigma\psi_1} + h_{\varphi_2\sigma\psi_2} = h_{\varphi_1\sigma\psi_1+\varphi_2\sigma\psi_2},\end{aligned}$$

where the inequality above is due to the joint concavity in [85, Theorem 3.5] (also [51, Corollary 3.2.3]).

Corresponding to (D.1) we have the integral expression of $\varphi\sigma\psi$ as follows:

Theorem D.7 *For every* $\varphi, \psi \in M_*^+$,

$$(\varphi\sigma\psi)(x) = \alpha\varphi(x) + \beta\psi(x) + \int_{(0,\infty)} \frac{1+s}{s}((s\varphi) : \psi)(x)\, d\mu(s), \qquad x \in M.$$

Proof Let $\omega := \varphi + \psi$. By applying (D.2) to $T_{\varphi/\omega}$ and $T_{\psi/\omega}$ we have

$$\begin{aligned}(\varphi\sigma\psi)(x) &= \mathrm{tr}(h_\omega^{1/2} x h_\omega^{1/2})(T_{\varphi/\omega}\sigma T_{\psi/\omega}) \\ &= \alpha \mathrm{tr}(h_\omega^{1/2} x h_\omega^{1/2}) T_{\varphi/\omega} + \beta \mathrm{tr}(h_\omega^{1/2} x h_\omega^{1/2}) T_{\psi/\omega} \\ &\quad + \int_{(0,\infty)} \frac{1+s}{s} \mathrm{tr}(h_\omega^{1/2} x h_\omega^{1/2})((sT_{\varphi/\omega}) : T_{\psi/\omega})\, d\mu(s) \\ &= \alpha\varphi(x) + \beta\psi(x) + \int_{(0,\infty)} \frac{1+s}{s}((s\varphi) : \psi)(x)\, d\mu(s),\end{aligned}$$

as asserted.

The following variational expression is the variant of (D.4) for the parallel sum $\varphi : \psi$. This expression and the integral expression in Theorem D.7 together may serve as a definition of $\varphi\sigma\psi$.

Theorem D.8 *For every* $\varphi, \psi \in M_*^+$ *and any* $x \in M$,

$$(\varphi : \psi)(x^*x) = \inf\{\varphi(y^*y) + \psi(z^*z) : y, z \in M,\ y + z = x\}. \tag{D.7}$$

Proof Let $\omega := \varphi + \psi$. Note that

$$\begin{aligned}\inf_{y+z=x} \{\varphi(y^*y) + \psi(z^*z)\} &= \inf_{y+z=x} \inf_{\varepsilon>0} \{(\varphi + \varepsilon\omega)(y^*y) + (\psi + \varepsilon\omega)(z^*z)\} \\ &= \inf_{\varepsilon>0} \inf_{y+z=x} \{(\varphi + \varepsilon\omega)(y^*y) + (\psi + \varepsilon\omega)(z^*z)\}.\end{aligned}$$

From this and (D.6) it suffices to prove (D.7) in the case $\varphi \sim \psi$. Let $e := s(\varphi) = s(\omega)$ and choose an $A \in eMe$ such that $h_\varphi^{1/2} = Ah_\omega^{1/2} = h_\omega^{1/2}A^*$. Then A is invertible in eMe and $T_{\varphi/\omega} = A^*A$. Since $T_{\varphi/\omega} + T_{\psi/\omega} = e$, we find that

$$\begin{aligned}h_{\varphi:\psi} &= h_\omega^{1/2}(T_{\varphi/\omega} : T_{\psi/\omega})h_\omega^{1/2} \\ &= h_\omega^{1/2} T_{\varphi/\omega}(T_{\varphi/\omega} + T_{\psi/\omega})^{-1} T_{\psi/\omega} h_\omega^{1/2} \\ &= h_\omega^{1/2} A^*A(e - A^*A) h_\omega^{1/2} \\ &= h_\omega^{1/2} A^*(e - AA^*)A h_\omega^{1/2} = h_\varphi^{1/2}(e - AA^*)h_\varphi^{1/2}.\end{aligned}$$

When $y, z \in M$ and $y + z = x$, one has

$$\begin{aligned}
&\varphi(y^*y) + \psi(z^*z) - (\varphi : \psi)(x^*x) \\
&\quad = \mathrm{tr}(x - z)^*(x - z)h_\varphi + \mathrm{tr}\, z^*zh_\psi - \mathrm{tr}\, x^*x(h_\varphi - h_\varphi^{1/2}AA^*h_\varphi^{1/2}) \\
&\quad = \mathrm{tr}\, z^*zh_\varphi + \mathrm{tr}\, z^*zh_\psi - \mathrm{tr}\, x^*zh_\varphi - \mathrm{tr}\, z^*xh_\varphi + \mathrm{tr}\, x^*xh_\varphi^{1/2}AA^*h_\varphi^{1/2} \\
&\quad = \mathrm{tr}\, z^*zh_\omega + \mathrm{tr}\, A^*h_\varphi^{1/2}x^*xh_\varphi^{1/2}A - 2\mathrm{Re}\,\mathrm{tr}\, x^*zh_\varphi \\
&\quad = \|zh_\omega^{1/2}\|_2^2 + \|xh_\varphi^{1/2}A\|_2^2 - 2\mathrm{Re}\,\mathrm{tr}\, A^*h_\varphi^{1/2}x^*zh_\varphi^{1/2}A^{*-1} \\
&\quad = \|zh_\omega^{1/2}\|_2^2 + \|xh_\varphi^{1/2}A\|_2^2 - 2\mathrm{Re}\,\langle xh_\varphi^{1/2}A, zh_\omega^{1/2}\rangle
\end{aligned}$$

thanks to $h_\omega^{1/2} = h_\varphi^{1/2}A^{*-1}$. Therefore, the Schwarz inequality gives

$$\varphi(y^*y) + \psi(z^*z) - (\varphi : \psi)(x^*x) \geq 0.$$

Moreover, since $\overline{Mh_\omega^{1/2}} = L^2(M)e$ and $xh_\varphi^{1/2}A \in L^2(M)e$, one can choose a sequence $\{z_n\} \subset M$ such that $\|z_nh_\omega^{1/2} - xh_\varphi^{1/2}A\| \to 0$. Letting $y_n := x - z_n$ one has

$$\begin{aligned}
&\varphi(y_n^*y_n) + \psi(z_n^*z_n) - (\varphi : \psi)(x^*x) \\
&\quad = \|z_nh_\omega^{1/2}\|_2^2 + \|xh_\varphi^{1/2}A\|_2^2 - 2\mathrm{Re}\,\langle xh_\varphi^{1/2}A, z_nh_\omega^{1/2}\rangle \longrightarrow 0,
\end{aligned}$$

so that (D.7) follows.

Now we are in a position to prove Theorem D.6 (ii).

Proof (Theorem D.6 (ii)) For the transformer inequality, by Theorem D.7 it suffices to prove (D.5) for the parallel sum $\varphi : \psi$. For every $x, a \in M$, using (D.7) twice we have

$$\begin{aligned}
(a^*(\varphi : \psi)a)(x^*x) &= (\varphi : \psi)((xa^*)^*(xa^*)) \\
&= \inf\{\varphi(y^*y) + \psi(z^*z) : y, z \in M,\ y + z = xa^*\} \\
&\leq \inf\{\varphi((ya^*)^*(ya^*)) + \psi((za^*)^*(za^*) : y, z \in M,\ y + z = x\} \\
&= \inf\{(a^*\varphi a)(y^*y) + (a^*\psi a)(z^*z) : y, z \in M,\ y + z = x\} \\
&= ((a^*\varphi a) : (a^*\psi a))(x^*x),
\end{aligned}$$

showing inequality (D.5).

Next, assume that $s(\varphi + \psi) \leq s(aa^*)$, and prove that the equality holds in (D.5). Let $\omega := \varphi + \psi$, $h := h_\omega$, and $a^*h^{1/2} = vk^{1/2}$ be the polar decomposition with $k^{1/2} = |a^*h^{1/2}|$, so that $k = h^{1/2}aa^*h^{1/2} \in L^1(M)_+$ and $s(k) = s(h)$ since $s(h) = s(\omega) \leq s(aa^*)$. Further, let $\widetilde{\omega}, \widetilde{\varphi}, \widetilde{\psi} \in M_*^+$ be such that $h_{\widetilde{\omega}} = k$, $h_{\widetilde{\varphi}} = k^{1/2}T_{\varphi/\omega}k^{1/2}$

and $h_{\widetilde{\psi}} = k^{1/2}T_{\psi/\omega}k^{1/2}$. Since $h^{1/2}(T_{\varphi/\omega} + T_{\psi/\omega})h^{1/2} = h_\varphi + h_\psi = h$, it follows that $T_{\varphi/\omega} + T_{\psi/\omega} = s(h)$ and $h_{\widetilde{\varphi}} + h_{\widetilde{\psi}} = k^{1/2}s(h)k^{1/2} = k$, so that

$$\widetilde{\omega} = \widetilde{\varphi} + \widetilde{\psi}, \qquad T_{\widetilde{\varphi}/\widetilde{\omega}} = T_{\varphi/\omega}, \qquad T_{\widetilde{\psi}/\widetilde{\omega}} = T_{\psi/\omega}.$$

One has

$$vh_{\widetilde{\varphi}}v^* = vk^{1/2}T_{\widetilde{\varphi}/\widetilde{\omega}}k^{1/2}v^* = a^*h^{1/2}T_{\varphi/\omega}h^{1/2}a = a^*h_\varphi a$$

and similarly $vh_{\widetilde{\psi}}v^* = a^*h_\psi a$. Moreover, one has

$$\begin{aligned} vh_{\widetilde{\varphi}\sigma\widetilde{\psi}}v^* &= vk^{1/2}(T_{\widetilde{\varphi}/\widetilde{\omega}}\sigma T_{\widetilde{\psi}/\widetilde{\omega}})k^{1/2}v^* \\ &= a^*h^{1/2}(T_{\varphi/\omega}\sigma T_{\psi/\omega})h^{1/2}a = a^*h_{\varphi\sigma\psi}a. \end{aligned}$$

From these with Theorem A.33,

$$a^*\varphi a = v\widetilde{\varphi}v^*, \qquad a^*\psi a = v\widetilde{\psi}v^*, \qquad a^*(\varphi\sigma\psi)a = v(\widetilde{\varphi}\sigma\widetilde{\psi})v^*.$$

Hence it suffices to show that

$$v(\widetilde{\varphi}\sigma\widetilde{\psi})v^* = (v\widetilde{\varphi}v^*)\sigma(v\widetilde{\psi}v^*). \tag{D.8}$$

Note that $v\widetilde{\varphi}v^* + v\widetilde{\psi}v^* = v\widetilde{\omega}v^*$ and $h_{v\widetilde{\omega}v^*}^{1/2} = (vkv^*)^{1/2} = vk^{1/2}v^*$, and it is immediate to see that

$$T_{v\widetilde{\varphi}v^*/v\widetilde{\omega}v^*} = vT_{\widetilde{\varphi}/\widetilde{\omega}}v^*, \qquad T_{v\widetilde{\psi}v^*/v\widetilde{\omega}v^*} = vT_{\widetilde{\psi}/\widetilde{\omega}}v^*.$$

Therefore, we find that

$$\begin{aligned} h_{(v\widetilde{\varphi}v^*)\sigma(v\widetilde{\psi}v^*)} &= h_{v\widetilde{\omega}v^*}^{1/2}(T_{v\widetilde{\varphi}v^*/v\widetilde{\omega}v^*}\sigma T_{v\widetilde{\psi}v^*/v\widetilde{\omega}v^*})h_{v\widetilde{\omega}v^*}^{1/2} \\ &= vk^{1/2}v^*((vT_{\widetilde{\varphi}/\widetilde{\omega}}v^*)\sigma(vT_{\widetilde{\psi}/\widetilde{\omega}}v^*))vk^{1/2}v^* \\ &= vk^{1/2}v^*v(T_{\widetilde{\varphi}/\widetilde{\omega}}\sigma T_{\widetilde{\psi}/\widetilde{\omega}})v^*vk^{1/2}v^* \\ &= vk^{1/2}(T_{\widetilde{\varphi}/\widetilde{\omega}}\sigma T_{\widetilde{\psi}/\widetilde{\omega}})k^{1/2}v^* = vh_{\widetilde{\varphi}\sigma\widetilde{\psi}}v^*, \end{aligned}$$

where the second equality above follows from Lemma D.3. Thus, (D.8) has been shown.

Let $\mathcal{H}$ and $\mathcal{K}$ be Hilbert spaces. For any operator connection σ and any positive linear map $\alpha : B(\mathcal{H}) \to B(\mathcal{K})$, the following inequality is well-known:

$$\alpha(A\sigma B) \le \alpha(A)\sigma\alpha(B), \qquad A, B \in B(\mathcal{H}), \tag{D.9}$$

which is due to Ando [4] though stated only for the geometric mean and the parallel sum. Inequality (D.9) for general σ can be seen, for instance, by using the integral expression in (D.2), which is considered as an extension of the transformer inequality.

Proposition D.9 *Let $\gamma : N \to M$ be a normal positive linear map between von Neumann algebras. Then for every $\varphi, \psi \in M_*^+$,*

$$(\varphi\sigma\psi) \circ \gamma \le (\varphi \circ \gamma)\sigma(\psi \circ \gamma).$$

Proof Although γ is not necessarily unital here, with $\omega := \varphi + \psi$ one can define a normal positive map $\gamma_\omega^* : M \to e_0 M e_0$ (where $e_0 := s(\omega \circ \gamma)$) as in Lemma 8.3 (also Proposition 6.6). Then the result can be shown as in Lemma 8.6. Indeed, we have

$$\begin{aligned} h_{(\varphi\sigma\psi)\circ\gamma} &= \gamma_*(h_{\varphi\sigma\psi}) \\ &= \gamma_*(h_\omega^{1/2}(T_{\varphi/\omega}\sigma T_{\psi/\omega})h_\omega^{1/2}) \\ &= h_{\omega\circ\gamma}^{1/2}\gamma_\omega^*(T_{\varphi/\omega}\sigma T_{\psi/\omega})h_{\omega\circ\gamma}^{1/2} \\ &\le h_{\omega\circ\gamma}^{1/2}(\gamma_\omega^*(T_{\varphi/\omega})\sigma\gamma_\omega^*(T_{\psi/\omega}))h_{\omega\circ\gamma}^{1/2} \\ &= h_{\omega\circ\gamma}^{1/2}(T_{\varphi\circ\gamma/\omega\circ\gamma}\sigma T_{\psi\circ\gamma/\omega\circ\gamma})h_{\omega\circ\gamma}^{1/2} \\ &= h_{(\varphi\circ\gamma)\sigma(\psi\circ\gamma)}. \end{aligned}$$

In the above, the third and the fourth equalities are due to Lemmas 8.3 and 8.5, respectively, which hold without the unitality assumption of γ, and the inequality above follows from (D.9).

The last result is the relation between the connection $\varphi\sigma\psi$ and the maximal f-divergence, which is useful in Sect. 8.2.

Proposition D.10 *For every operator connection σ and every $\varphi, \psi \in M_*^+$,*

$$\widehat{S}_{-k_\sigma}(\psi\|\varphi) = -(\varphi\sigma\psi)(1).$$

Proof When $\varphi \sim \psi$, by Definition 4.1 and Lemma D.4,

$$\widehat{S}_{-k_\sigma}(\psi\|\varphi) = \mathrm{tr}\big(h_\varphi^{1/2}(-k_\sigma)(T_{\psi/\varphi})h_\varphi^{1/2}\big) = -(\varphi\sigma\psi)(1).$$

For general $\varphi, \psi \in M_*^+$, by Definition 4.3 and (D.6), with $\omega := \varphi + \psi$ we have

$$\begin{aligned} \widehat{S}_{-k_\sigma}(\psi\|\varphi) &= \lim_{\varepsilon\searrow 0} \widehat{S}_{-k_\sigma}(\psi + \varepsilon\omega\|\varphi + \varepsilon\omega) \\ &= -\lim_{\varepsilon\searrow 0}((\varphi + \varepsilon\omega)\sigma(\psi + \varepsilon\omega))(1) = -(\varphi\sigma\psi)(1), \end{aligned}$$

as asserted.

References

1. Accardi, L., Cecchini, C.: Conditional expectations in von Neumann algebras and a theorem of Takesaki. J. Funct. Anal. **45**, 245–273 (1982)
2. Alberti, P.M.: A note on the transition probability over C^*-algebras. Lett. Math. Phys. **7**, 25–32 (1983)
3. Alberti, P.M., Uhlmann, A.: On Bures distance and $*$-algebraic transition probability between inner derived positive linear forms over W^*-algebras. Acta Appl. Math. **60**, 1–37 (2000)
4. Ando, T.: Concavity of certain maps on positive definite matrices and applications to Hadamard products. Linear Algebra Appl. **26**, 203–241 (1979)
5. Ando, T., Hiai, F.: Log majorization and complementary Golden-Thompson type inequalities. Linear Algebra Appl. **197/198**, 113–131 (1994)
6. Ando, T., Hiai, F.: Operator log-convex functions and operator means. Math. Ann. **350**, 611–630 (2011)
7. Araki, H.: Relative Hamiltonian for faithful normal states of a von Neumann algebra. Publ. Res. Inst. Math. Sci. **9**, 165–209 (1973)
8. Araki, H.: Golden-Thompson and Peierls-Bogolubov inequalities for a general von Neumann algebras. Commun. Math. Phys. **34**, 167–178 (1973)
9. Araki, H.: Some properties of modular conjugation operator of von Neumann algebras and a non-commutative Radon-Nikodym theorem with a chain rule. Pacif. J. Math. **50**, 309–354 (1974)
10. Araki, H.: Relative entropy of states of von Neumann algebras. Publ. Res. Inst. Math. Sci. **11**, 809–833 (1976)
11. Araki, H.: Relative entropy for states of von Neumann algebras II. Publ. Res. Inst. Math. Sci. **13**, 173–192 (1977)
12. Araki, H.: On an inequality of Lieb and Thirring. Lett. Math. Phys. **19**, 167–170 (1990)
13. Araki, H., Masuda, T.: Positive cones and L_p-spaces for von Neumann algebras. Publ. Res. Inst. Math. Sci. **18**, 339–411 (1982)
14. Arias, A., Gheondea, A., Gudder, S.: Fixed points of quantum operations. J. Math. Phys. **43**, 5872–5881 (2002)
15. Audenaert, K.M.R., Datta, N.: α-z-relative entropies. J. Math. Phys. **56**, 022202, 16 pp. (2015)
16. Beigi, S.: Sandwiched Rényi divergence satisfies data processing inequality. J. Math. Phys. **54**, 122202, 11 pp. (2013)
17. Belavkin, V.P., Staszewski, P.: C^*-algebraic generalization of relative entropy and entropy. Ann. Inst. H. Poincaré Phys. Théor. **37**, 51–58 (1982)
18. Bergh, J., Löfström, J.: Interpolation Spaces: An Introduction. Springer, Berlin/Heidelberg/New York (1976)

F. Hiai, *Quantum f-Divergences in von Neumann Algebras*,
Mathematical Physics Studies, https://doi.org/10.1007/978-981-33-4199-9

19. Berta, M., Fawzi, O., Tomamichel, M.: On variational expressions for quantum relative entropies. Lett. Math. Phys. **107**, 2239–2265 (2017)
20. Berta, M., Scholz, V.B., Tomamichel, M.: Rényi divergences as weighted non-commutative vector valued L_p-spaces. Ann. Henri Poincaré **19**, 1843–1867 (2018)
21. Bhatia, R.: Matrix Analysis. Springer, New York (1996)
22. Bratteli, O., Robinson, D.W.: Operator Algebras and Quantum Statistical Mechanics 1, 2nd edn. Springer, New York (1987)
23. Bratteli, O., Robinson, D.W.: Operator Algebras and Quantum Statistical Mechanics, vol. 2, 2nd edn. Springer, Berlin (1997)
24. Bratteli, O., Jorgensen, P.E.T., Kishimoto, A., Werner, R.F.: Pure states on O_d. J. Oper. Theory **43**, 97–143 (2000)
25. Choi, M.-D.: A Schwarz inequality for positive linear maps on C^*-algebras. Illinois J. Math. **18**, 565–574 (1974)
26. Connes, A.: Sur le théoème de Radon-Nikodym pour les poids normaux fidéles semi-finis. Bull. Sci. Math. (2) **97**, 253–258 (1973)
27. Connes, A.: Une classification des facteurs de type III. Ann. Sci. École Norm. Sup. (4) **6**, 133–252 (1973)
28. Connes, A.: Caractérisation des espaces vectoriels ordonnés sous-jacents aux algèbres de von Neumann. Ann. Inst. Fourier (Grenoble) **24**, 121–155 (1974)
29. Connes, A.: Classification of injective factors, Cases II_1, II_∞, III_λ, $\lambda \neq 1$. Ann. Math. (2) **104**, 73–115 (1976)
30. Connes, A., Takesaki, M.: The flow of weights on factors of type III, Tôhoku Math. J. **29**, 473–575 (1977)
31. Datta, N.: Min- and max-relative entropies and a new entanglement monotone. IEEE Trans. Inf. Theory **55**, 2816–2826 (2009)
32. Dixmier, J.: Formes linéaires sur un anneau d'opérateurs. Bull. Soc. Math. France **81**, 9–39 (1953)
33. Donald, M.J.: Relative hamiltonians which are not bounded from above. J. Funct. Anal. **91**, 143–173 (1990)
34. Douglas, R.G.: On majorization, factorization, and range inclusion of operators on Hilbert space. Proc. Am. Math. Soc. **17**, 413–415 (1966)
35. Ebadian, A., Nikoufar, I., Gordji, M.E.: Perspectives of matrix convex functions. Proc. Natl. Acad. Sci. U. S. A **108**, 7313–7314 (2011)
36. Effros, E.G.: A matrix convexity approach to some celebrated quantum inequalities. Proc. Natl. Acad. Sci. U. S. A. **106**, 1006–1008 (2009)
37. Fack, T., Kosaki, H.: Generalized s-numbers of σ-measurable operators. Pacif. J. Math. **123**, 269–300 (1986)
38. Frank, R.L., Lieb, E.H.: Monotonicity of a relative Rényi entropy. J. Math. Phys. **54**, 122201, 5 pp. (2013)
39. Franz, U., Hiai, F., Ricard, É.: Higher order extension of Löwner's theory: Operator k-tone functions. Trans. Am. Math. Soc. **366**, 3043–3074 (2014)
40. Fujii, J.: On Izumino's view of operator means. Math. Japon. **33**, 671–675 (1988)
41. Furuta, T.: $A \geq B \geq 0$ assures $(B^r A^p B^r)^{1/q} \geq B^{(p+2r)/q}$ for $r \geq 0$, $p \geq 0$, $q \geq 1$ with $(1+2r)q \geq p+2r$. Proc. Am. Math. Soc. **101**, 85–88 (1987)
42. Haagerup, U.: The standard form of von Neumann algebras, Notes, Copenhagen University (1973)
43. Haagerup, U.: The standard form of von Neumann algebras. Math. Scand. **37**, 271–283 (1975)
44. Haagerup, U.: On the dual weights for crossed products of von Neumann algebras I. Removing separability conditions. Math. Scand. **43**, 99–118 (1978/1979)
45. Haagerup, U.: On the dual weights for crossed products of von Neumann algebras II. Application of operator-valued weights. Math. Scand. **43**, 119–140 (1978/1979)
46. Haagerup, U.: L^p-spaces associated with an arbitrary von Neumann algebra. In: Colloquium Internaternational CNRS, vol. 274, pp. 175–184. CNRS, Paris (1979)

47. Haagerup, U.: On the uniqueness of injective III_1 factor (1985). Preprint. arXiv:1606.03156 [math.OA]
48. Haagerup, U.: Connes' bicentralizer problem and uniqueness of the injective factor of type III_1. Acta Math. **158**, 95–148 (1987)
49. Hansen, F.: An operator inequality. Math. Ann. **246**, 249–250 (1980)
50. Hayashi, M.: Quantum Information. An Introduction. Springer, Berlin (2006)
51. Hiai, F.: Matrix analysis: matrix monotone functions, matrix means, and majorization. Interdiscipl. Inf. Sci. **16**, 139–248 (2010)
52. Hiai, F.: Concavity of certain matrix trace and norm functions. Linear Algebra Appl. **439**, 1568–1589 (2013)
53. Hiai, F.: Concavity of certain matrix trace and norm functions. II. Linear Algebra Appl. **496**, 193–220 (2016)
54. Hiai, F.: Quantum f-divergences in von Neumann algebras I. Standard f-divergences. J. Math. Phys. **59**, 102202, 27 pp. (2018)
55. Hiai, F.: Quantum f-divergences in von Neumann algebras II. Maximal f-divergences. J. Math. Phys. **60**, 012203, 30 pp. (2019)
56. Hiai, F.: Lectures on Selected Topics of von Neumann Algebras. EMS Series of Lectures in Mathematics, European Mathematical Society (EMS) (2021)
57. Hiai, F., Kosaki, H.: Connections of unbounded operators and some related topics: von Neumann algebra case (2020). Preprint. arXiv:2101.01176 [math-OA]
58. Hiai, F., Mosonyi, M.: Different quantum f-divergences and the reversibility of quantum operations. Rev. Math. Phys. **29**, 1750023, 80 pp. (2017)
59. Hiai, F., Nakamura, Y.: Distance between unitary orbits in von Neumann algebras. Pacif. J. Math. **138**, 259–294 (1989)
60. Hiai, F., Petz, D.: The Golden-Thompson trace inequality is complemented. Linear Algebra Appl. **181**, 153–185 (1993)
61. Hiai, F., Ohya, M., Tsukada, M.: Sufficiency, KMS condition and relative entropy in von Neumann algebras. Pacif. J. Math. **96**, 99–109 (1981)
62. Hiai, F., Ohya, M., Tsukada, M.: Sufficiency and relative entropy in *-algebras with applications in quantum systems. Pacif. J. Math. **107**, 117–140 (1983)
63. Hiai, F., Mosonyi, M., Petz, D., Bény, C.: Quantum f-divergences and error correction. Rev. Math. Phys. **23**, 691–747 (2011); Erratum: Quantum f-divergences and error correction. Rev. Math. Phys. **29**, 1792001, 2 pp. (2017)
64. Jaksic, V., Ogata, Y., Pautrat, Y., Pillet, C.-A.: Entropic fluctuations in quantum statistical mechanics. An introduction. In: Quantum Theory from Small to Large Scales, August 2010. Lecture Notes of the Les Houches Summer School, vol. 95 (Oxford University Press, Oxford, 2012), pp. 213–410. arXiv:1106.3786
65. Jenčová, A.: Reversibility conditions for quantum operations. Rev. Math. Phys. **24**, 1250016, 26 pp. (2012)
66. Jenčová, A.: Rényi relative entropies and noncommutative L_p-spaces. Ann. Henri Poincaré **19**, 2513–2542 (2018)
67. Jenčová, A.: Rényi relative entropies and noncommutative L_p-spaces II (2020). arXiv:1707.00047v2 [quant-ph]
68. Jenčová, A., Petz, D.: Sufficiency in quantum statistical inference. Commun. Math. Phys. **263**, 259–276 (2006)
69. Jenčová, A., Petz, D.: Sufficiency in quantum statistical inference. A survey with examples. Infin. Dimens. Anal. Quantum Probab. Relat. Top. **9**, 331–351 (2006)
70. Jenčová, A., Petz, D., Pitrik, J.: Markov triplets on CCR-algebras. Acta Sci. Math. (Szeged) **76**, 111–134 (2010)
71. Kadison, R.V.: Strong continuity of operator functions. Pacif. J. Math. **26**, 121–129 (1968)

72. Kadison, R.V., Ringrose, J.R.: Fundamentals of the Theory of Operator Algebras, I. Elementary Theory, II. Advanced Theory. Pure and Applied Mathematics, vol. 100. Academic, New York (1983, 1986)
73. Kato, T.: Trotter's product formula for an arbitrary pair of self-adjoint contraction semigroups. In: Topics in Functional Analysis (essays dedicated to M. G. Kreĭn on the occasion of his 70th birthday). Advances in Mathematics Supplementary Studies, vol. 3, pp. 185–195. Academic, New York/London (1978)
74. Kato, T.: Perturbation Theory for Linear Operators. Reprint of the 1980 edition. Classics in Mathematics. Springer, Berlin (1995)
75. Kosaki, H.: Positive cones associated with a von Neumann algebra. Math. Scand. **47**, 295–307 (1980)
76. Kosaki, H.: Linear Radon-Nikodym theorem for states on a von Neumann algebra. Publ. Res. Inst. Math. Sci. **18**, 379–386 (1981)
77. Kosaki, H.: Interpolation theory and the Wigner-Yanase-Dyson-Lieb concavity. Commun. Math. Phys. **87**, 315–329 (1982)
78. Kosaki, H.: Applications of the complex interpolation method to a von Neumann algebra: non-commutative L^p-spaces. J. Funct. Anal. **56**, 29–78 (1984)
79. Kosaki, H.: Relative entropy of states: a variational expression. J. Oper. Theory **16**, 335–348 (1986)
80. Kosaki, H.: An inequality of Araki-Lieb-Thirring (von Neumann algebra case). Proc. Am. Math. Soc. **114**, 477–481 (1992)
81. Kosaki, H.: A remark on Sakai's quadratic Radon-Nikodym theorem. Proc. Am. Math. Soc. **116**, 783–786 (1992)
82. Kosaki, H.: Parallel sum of unbounded positive operators. Kyushu J. Math. **71**, 387–405 (2017)
83. Kosaki, H.: Absolute continuity for unbounded positive self-adjoint operators. Kyushu J. Math. **72**, 407–421 (2018)
84. Krieger, W.: On ergodic flows and the isomorphism of factors. Math. Ann. **223**, 19–70 (1976)
85. Kubo, F., Ando, T.: Means of positive linear operators. Math. Ann. **246**, 205–224 (1980)
86. Kümmerer, B., Nagel, R.: Mean ergodic semigroups on W^*-algebras. Acta Sci. Math. (Szeged) **41**, 151–159 (1979)
87. Lesniewski, A., Ruskai, M.B.: Monotone Riemannian metrics and relative entropy on noncommutative probability spaces. J. Math. Phys. **40**, 5702–5724 (1999)
88. Lieb, E.: Convex trace functions and the Wigner-Yanase-Dyson conjecture. Adv. Math. **11**, 267–288 (1973)
89. Łuczak, A.: Quantum sufficiency in the operator algebra framework. Int. J. Theor. Phys. **53**, 3423–3433 (2014)
90. Łuczak, A.: On a general concept of sufficiency in von Neumann algebras. Probab. Math. Stat. **35**, 313–324 (2015)
91. Matsumoto, K.: On single-copy maximization of measured f-divergences between a given pair of quantum states (2016). Preprint. arXiv:1412.3676v5 [quant-ph]
92. Matsumoto, K.: A new quantum version of f-divergence. In: Reality and Measurement in Algebraic Quantum Theory. Springer Proceedings in Mathematics & Statistics, vol. 261, pp. 229–273. Springer, Singapore (2018)
93. McCarthy, J.E.: Geometric interpolation between Hilbert spaces. Ark. Mat. **30**, 321–330 (1992)
94. Mosonyi, M.: Convexity properties of the quantum Rényi divergences, with applications to the quantum Stein's lemma. In: 9th Conference on the Theory of Quantum Computation, Communication and Cryptography (TQC 2014). LIPIcs. Leibniz International Proceedings in Informatics, vol. 27, pp. 88–98 (2014). arXiv:1407.1067 [quant-ph]
95. Mosonyi, M., Hiai, F.: On the quantum Rényi relative entropies and related capacity formulas. IEEE Trans. Inf. Theory **57**, 2474–2487 (2011)
96. Mosonyi, M., Ogawa, T.: Quantum hypothesis testing and the operational interpretation of the quantum Rényi relative entropies. Commun. Math. Phys. **334**, 1617–1648 (2015)

97. Mosonyi, M., Ogawa, T.: Strong converse exponent for classical-quantum channel coding. Commun. Math. Phys. **355**, 373–426 (2017)
98. Müller-Lennert, M., Dupuis, F., Szehr, O., Fehr, S., Tomamichel, M.: On quantum Rényi entropies: a new generalization and some properties. J. Math. Phys. **54**, 122203, 20 pp. (2013)
99. Nelson, E.: Notes on non-commutative integration. J. Funct. Anal. **15**, 103–116 (1974)
100. Nielsen, M.A., Chuang, I.L.: Quantum Computation and Quantum Information. Cambridge University Press, Cambridge (2000)
101. Ohya, M., Petz, D.: Quantum Entropy and Its Use. Springer, Berlin (1993); 2nd edn. (2004)
102. Pedersen, G.K.: C^*-Algebras and Their Automorphism Groups. London Mathematical Society Monographs, vol. 14. Academic, London/New York (1979)
103. Petz, D.: Quasi-entropies for states of a von Neumann algebra. Publ. Res. Inst. Math. Sci. **21**, 787–800 (1985)
104. Petz, D.: Quasi-entropies for finite quantum systems. Rep. Math. Phys. **23**, 57–65 (1986)
105. Petz, D.: Sufficient subalgebras and the relative entropy of states of a von Neumann algebra. Commun. Math. Phys. **105**, 123–131 (1986)
106. Petz, D.: Sufficiency of channels over von Neumann algebras. Quart. J. Math. Oxford Ser. (2) **39**, 97–108 (1988)
107. Petz, D.: A variational expression for the relative entropy. Commun. Math. Phys. **114**, 345–349 (1988)
108. Petz, D.: Discrimination between states of a quantum system by observations. J. Funct. Anal. **120**, 82–97 (1994)
109. Petz, D.: Quantum Information Theory and Quantum Statistics. Theoretical and Mathematical Physics. Springer, Berlin (2008)
110. Petz, D., Ruskai, M.B.: Contraction of generalized relative entropy under stochastic mappings on matrices. Infin. Dimens. Anal. Quantum Probab. Relat. Top. **1**, 83–89 (1998)
111. Pusz, W., Woronowicz, S.L.: Functional calculus for sesquilinear forms and the purification map. Rep. Math. Phys. **8**, 159–170 (1975)
112. Raggio, G.A.: Comparison of Uhlmann's transition probability with the one induced by the natural positive cone of a von Neumann algebra in standard form. Lett. Math. Phys. **6**, 223–236 (1982)
113. Reed, M., Simon, B.: Methods of Modern Mathematical Physics II. Fourier Analysis, Self-Adjointness. Academic, New York/London (1975)
114. Reed, M., Simon, B.: Methods of Modern Mathematical Physics I: Functional Analysis, 2nd edn. Academic, New York (1980)
115. Remmert, R.: Classical Topics in Complex Function Theory. Translated from the German. Graduate Texts in Mathematics, vol. 172. Springer, New York (1998)
116. Sakai, S.: C^*-Algebras and W^*-Algebras. Reprint of the 1971 edition. Classics in Mathematics. Springer, Berlin (1998)
117. Schmüdgen, K.: Unbounded Self-adjoint Operators on Hilbert Space. Graduate Texts in Mathematics, vol. 265. Springer, Dordrecht (2012)
118. Segal, I.E.: A non-commutative extension of abstract integration. Ann. Math. (2) **57**, 401–457 (1953); Correction to "A non-commutative extension of abstract integration" 58, 595–596 (1953)
119. Simon, B.: Lower semicontinuity of positive quadratic forms. Proc. R. Soc. Edinb. Sect. A **79**, 267–273 (1977/1978)
120. Simon, B.: Operator Theory, A Comprehensive Course in Analysis, Part 4. American Mathematical Society, Providence, RI (2015)
121. Stinespring, W.E.: Integration theorems for gages and duality for unimodular groups. Trans. Amer. Math. Soc. **90**, 15–56 (1959)
122. Strătilă, S.: Modular Theory in Operator Algebras. Editura Academiei and Abacus Press, Tunbridge Wells (1981)
123. Strătilă, S., Zsidó, L.: Lectures on von Neumann Algebras. Editura Academiei and Abacus Press, Tunbridge Wells (1979)

124. Takesaki, M.: Tomita's Theory of Modular Hilbert Algebras and Its Applications. Lecture Notes in Mathematics, vol. 128. Springer, Berlin/New York (1970)
125. Takesaki, M.: Conditional expectations in von Neumann algebras. J. Funct. Anal. **9**, 306–321 (1972)
126. Takesaki, M.: Duality for crossed products and the structure of von Neumann algebras of type III. Acta Math. **131**, 249–310 (1973)
127. Takesaki, M.: Theory of Operator Algebras I. Reprint of the first (1979) edition. Encyclopaedia of Mathematical Sciences, vol. 124. Springer, Berlin (2002)
128. Takesaki, M.: Theory of Operator Algebras II. Encyclopaedia of Mathematical Sciences, vol. 125. Springer, Berlin (2003)
129. Terp, M.: L^p spaces associated with von Neumann algebras. Notes, Copenhagen University (1981)
130. Tomamichel, M.: Quantum Information Processing with Finite Resources. Mathematical Foundations. SpringerBriefs in Mathematical Physics, vol. 5. Springer, Cham (2016)
131. Tomiyama, J.: On the projection of norm one in W^*-algebras. Proc. Jpn. Acad. **33**, 608–612 (1957)
132. Umegaki, H.: Conditional expectation in an operator algebra. Tôhoku Math. J. (2) **6**, 177–181 (1954)
133. Umegaki, H.: Conditional expectation in an operator algebra, III. Kōdai Math. Sem. Rep. **11**, 51–64 (1959)
134. Umegaki, H.: Conditional expectation in an operator algebra, IV (entropy and information). Kōdai Math. Sem. Rep. **14**, 59–85 (1962)
135. Wilde, M.M.: Quantum Information Theory. Cambridge University Press, Cambridge (2013); 2nd edn. (2017)
136. Wilde, M.M., Winter, A., Yang, D.: Strong converse for the classical capacity of entanglement-breaking and Hadamard channels via a sandwiched Rényi relative entropy. Commun. Math. Phys. **331**, 593–622 (2014)
137. Yeadon, F.J.: Non-commutative L^p-spaces. Math. Proc. Cambridge Philos. Soc. **77**, 91–102 (1975)

Index

F. Hiai, *Quantum f-Divergences in von Neumann Algebras*,
Mathematical Physics Studies, https://doi.org/10.1007/978-981-33-4199-9

Printed in the United States
by Baker & Taylor Publisher Services

Printed in the United States
by Baker & Taylor Publisher Services